D0341458

TOWARD A GENERAL SCIENCE OF VIABLE SYSTEMS

TOWARD A GENERAL SCIENCE OF VIABLE SYSTEMS

Arthur S. Iberall
Chief Scientist
General Technical Services, Inc.
Upper Darby, Pennsylvania

McGraw-Hill Book Company

New York
St. Louis
San Francisco
Düsseldorf
Johannesburg
Kuala Lumpur
London
Mexico
Montreal
New Delhi
Panama
Rio de Janeiro
Singapore
Sydney
Toronto

Toward a General Science of Viable Systems

Library of Congress Catalog Card Number 77-128790
07-031672-4

1234567890 MAMM 7987654321

This book was set in Times, and was printed and bound by The Maple Press Company. The designer was M. Haber; the drawings were done by Alan Chapman. The editors were Jeremy Robinson and Maureen McMahon. Les Kaplan supervised production.

To Professor George Gamow, whose
teaching was most inspirational at,
and to, the right impressionable age.
To Warren McCulloch, truly gentle
and wise, whose cultivation and
inspiration of youth, by chance, also
touched me.

Contents

Preface

This book attempts to provide a unifying introduction on the general nature of systems to those students of science who are interested in systems' science and engineering. It offers a background of ideas and tools, not specific methods or detailed information. However, it is not written in popular form for its less technically trained readers. Instead, it is a challenging introductory working document, a textbook to be developed by the reader as it was developed by me. I can only attempt, therefore, to share the insights that I have acquired, and also the confusions and uncertainties with which I am left. I cannot pretend to be an expert teacher, but can be only a guide. Further, because of its scope and point of view, my presentation has no chance of satisfying all my technical readers. The specialist in his field who thinks that the biological system can be understood only in terms of detailed biochemical reactions at the molecular level, or that the social system can be understood only in terms of forces that make up the daily economic market, or that physics is based only on the theory of nuclear particles is likely to be out of sympathy with this effort. Yet there is another level to science that seeks lesser precision in detailed focus (assuming as it does that the specialist is much better able to interpret the localized details) but broader and grosser insight. The preparation that went into this study was devoted from the start to those technically oriented people who are willing to be exposed to such an overall view.

With regard to my peers in the physical sciences, I have added reason for being very much concerned about obtaining their ear and their suspension of judgment until the overall design of my major theses becomes apparent.

A particular point of view concerning quantum mechanics may be quite trying to physicist colleagues. Although I accept modern quantum mechanics as a working tool and am satisfied that it is useful at the nuclear, atomic, and molecular level, I am not satisfied with the foundations that it suggests for the logic of science. The difficulty is that it is a theory that accounts for quantization as an eigenvalue problem which I regard as being particularly associated with linear theory. From a logical-esthetic point of view, I would prefer to see quantization emerge from a nonlinear field process, preferably in the form of limit cycles. To my mind, this questioning falls into the same camp as the hidden-variable criticism of David Bohm.

In this document, a thesis is tentatively put forth on how to take an overview of the hierarchical organization of systems and yet keep the detailed logic or change in logic at any systems' level from toppling the overall structure of systems' science. Actually, it appears that only two systems' levels exhibit a character different from the one fairly unified theme in all others. The fundamental character of explanation identified with quantum mechanics is one case in point. I can keep a unitary point of view for all systems' levels if physicist colleagues will tolerate the following idea:

Regard quantum mechanics (as a linear theory that seeks to describe the field phenomena by eigenvectors, with eigenvalues arising from boundary-value problems) as a description of the stable "contours" or boundaries permitted within an abstract space that describes the possible field solutions. It represents those stable solutions that may persist, or transformations between stable solutions that may persist. It is not strange, therefore, that these contours are expressed as probabilities. However, the underlying field ought to be described by continuous nonlinear field equations, possibly partly classical but certainly not completely discovered today. These equations should describe hidden continuous variables. In imagination, it is perhaps by the use of Poincaré's characteristic functions applied to these nonlinear equations that the linear set that is quantum mechanics might emerge essentially as the secular equations.

I do not assert that this is true, only that I would prefer it to be true or can imagine it to be true. It awaits a process of discovery which may be infinitely more difficult than discovering the "linear contours" that are already given by quantum mechanics. Thus I can say very little about it, but I have taken the liberty, every time the issue presents itself, to suggest this conceptual alternative. Although this may be annoying to the physicist, it was done so that the nonphysicist can understand that there may be a foundational issue that continues to arise.

The second issue, less trying but related, is the description of a conservative system, such as the three-body gravitational problem or the solar system. Again for consistency in describing all systems' levels, I am concerned with classifying complex systems and their stationary or near-stationary constellations, and attempt to use a purist doctrine that limit cycles can emerge only from nonconservative systems. Can one not disregard losses and make use of near-periodic phenomena such as 5 billion years of solar-system organization? Certainly, but there appears a persistently annoying stumbling block. Descriptions of the three-body problem and other conservative systems furnish no classification of periodic and nonperiodic orbits, although they enable one to compute future epochs with precision from any given start. Only the orbits cannot be well classified. However, I prefer in this present line of thought to deal with systems with well-classifiable orbits—in fact with limit cycles. Must the solar system be eliminated from our thinking? Not necessarily, if we take a larger view in which the solar system is part of, in fact a phase of, cosmological formation and galactic formation, or else consider it as another exceptional case at one level, like quantum mechanics, in which persistence of orbital form seems to take place for a sufficiently long time without requiring that the periodic orbits necessarily be nonlinear limit cycles that emerge from a system with nonconservative nature.

My position is that periodic orbits should invariably emerge in the cosmos, essentially from systems with a nonconservative nature, but even if they do not, the general description of hierarchical systems will still be a useful construct. If the reader does not hang up on such points, or similar points in which the author errs in some particular or even significant detail, perhaps the overall thesis will have some merit. If it does not appeal to him in whole or in part, I can only, with hearty applause, welcome not only his detailed dissent but also his willingness to offer, either in whole or part, substitute theses.

Above all, I am not expert in all the subjects I cover. I will be wrong in many details. I invite vigorous argument and correction. By a discussion of whether or how all systems can be illuminated by principles of a science of systems, the detailed faults and illustrations may be corrected, or perhaps other more insightful views may be formed.

As far as my general theses are concerned, I will anticipate a general criticism that perhaps only analogies are being proposed, rather than real principles of science, with the words of E. T. Whittaker, in his obituary to V. Volterra (1):

"Biologists have been apt to criticize Volterra for preoccupying himself so elaborately with abstract mathematical models based on simplifying assumptions remote from the complexities of nature. Yet this, after all, is the procedure on which the triumphs of physical science have been founded. It would be rash to say whether the analogies with physical science which he unearthed will remain what they appear to be at first, and certainly are, at least, a clever and remarkable tour de force—or whether they will eventually be seen as the gems of a profound biodynamics, essential to the theoretical and economic biology of the future: what is beyond dispute is that his contributions to pure mathematics will be in demand more and more inescapably as mathematical biology develops."

The abstract ideas about general systems in this book should be similarly blessed!

Finally, in a willingness, perhaps even an unseeming eagerness, to put forth many all-embracing theses, I have also found encouragement in the words of Toynbee (2) that indicate that others have attempted and have perhaps stumbled on such grand tours:

"There is a strong tendency to depreciate works . . . which are created by single minds, and the depreciation becomes the more emphatic the nearer such works approximate to being 'Universal Histories'. For example, Mr. H. G. Wells', *The Outline of History* was received with unmistakable hostility by a number of historical specialists. They criticized severely the errors which they discovered at the points where the writer, in his long journey through time and space, happened to traverse their tiny allotments. They seemed not to realize that, in re-living the entire life of Mankind as a single imaginative experience, Mr. Wells was achieving something which they themselves would hardly have dared to attempt—something, perhaps, of which they had never conceived the possibility. In fact, the purpose and value of Mr. Wells' book seem to have been better appreciated by the general public than by the professional historians of the day."

May Fortune so smile on us!

Further, for any who wonder at the path taken, in which life and man are only bypasses to the larger world, I may offer the words of Lévi-Strauss, whose lesson is drawn from both more and less savage cultures (3).

"In this century . . . it has perhaps never been more urgent to point out that a well-ordered humanism does not begin with oneself, but places the world above life, life above man, respect for others above egotism. And that, even a residence of one or two thousand years . . . cannot be considered as license for any species, including our own."

The reader who wishes some idea and measure of what I am attempting both to challenge and to reconstruct as a foundation for a working interdisciplinary science may turn to Kuhn (4) for guidance. Perhaps by the end of my "novel," a fragmentary answer will be evident for the questions that Kuhn poses: ". . . why the evolutionary process should work. What must nature, including man, be like in order that science be possible at all?"

I am grateful to all the readers who graciously plowed through this "eruption," and offered me their insights, thoughts, and wisdoms. I have tried to respond to each clue they proferred me. In age and experience, my readers are my peers. I must also thank younger members of the scientific community who, in measure of equal importance with wisdom, gave me their feelings. When they are famous, or at least well known, I will then say that they did me the great honor of reviewing this book.

Finally, I am also grateful to the United States Army, whose support of our work in systems science ultimately made this book possible.

ARTHUR S. IBERALL

Acknowledgments

The author has made every effort to determine and credit the holders of copyright of material quoted in this book. The following special credits are added to fulfill the requirements of the copyright holder.

American Association for the Advancement of Science. The quotation on page 348, by Unsöld, is taken from *Science*, vol. 163, no. 3871, page 1015, 1969. Copyright 1969 by the American Association for the Advancement of Science.

Cornell University Press. The quotations on pages 128 and 129 are reprinted from Linus Pauling, *The Nature of the Chemical Bond*. Copyright 1939 and 1940 by Cornell University. Third edition, copyright 1960 by Cornell University. Used by permission of Cornell University Press.

Dover Publications, Inc. The quotation by E. Whittaker (Preface, page xii) is from *Theory of Functionals and Integral and Integro Differential Equations* by Vito Volterra, Dover Publications, Inc., New York. The quotation on page 31, by Dziobeck, is from *Mathematical Theories of Planetary Motions* by Otto Dziobeck, Dover Publications, Inc., New York. The quotation on page 126, by Hume-Rothery, is from *Electrons, Atoms, Metals and Alloys* by Wm. Hume-Rothery, Dover Publications, Inc., New York. These three quotations are reprinted through permission of the publisher.

New York Times. The quotation in the footnote on page 204, by N. Hentoff, was copyrighted in 1968 by the New York Times Company. Reprinted by permission.

Reinhold Publishing Corp. The quotations on pages 76 and 93 by R. Tolman, and referenced on page 106 as published in the 1927 Chemical Engineering Catalogue Co., is copyrighted by Reinhold Publishing Corp. and reprinted by their permission.

Scientific American, Inc. The quotation on page 126 by A. H. Cottrell was taken from a chapter in *Materials, A Scientific American Book*, W. H. Freeman and Co. The chapter first appeared as an article in the September 1967 issue of *Scientific American*: A. H. Cottrell, The Nature of Metals, page 90.

University of California Press. The quotation on page 195 by W. Thorpe originally appeared in *Perspectives in Marine Biology*, A. Buzzati-Traverso (ed.), University of California Press, Berkeley, 1965. It is reprinted with their permission.

As a novice at getting books out, the author wishes to acknowledge, with praise and gratitude, the secretarial skills of Sonya Bulkey. She has an uncanny ability to type rapidly, calmly, intelligently, all documents—including books—and sort out the material, no matter how complex or ambiguous the writing, the changes, or the errors.

BIBLIOGRAPHY

1. Volterra, V.: "Theory of Functionals and Integral and Integro Differential Equations," Dover Publications, Inc., New York, 1959.
2. Toynbee, A.: "A Study of History, Geneses of Civilization," part 1, Oxford University Press, New York, 1962.
3. Lacroix, J.: Lévi-Strauss and the "Primitives"—On With His Mythologigues, *Atlas*, vol. 17, p. 64, translation of book review of Lévi-Strauss, "The Origin of Table Manners," Jan., 1969.
4. Kuhn, T.: "The Structure of Scientific Revolutions," University of Chicago Press, Chicago, 1962.

part one

Physical Preliminaries

Introduction

This study, admittedly speculative, attempts to present the foundations for a central structure of natural and human sciences of interest to society under the cover of a few unifying principles. The material is directed primarily at a technical audience. Although the specific end desired may be too ambitious, the study will still have succeeded if it stimulates the reader to think about the community of some technical notions both in and out of his own field of specialization.

Since such a study is obviously an attempt at a highly organized endeavor, it is useful to summarize the tools of rational study in that form commonly referred to as the scientific method. Science here is viewed as rational "explanation" of natural phenomena observed in time. Explanation is viewed as the establishment of one-to-one correspondence between patterns of phenomena and a notable condensation in patterns of words—spoken or written—that represent explanation. However, what any one man will accept as explanation is left unexplored.

Furthermore, scientific explanation can be viewed within a first category of *formal explanation* at three possible levels: heuristic, as the level at which the elements and relations that are discovered in a field of phenomena are named and ordered; phenomenological, as the level at which the functional transformations that may occur in space and time are described; and analytic, as the level at which the elements, relations, and functions are condensed into abstract form, which may then undergo manipulations without reference back to the physical field of phenomena. There is also a prescientific level, the human clerical function, that names and catalogues things. As illustrations, biological taxonomy is heuristic; metallurgy is commonly phenomenological; wave guide design is analytic; whereas most collecting is often only a clerical routine.

A second possible category of explanation is *isomorphic explanation* (implying identical structure). It may be that this category has no content other than the formal.[1]

The types of scientific method are:

Inductive   From sequences of observations a central principle is ultimately inferred. [See, for example, Darwin (1)].

Deductive   From a small structure of principles, sequences of observations are described (illustrated by Euclid or Aristotle).

Dialectic   Arguments are assembled first as a thesis for explaining some class of phenomena; arguments are then marshaled as an antithesis; resolution is then effected in a new thesis by some weighing of the dual structure of ideas thus uncovered (illustrated by Socrates, or Marx).[2]

Mystical   A central principle, essentially metaphysical, is selected by the systematizer. (Illustrative of such principles have been: Occam's razor, the parsimony of assumption—explanation is best which is simplest; Einstein's philosophy—the universe is not capriciously organized and in fact should be describable with simple mathematical elegance; Bridgman's operational philosophy—physical entities are to be defined within the operational contexts in which they are measured. Bridgman's principle is touched on later in a section on metaphysics.

---

[1] Two structures would be viewed as isomorphic if they had a considerable similarity in form. By extension, the description of two systems may be viewed as isomorphic if both form and function of the systems were quite similar, for all their likely functional transformations. To assure isomorphic similarity may require much more experience with these systems than most of us possess.

[2] For the reader who wishes to take the study of this subject of general systems' science quite seriously, there is an excellent debate that has recently surfaced, which provides good study material for attempting to evaluate the validity of technical argument and at the same time to see the difficulties when the technical subject is the history of past events involving men. The subject is the origin of the cold war (2).

Pragmatic   The organized empirical facts stand in their own right as the system, so that their summarization is more descriptive and eclectic than principled.

Abductive   By questioning whether incomplete information is a case under one known rule or another, the observer makes executive commitment as a probabilistic decision. [Discussed by Aristotle, and by Peirce (3), it is logically fallible but useful. It is the common logical method used in medical diagnosis.][1]

The principal systems that will be used for development here are mystical (a central principle has been selected) and dialectic. As far as possible, the views of major protagonists, generally historical innovators, will ultimately be referenced for the central theme in each science. More modern sources, either favorable or unfavorable, but presented to bring the field into more modern view, will then be referenced, and an argument will be pursued to bring it within the central frame.

It must be recognized that the dialectic method is no more capable of discovering "truth" than any other method. The protagonists on one side or the other (or the ones selected by the author) may both be beside the point. As a result, the resolution may be just as far away from truth as if no argument at all had gone on. The methodology thus tends to be both an individual preference and one that seems to fit the particular field during the particular era. After Euclid, for example, the facts common to geometry would seldom again be explored by dialectic argument—until a non-Euclidean geometry was required.

As a physical scientist, the author has always sought theories that are isomorphic, i.e., their explanations and predictions are in one-to-one correspondence with the phenomena they purport to describe. However, in the case of a general science of systems this may be too ambitious. The central principle to be pursued is thus regarded tentatively as a formal guide, rather than an isomorphic guide. To propose a metaphor, it is a scaffolding which —hopefully—outlines the form ; that is, it is close to the real structure though it is not quite the structure itself. Yet Peirce's comments (3) that this may be misleading should be noted.

The key thesis is that at every hierarchical level of systems there are fluctuating, or oscillatory, active "atomistic" elements[2] which are capable of

---

[1] To these six, the temptation is great to add a seventh, seductive. Often used currently, as a freewheeling empiricism which combines all methods, quite ruthlessly, in an attempt to squeeze conclusions out of observations. The better scientific results tend to withstand the rough handling of more than one method.

[2] The minimal elemental system retaining function, atomistic elements are compact autonomous oscillators, auto-oscillators for short. In electrical engineering, active networks would be those which contain power sources, and ac active networks are those whose power sources appear periodically varying ("alternating current" in engineering terminology or variable flux sources, as compared with dc "direct current," or constant flux sources).

emitting and absorbing energy by discrete steps and capable of interacting among themselves; that, as the geometric field of such elements grows and the focus of attention—in time and space—becomes grosser, the ensemble of such elements permits description as a continuum field; that as the field continues to grow further in space and time, disordering relations that are implicit in the atomistic sublevel ultimately destroy the interacting ordering, and the large continuum field again forms, discretely, one or more superatoms (it winds up, as a vortex or helical structure, or fragments, as an elastically stressed crystallite lattice, forming a gross localized structure) in which now the material of the superatom is itself a portion of a continuum field; and that this process extends indefinitely (not necessarily without end, but not closed in the sense of knowing whether it ends) toward the small, and toward the large. This act of absorbing energy locally, by rotational momentum or momentum associated with internal degrees of freedom, as will be emphasized later, can keep the field structure of these nested levels compact.

The intent of this study is not to produce novelty, but unity. It proposes, by selection and arrangement and by bringing in the least possible amount of new "revelation," to show how much of a general science of systems may be ordered out of existing knowledge. On the other hand, to some it may seem to consist essentially of a reworking of a nineteenth-century doctrine, energetics, which had received strong support at certain times and was later dropped with the advent of atomic and quantum physics. One might judge, in reviewing history on a more extended time scale, that energetics, as a doctrine that always seemed to have a quasistatic cast, was superseded in each instance by a doctrine of science that was dynamic and dealt with quantized entities.[1] It is hoped by embedding energetics within a suitable nonlinear mathematical framework, structured both by a combination of statistical and deterministic mechanical ideas and Gibbs' ergodic theory, that a primitive twentieth-century general science of systems might be begun. (Henceforth to be referred to as general systems' science or GSS.)

For present purposes, a system may be viewed both as a field of physical phenomena in which a class of elements exhibits its functions or behavior in space and time, and as an abstract description which presumably may be

---

[1] A more literary statement of the theme of this study might very well be the aphorism, "Plus ça change, plus c'est la même chose!"

In energetics, attention is devoted to accounting for all of the energies and their transformations in a system. (In the present era, compartmental analysis, as used in biomedical engineering to analyze internal biological organ systems, is tending toward energetics. Compartmental analysis is a more rigorous accounting of biochemical pools than the biologist has become accustomed to think of.)

On the other hand, the alternate doctrine invokes the existence of active discrete particles of similar nature constantly in motion. It is those repeated entities that carry both the energy and structure of the system. The molecular biologist's description of the biological system conforms to this view.

isomorphic with the physical field. However, a meaningful systems' science which is tied to reality must always seek to be related to the physical field rather than to the abstract, formal, purely logical description. However, to provide an abstract foundation for GSS, two systems will be viewed as functionally isomorphic over a dynamic range of behavior if they have the same singularities of motion, in a stability sense, over that range.[1]

Before beginning the organized content of this study, a very rudimentary example of how a hierarchical structure of atom-continuum-atom is viewed will be given.

Consider a river flowing downstream as a constant current source. (The velocity field that makes up the river is derived from a gravitational potential and from ground waters.) Consider small particles carried by the stream. As a result of gravity, they fall vertically. Yet, upon their striking the bed of the stream, there is an interaction via hydrodynamic forces. Instead of coming to rest, the particles bounce up in a motion described as saltation (a nearly cycloidal motion). The interaction of many such particles with the stream's boundary layer creates an equilibrium distribution, described as a "law of atmospheres" for the bed load carried by the stream. The collection of these particles may now be characterized, by a continuum description of the bed load of the river, as a dynamic entity that will be subject to gross dynamic laws as the bed load interacts with elements in the macroscopic river-field (such as bends, changes in river shape or cross-sectional area, or changes in the character of the soil).

The purpose of offering this example is to suggest a richer texture of possible examples of systems, beyond the more commonly used example of a collection of molecules obeying the Maxwellian statistics of elementary molecular theory, that can give rise to a variety of equilibrium distributions.

The line of sciences that will be discussed in the book is covered in the table of contents.

Of course, a book on systems' science may not proceed indefinitely without a more specific definition of a viable system.[2] A system is not simply a network of one or more loops involving equipollent elements (e.g., a network involving the elements of resistance, capacitance, inductance, and power

---

[1] This view attempts to make systems topologically similar in phase space. What is involved may become clearer after the chapter on nonlinear mechanics.

[2] *System* A complex unity formed of many, often diverse, parts subject to a common plan or serving a common purpose; a set of units combined by nature or art to form an integral, organic, or organized whole, an orderly working totality; a group of bodies moving together in an inter-related pattern or under the influence of related forces or attractions. (Synonyms: scheme, network, complex, organism.)

Viable Having attained such form and development of organs (i.e., internal parts) as to be normally capable of living (i.e., undergoing complex transformations of energy) outside the uterus (i.e., outside of the "laboratory," "crucible," or environment, in which the system had its start-up).

sources as an electrical engineer might visualize),[1] although these may be regarded as *simple systems*. The real concern, with what may be viewed as *large systems*, is with systems that involve not just elements of one hierarchical level but complex couplings with imperfectly definable measures of equipollency among elements and imperfectly definable spatial links. (A new animal species suddenly injected into a complex ecology, which itself is only recently formed, is an extreme illustration. It is not clear a priori whether the new species will "take." If it does, then one more element of complexity has been added to the system.) The qualification of viability is to indicate that we will be concerned with systems that continue to undergo energy transformations, rather than static systems, typically those that are written on pieces of paper (e.g., a philosophic system, a system of aesthetics).

This document, in an attempt to lay a foundation for systems engineering, does not deal in any operational detail with large systems, but attempts to abstract the principles common to all systems' levels, by structuring many simpler systems, each at its hierarchical level.

The following outlines may help diagram the exposition.

The tools assumed

Metaphysics
Existence
Reality
Communication
  ("The man who says he has no metaphysics has a bad one."
                                            —EILHARD VON DOMARUS)

Physical Observation
Sensory
Extensional
  ("Observations are by the methods of differences. Elephants are sometimes
  present and sometimes absent; but the air is always present."
                                            —ALFRED NORTH WHITEHEAD)

Language
First—a signaling system (internal language)
Second—an exchange system
  symbols (names of things)
  relations
  functions

Mathematics
Common language
Logic
The fields of mathematics

---

[1] Equipollence may be a somewhat strange word. However, it is used to put forth the idea of "forces" or "reactions" of relatable strength (e.g., $\Delta V = L\ddot{q}$, $\Delta V = R\dot{q}$, $\Delta V = q/c$; or 1 wife = 20 cows = 60 pigs, etc.).

Philosophic Synthesis
Chains of causality

Physical Tools
Mechanics
Gravitation
Electromagnetism
Nonlinear mechanics
Statistical mechanics
Quantum mechanics

Although these tools are assumed to be available to the reader, they are discussed to the limited extent that appears to be necessary to carry on the search for a GSS.

Some philosophic introduction to the background of purely kinematic ideas needed to discuss so many diverse fields is in order. It is hoped that by alluding to quantization, the formation of orbits and of constellations, and the occurrence of discrete jumps from one orbit to another orbit both in the atomic field and in the atomistic clusters that appear in the social behavior of humans, that the possible common character of a science to cover both areas may finally reach the reader.

By "kinematic" ideas are meant those that describe systems' motions in space and time without recourse to the specific dynamic "forces" that may be responsible. They relate to motion in the abstract. By pursuing this idea in some depth, we hope to be better able to focus on the common phenomenological character. It is the development of trajectories of motion and of certain closed paths that may be regarded as orbits and the essential near-constancy of these orbital periodicities that we wish to stress. For a literary objective, we wish to reconcile the cyclicality of change that many see in history with the appearance in the mind of the individual of various new insights into the perceived field of phenomena that historically unrolls before him, and bring it within the same ideological structure of quantum mechanics —that is, of its quantization of the action of systems. Yet, as we will stress later, the action of both classes of systems is sufficiently complex than an emergent evolution arises in sufficiently long time scale.

SUMMARY

1. The major line of systems of concern to man can be described by a few unifying scientific principles.

2. Science consists of a description of phenomena by abstraction and condensation.

3. The tools of scientific method by which description is developed are matters of individual style. Induction, deduction, dialectics, mysticism, pragmatism, and abduction illustrate such tools.

4. The first principle of general systems science (GSS) is that at every hierarchical level at which phenomena are to be described there are fluctuating or oscillatory, active, atomistic elements

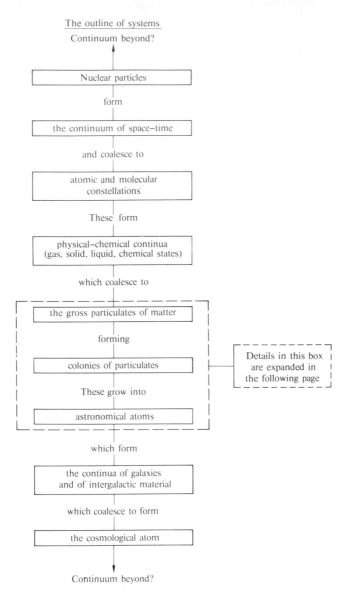

Fig. 1 The main line of general systems' science.

capable of emitting and absorbing energy by discrete steps and capable of interacting among themselves; that as the geometric field of such elements increases in scale and the focus of attention—in time and space—becomes grosser, the system of such elements may be described by continuum field equations; that as the field continues to grow further in space and time, disordering relations that are implicit in the atomistic sublevel ultimately destroy the interacting ordering, and the large continuum field again forms, discretely, one or more superatoms (it winds up, as in a vortex or helix, or elastically stores energy in a stressed crystallite so as to form a gross

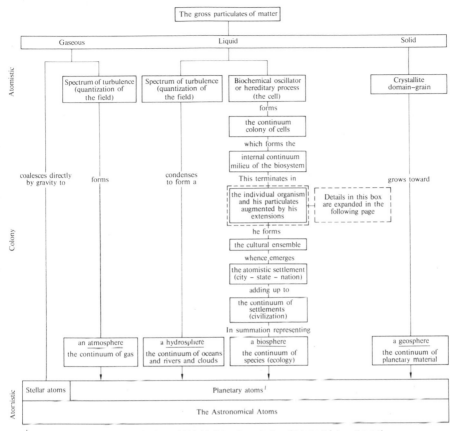

Fig. 2   The gross particulates of matter.

localized structure) in which now the material of the superatom is itself a portion of a continuum field; and that this process extends indefinitely (not necessarily without end, but not closed in the sense of knowing whether it ends) toward the small, and toward the large. This act of absorbing energy locally, by rotational momentum or vibrational momentum associated with internal degrees of freedom, can keep the structure of fields compact.[1]

5. The tools of science are metaphysics, observation, language, mathematics, philosophy, and physical science.

6. The main line of systems is nuclear-physical systems, atomic and molecular systems, gross physical-chemical states of matter, and gross material structures—of planets and stars, galaxies, and the cosmos.

[1] Is the coalescence into local systems random or determinate? We do not really know until or unless we have a detailed global view. Provisionally we would conjecture that the local coalescence is accidental, but we can't tell whether a similar coalescence might not have taken place if a somewhat changed collection of elements had come together at some other point in the field. The reader may rejudge the issue when he comes to the last chapters.

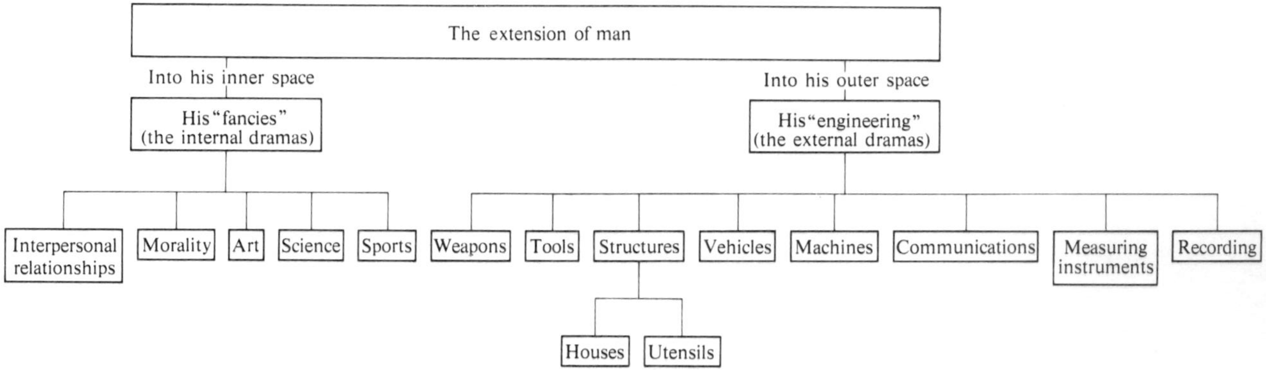

Fig. 3 The extension of man.

7. The gross physical-chemical states of matter coalesce into stars at one extreme, and—on planets—coalesce into the geosphere, hydrosphere, atmosphere, and biosphere.

8. The line of biochemical oscillators emerges from the liquid state and develops hierarchically into the protocell, the cell, the coordinated cellular colony, the individual complex organism, the cultural ensemble, the social settlement, the canvas of civilization, up to the total ecology.

9. From the complex biochemical individual—man—arise his extensions. Extended into the outer space, they represent his engineering and technological achievements; extended into the inner space, his creative dramas.

QUESTIONS

(Note to teachers or mentors: This book is perhaps a useful 1-year course for such Ph.D. candidates near the end of their formal school work as are interested in a broadened interdisciplinary view of science. These questions are intended to be, or to suggest, a yearlong project—perhaps carried on among a few students of different disciplines. A preliminary report could be handed in at the end of 6 months, for approval, and developed to final form for the next 6 months. A mature thesislike study would be expected.)

1. Illustrate in detail the difference between a formal scaffold for a particular branch of science and a fully isomorphic structure of explanation for some other branch of science. Trace this through its most definitive statements.

2. Pursue the structure, in terse form, of some great dialectic that has gone on. Typical ones to consider might be the cold war, the current dialectic between youth and society, or Marxian socialism versus capitalism. (Do not let the argument hang upon the observation that the dictatorship of the proletariat is not likely to emerge. The issue proposed is their relative success as socioeconomic systems.)

BIBLIOGRAPHY

1. Eiseley, L.: "Darwin's Century," Anchor Books, Doubleday & Company, Inc., Garden City, N.Y., 1961.
2. Alperovitz, G.: How Did the Cold War Begin?, *N.Y. Rev.*, p. 6, March 23, 1967; The Trump Card, *N.Y. Rev.*, p. 6, June 15, 1967.
   Gareau, F.: The Cold War 1947 to 1967: A Quantitative Study, *University of Denver Monograph Series in World Affairs*, vol. 6, no. 1, University of Denver, Denver, Colo., 1969.
   Halle, L.: "The Cold War as History," Harper & Row, Publishers, Incorporated, New York, 1967.
   Lasch, C.: The Cold War, Revisited and Re-visioned, *N.Y. Times*, January 14, 1968.
   Leo, J.: "Revisionist" Historians Blame U.S. for Cold War, *N.Y. Times*, September 24, 1967.
   Morgenthau, H.: Historical Justice and the Cold War, *N.Y. Rev.*, vol. 13, no. 1, p. 10, 1969.
   Schlesinger, A.: Origins of the Cold War, *Foreign Affairs*, vol. 46, p. 22, 1967.
   Starobin, J.: Origins of the Cold War: The Communist Dimension, *Foreign Affairs*, vol. 47, no. 4, p. 680, 1969.
3. Peirce, C.: "Selected Writings," Dover Publications, Inc., New York, 1958.

1

Dynamics of Historical Speculation and of Physical Systems

In viable systems we tend to be concerned with two apparently disparate kinematic processes; one, the emergence of new paths, new forms, new ideas from the past states, i.e., the appearance of new transients that take us from one state to another; two, the endless repetition of what went on before, i.e., the emergence of repetitive cycles of phenomena. How may one proceed to tie together the changing course of man's thoughts about himself and nature and the dynamics of nature with a common systematic science? It is with such questions that we intend to deal.

We will attempt to illustrate some dynamic changes in human thinking and shall suggest some physical ideas for their interpretation. Mach points out in *The Science of Mechanics* (1) that mechanics developed in an age of predominant theological cast (Chap. 4, Sec. 2). Theological questions were excited and modified by every line of thought. On the technical side, for example, Galileo became occupied with the strength of materials in nature. The forms of various natural structures—bones, feathers, stalks—made a profound impression on the thinking beholder apparently as it did, much later, to D'Arcy Thompson. Such complexity has been used over and over as evidence for design, or a designer, in nature. It requires considerable technical knowledge even to imitate or understand such complex structures (e.g., the design of a pinion feather), let alone invent them. However, investigation —not admiration—is the office of science. For the appearance of such new and technically complex structures, in the case of living systems, it was Darwin who later proposed an explanation by the theory of natural selection, which, even if not complete, was a first significant attempt to replace mere admiration of the process of adaptive development by serious inquiry into its mode of operation.

> [1] In an earlier study on technological forecasting (2), a time constant of 200 years was suggested for the emergence of ideas in society. It was proposed that 10 to 100 men, very loosely coupled to each other but coupled to the milieu of existing ideas, contribute to the emergence of new ideas as an adaptive process. A major question that shall concern us is the nature of the processes by which any well-established system, whether biological or inanimate, yet possessing both massive inertial characteristics and also statistical characteristics, changes in time.
>
> An ubiquitous impression in examining such systems is always of the existence of "noise" as a background to the system. Noise seems always to be present, leaking in at the boundary of systems. What is at issue is whether the conditions within the milieu are such that the noise is persistent and becomes organized rather than dissipated. If the noise can roll up or be entrained within any particular system into a local vortex or into a repetitive chain that will be propagated at least for some distance or for sufficient time, then it may cohere and develop into a new atomistic entity. The two kinds of propagation that can exist in systems are diffusive and wavelike. If the entity can lock itself up into a coherent structure because of the overall energy associated with the individuals involved, not because of the "force" of the concept or entity but because of its reinforcing phasing, then the strength of the packet—for example the summation of human thought in the human case—can develop a propagating organizational structure.[2] That group structure, representing or expounding the concept or entity, begins to be a minor "elite" element. What is characteristic of an elite entity or structure, as our working concept, is that it ties up more power density than neighboring field regions. The Hebrews, Christians, Muhammedans, Protestants, Nazis, Newtonians, relativists, quantum mechanists, and Black Panthers are all examples of such human groups.
>
> Since order seems to emerge from disorder, the organized tie-up of energy might be viewed, in physical terms, as a bootstrap process by which living systems counter the tide of entropic degradation. It is not really. The fact is that quantized absorption and emission, not always by radiation but generally by lower mechanical atomistic processes, provide exchange energy as a transfer coin of the realm in the physical milieu. "Kinetic" energy, or energy in transit, is the most elementary form. (For example, energy in the hydrodynamic field passes from potential energy

---

[1] In order to permit a dialectic and yet not confuse the reader, in this volume the author's comments will be slightly indented and in small print whenever extensive.

[2] An interesting reference for change in the human social system is Rogers (3).

of pressure to kinetic energy to energy entrained in the form of vortices before it finally degrades to thermal energy.) The lack of long-range stability in the large space-time domain of the entire society of biological or physical particles permits them to fracture into more restricted regimes. As a result, a pattern of large-scale "atoms" or oscillators arises, each of them made up of a portion of the social domain. Cutting across disciplines, whether drawn as physical illustrations or human illustrations, the basic problem of stability, of stabilizing new structures and functions, is thereby introduced.

The specific illustration that Mach illuminates [(1), p. 551] is really the 200-year transition from theology to physics, which he highlights with the names of such men as Galileo, Descartes, Fermat, Leibnitz, and Euler.

During the entire sixteenth and seventeenth centuries, down to the close of the eighteenth, the prevailing inclinations of inquirers was to find in all physical laws some particular disposition of the Creator.

Towards the close of the eighteenth century a remarkable change took place—change which was apparently an abrupt departure from the current trend of thought, but in reality was the logical outcome of the development indicated. After an attempt in a youthful work to found mechanics on Euler's principle of least action, Lagrange, in a subsequent treatment of the subject, declared his intention of utterly disregarding theological and metaphysical speculation, as in their nature precarious and foreign to science.

The idea that theology and physics are two distinct branches of knowledge thus took, from its first germination in Copernicus till its final promulgation by Lagrange, almost two centuries to attain clarity in the minds of investigators.[1] At the same time, it cannot be denied that this truth was always clear to the greatest minds, like Newton. Newton never, despite his profound religiosity, mingled theology with the questions of science (p. 552).

It is fascinating, but tangential, to pursue Mach's remarks further. To explain the transition he points out that the general views and tendencies of an age must not be judged by its greatest but by its average minds. In the pre-sixteenth century age, religion was the sole education. Techniques were primitive. Freedom and enlightenment were for very few in the Middle Ages, mostly poets and scientists, with very few connections to the views of the people at large. "Rationalism does not seem to have gained a broad theater of action till the literature of the eighteenth century. Humanistic,

---

[1] It is not inappropriate to note Mach's remark from Vailati, Atti della R. Acad. di Torino 33, 1898, "The rapid spread of firearms in the fourteenth century gave a distinct impulse to the study of the motion of projectiles and indirectly to that of mechanics generally." Thus another 200 years was the period from technological innovation to causal thinking in the theological terms of the day.

philosophical, historical and physical science here met and gave each other mutual encouragement."[1] (p. 553).

> These historical themes of the past are illustrative of the chain reaction, the formation of vortices and orbits, and local entrainment by angular momentum or within internal modes.

On the technical side, the marvels of such theological metaphysical principles as the principle of shortest time, least action, and other maximum or minimum principles gradually became understood as imprecise (Mach points out that light does not take the shortest path, but all paths, etc.) or as more representative of states reached because causes of all further alteration are removed.

> However, in accordance with the ideas of nonlinear mechanics, a foundation for which had already been laid down by Poincaré at Mach's time, we choose to stress the added state of a limit cycle capable of drawing from "potential" energy in the milieu to form a closed orbit. A limit cycle is the closed, isolated path in a phase space of velocity and displacement, which a dynamic system may approach from any starting motion within or without the closed path. For example, a watch has a limit cycle whether it is permitted to start from rest or from an accelerated shake.

By the eighteenth century, in French Encyclopedism, the view existed of a nearly complete system explaining the world in final physical and mechanical form. Laplace conceived a mind competent to foretell the progress of nature for all eternity, "if but the masses, their positions and initial velocities were given" (p. 558). Whereas Mach states he can sympathize with this expression of intellectual joy, 100 years later this has appeared to be a mechanical myth—as mythical as the animistic, older religious view. Yet biology, sensation, etc., are to be explained in physical terms. However, science no longer pretends to be a complete view of the world. The "modern" scientist tolerates an incomplete conception and view of the world and makes no investigation into things that are yet inaccessible. But when new areas open to research, no well-organized researcher will hesitate to consider inquiry.

Our thoughts today are still so close to these that one must simply draw the conclusion that we (1970s) are embedded in the same scientific milieu as Mach (1901). Thus, the change in many technical-philosophic ideas over 70 years is not so great. Another conclusion that one may draw is that the period of the eighteenth to the twentieth century has been the transition from the idea of individual certainty to the idea of statistical certainty but individual indeterminacy.

---

[1] On the other hand it would burden the nonsocial-scientist reader to examine Saint-Simon's *Historical Memoirs, 1691–1709* [L. Norton (ed.), vols. 1 and 2, McGraw-Hill, 1969] for the quality of the age and court of Louis XIV. Suffice it that he knows such documentation exists, if and when he becomes curious.

Yet one may note in one volume of a series that has nearly swept the field of rudimentary graduate training in physics, both here and abroad, Landau and Lifshitz' *Statistical Physics* (1958), "If we write down the equations of motion of a mechanical system (of which there are as many as there are degrees of freedom of the system) and integrate them, we can in principle completely determine the motion of the system. If, however, we are dealing with a classical system with a very large number of degrees of freedom, then the actual application of classical methods requires the statement and solution of an equally large number of differential equations which is usually impracticable."

Quite on the contrary, in our view of complex systems, each epoch develops its own limit-cycle pattern of behavior and thought, which although related to and drawn from past elements, no longer is the temporal projection of past velocities and directions.

> This brings the argument up to another part of the central thesis of the present study, which is very consciously based on a premise that fits twentieth-century thought. No system can be so adequately isolated that noise cannot creep in. Thus, a science of open systems—at any level—must include a description of both the noise and the systems' dynamics. It is not solely a matter of the "practicality" of the computation, nor simply the problem, as Landau and Lifshitz state, "of the time and paper it would consume."

A basic premise is that at every level there are noise sources: degrading friction at an ordinary macroscopic scale; Brownian motion at a molecular scale; the uncertainty principle at nuclear scale; or pair production, new star production and other cosmic radiation phenomena at the cosmological scale. Demonstrable theoretical prototypes are the Einstein theory of Brownian motion, the theory of Johnson-Nyquist noise, and (hopefully to be accepted in time) some work herein referenced on the solution of hydrodynamic equations for turbulence.

It is with these brief remarks that the stage is set for the transition from Mach's social history of mechanical ideas to Gibbs' statistical mechanics for all kinds of systems.

In learning that we propose to pursue a doctrine of "energetics," the modern physicist may tend to feel, "Oh, Lord, must we go over this again!" Yet our answer is, "Yes!" The physicist is faced with the following dilemma. After Newton had enunciated his laws of motion, mechanical motion seemed in hand and gravitational forces were encompassed. When the researches of the seventeenth, eighteenth, and nineteenth centuries were crowned by the summarizing electromagnetism of Maxwell, as well as his achievement of describing the statistical mechanics of the interaction of many bodies, it appeared that, except for the detail of the structure of mass in the small (its atomistic nature was also then clear) and the nature of the universe in the large, such fields as optics, radiation, mechanics, acoustics, the states of matter and electricity and magnetism were well accounted for and bounded, and various problems required only a great deal of detailing. (Certainly most departments of physics at universities are still convinced that there is very little unexplained content in classical physics.) There was a willingness, ambiguous perhaps, to do this on the basis either of continuum theory or a discrete theory of matter.

Then undergoing development and change from Planck to Bohr, the atom and quantum concept of matter and radiation grew until it appeared that the continuum simply and intrinsically had to be made up of quanta. This was followed by quantum theory, emerging finally in the form of a continuous-wave theory. Once more the ambiguity was pursued as a dual theory of matter—of atoms and waves—and would have been acceptable if only two fundamental particles had to be accounted for, the proton and electron, or even an added odd element, the neutron. However, Dirac reached in with topological ingenuity and pulled out pairs from a negative ocean to provide an expectation of the positron. Since this, with regularity, modern Big Science has plucked particles out in very much increased number. At the same time, the philosophic foundations of physics have become increasingly confusing.

Although these few sentences may have not dealt adequately with the physics of "fundamental" particles, we are quite serious with regard to their underlying logical position. The foundations of physics today are far from compelling, and there is much in scientific argument regarding its fundamentals that is not compelling.

If one were to seek a solid unshakable foundation for science, one would not find it in physics, or any other field. No one level of organization can provide the basis for a solid foundation. What we are proposing is that all one can depend on is the hierarchical levels of organization being there for interpretation, not on the details of logic of any field.

Yet, we wish to pursue a path that does not have the problem that a fault in the fundamental logic of any field will destroy the logic of a superior hierarchical level. Therefore, we prefer to pattern science on the formal structure of . . . atomistic level transforms into continuum level transforms into atomistic level . . .

$$\cdots A_m \to C_m \to A_{m+1} \to C_{m+1} \to A_{m+2} \to C_{m+2} \cdots$$

and consider $A_m \to C_m$ as meaning that A_m's existence is a sufficient phenomenological basis for C_m. Then C_m's existence is sufficient phenomenological basis for A_{m+1}, etc., and there is no particular worry at which level there is a breakdown in the fundamental logic of explanation during any historical epoch. (Atomic and shorter-lived particles exist. Can we get on with molecular physics?) With such a chicken-egg, egg-chicken scheme, it does not matter where one starts.

It is to provide such a thread of continuity that we have chosen the thought of an (not "the") old-fashioned "energetics" doctrine and an old-fashioned statistical-mechanics doctrine and have made a key role for the mathematical physicist—the trade practiced by the author.

For purposes of capturing human fancy, it is unfortunate to be caught in the role of the advocate of the antithesis (here to quantum mechanics), for it is the resolver—and the subsequent thesis-maker—who usually succeeds in capturing the maid in most romances. Nevertheless, the role is necessary to set the story straight, perhaps for others to resolve.

The nineteenth century fostered a doctrine, energetics, with such adherents as Mach, Helm, Ostwald, and Duhem. With the rise of quantum theory, in which Planck represents the bridge from old concepts to new concepts, the doctrine was discredited. An introduction to, not an extensive review of, the ideas of some of the adherents may be found in the following: Duhem (4) and (5); Ostwald (6); quantum theory (7) and (8).

With full recognition that the modern physicist may regard this as retrogressive, the marriage of an energetics doctrine that might satisfy Ostwald and Mach with a nonlinear mechanical structure that might please Poincaré and Gibbs by its assumption of an extensive nonlinear doctrine will be attempted. Its justification will be, not that it may be so completely useful for the foundations of modern physical theory, but that it may be able to cover so many fields.

An energetics doctrine, as viewed here, basically attempts to encompass any new field of phenomena by a "calorimetric" determination of the energy changes in the system. It counts the number of physical degrees of freedom in the system by the following test. A generalized displacement q_i* is a degree of freedom if there exists a conjugate variable p_i, a generalized momentum, such that from their cooperative action

$$\Delta E = p_i \, \Delta q_i$$

there results energy changes ΔE, which can be detected calorimetrically. By such means, these newly discovered degrees of freedom can be brought within a structure of generalized thermodynamics (9). (The problem which is indeterminate at the moment will be the structure of the "calorimeter" at different hierarchical levels. The laboratory calorimeter for example cannot be used for the measurement of the universe or the single nuclear particle.)

It is instructive to examine Ostwald's paper (6), given as a Faraday lecture in 1904. (It is certainly worth reading for its comfortable modern tone—which any system scientist or engineer would appreciate.) He points out that a question that always fascinated Faraday was the mutual

---

* The displacement is assumed susceptible to continuous transformation through a set of values, so that the change of displacement Δq_i between two values exists.

transformations of energy in various kinds. In every case, Faraday put the question: "How can I change a given force into another?" (which should delight any input-output thinker). However, this extended to the question of even attempting the once absurd notion of transmutation, and thus even such problems could be now given to the chemist for solution.

"Let none start at the difficult task and think the means far beyond him; everything may be gained by energy and perseverance."

> Lest any say that Faraday could not have meant this in a modern sense, even though the modern nuclear "chemist" can do what Faraday wanted, we must reply that this misses the thought that philosophically Faraday meant this in a very modern sense. He simply did not have the apparatus, verbal or experimental, to achieve it. It is this possibility of discovering ideas philosophically before their time that scientists who have specialized their own efforts ought to try to see. There is most often a philosophic tendency in history and man's thought to alternate between two poles—of *A* and *B*—of thesis and antithesis, of atomism and continuum, of idealism and realism, of holism and reductionism, etc. (e.g., the swinging pendulum of history). However, each age develops its own specific style of thinking, its paradigm, and the precursor in the past often cannot be effectively connected to the view of the future. It is not looseness of thought which is being sought by such statements but a philosophic richness of speculative material. What is dead is dead. However, even today we are still under the influence of a Faraday.

Following the examples or the spirit of the great discoveries of his master, Davy, Faraday says that methods of reduction even more progressive and advanced may exist to "put at some favorable moment the basis of the metals in our hands."

Ostwald attempted to follow that hint into the question of the nature of the elements and their compounds, and pointed out that Dumas, 35 years earlier, had delivered a Faraday lecture on precisely the same subject as his:

> Nevertheless, I do not shrink from repetition. Each generation of chemists must form its own views regarding this fundamental problem of science. The progress of science shows itself in the way in which this is done. [Faraday was fully influenced by Humphry Davy's brilliant discoveries and sought for the solution of the problem in Davy's way. For Dumas, the most important achievement of his day was the systematization of organic chemistry, condensed into the concept of homologous series.]
>
> It is well known that these ideas finally developed into the great generalization we owe to Newlands, Lothar Meyer and Mendeleeff. Although the problem of the decomposition of the elements was not solved in this way, these ideas proved to be most efficient factors in the general development of science.

[He then continues in a delightful vein.] From what store of ideas will a modern chemist derive the new materials for a new answer to the old question? A physicist will have a ready answer: he will construct the elements in a *mechanical* way, or, if he is of the most modern type, he will use *electricity* as timber. [Today of course the modern physicist has even newer synthetic materials for construction.] The chemist will look on these structures with due respect indeed, but with some reserve. Long experience has convinced chemists (or at least some of them) that every hypothesis taken from another science ultimately proves insufficient. They are adapted to express certain sides of his, the chemist's, facts, but on other not less important sides they fail, and the end is inadequacy. Learning by this experience, he makes a rule to use only chemical material for this work, and according to this rule I propose to proceed.

[Ostwald continues:] Hence, like Dumas, I put the question: what are the most important achievements of the chemistry of our day? I do not hesitate to answer: *chemical dynamics* or the theory of the progress of chemical reactions and the theory of chemical equilibrium. What answer can chemical dynamics give to the old question about the nature of the chemical elements?

The answer to this question sounds most remarkable; and to impress you with the importance I ascribe to this investigation, I will mention the result at once: *It is possible, to deduce from the principles of chemical dynamics, all the stoichiometrical laws; the law of constant proportions; the law of multiple proportions and the law of combining weights.* You all know that up to the present time it has only been possible to deduce these laws by help of the atomic hypothesis. Chemical dynamics has, therefore, made the atomic hypothesis unnecessary for this purpose and has put the theory of the stoichiometrical laws on more secure ground than that furnished by a mere hypothesis.[1]

I am quite aware that in making this assertion I am stepping on somewhat volcanic ground. I may be permitted to guess that among this audience there are only very few who would not at once answer that they are quite satisfied with the atoms as they are, and that they do not in the least want to change them for any other conception.

Briefly, we will provide just a few additional quotations from Ostwald.

And now let us go into the matter:
The first concept we start from is *equilibrium*. In its original meaning,

---

[1] It is not essential to accept this argument. It is unfortunate that Ostwald came to the conclusion that atoms do not exist and thereby confused the issue. Nevertheless it is important that each field introduce its own working ideas. Concordance at two or more levels then becomes a most useful bonus.

this word expresses the state of a balance when two loads are of the *same weight*. Later, the conception was transferred to forces of all kinds, and designates the state when the forces neutralize one another in such a way that *no motion* occurs. As the result of the so-called chemical forces not showing itself as a motion, the use of the word has to be extended still further to mean that *no variation* occurs in the properties of the system. In its most general sense, *equilibrium denotes a state independent of time*.

In summation, the beautiful points that Ostwald makes, in our opinion, are the possible historical repetition of a scientific idea; the questioning of the foundations of central key theses; the value of analogues, yet their ultimate limitation, except within the experimental content of their own field; and the value of self-consistency for any particular hierarchical level in science. Finally, Ostwald indicates—with beautiful clarity—the idea of force balance systems.

> With all of these, we agree and stress again in particular that the *A-C-A-C* structure is simply a formal device—that is, we consider it to be the best analogue for all the fields of science, as a logical tool—with some additional formal details. The detailed -*A*-, or -*C*- nature of each field must arise from that field's content.

It is with this that Ostwald begins his own detailed thesis, which we are not concerned with at this moment.[1] It is time to return to our own theses. Before furthering our arguments, it is worthwhile to bear in mind the developmental line—classical mechanics, the rest of classical physics, statistical mechanics, and statistical thermodynamics. (By "classical" mechanics, we mean the motion of systems of particles embedded in a real universe, either by the laws of Newton or Einstein.) That is, one assumes something about matter, energy, and a framework of space, and interrelation, and therefore of motion, and therefrom of time. Galilean transformations or Lorentz-Einstein transformations may be viewed as being selected for experimental reasons. Newtonian or Einsteinian mechanics may then be accepted as a general dynamics.

> The following whimsy may be suggestive. The author, during graduate studies (in a seminar on tensor analysis), gave a number of lectures on the discovery of the laws of physics by two little gnomes, one named "Newt," the other "Einst." Whilst tumbling through space, they could achieve a comfort state by using their stomachs as accelerometers and by whirling their arms and legs as

---

[1] We have learnt that if certain results, typically a phenomenological result at a particular hierarchical level and an analytic result from a lower hierarchical level, both agree, the confidence in going forward is much greater. The validity of transport coefficients postulated phenomenologically and derived from molecular theory is a case in point. Thus Ostwald's thesis, if it could be independently derived, would really become added support for the atomic hypothesis.

rotational momentum paddles. They could measure distance with their rolled-up arms and legs. By assuming classical ballet positions they could form their own local coordinate systems to measure events in space and in time using their hearts as clocks. (These metro-gnomes measured inches with their feet and time by their heartbeats and used their own internal cosmic juices as reaction motor elements.) Hence, they were now ready to observe the passing scene and discover the tensor laws of physics. It is with such a useful construct, or with other "thought" experiments, that one may start to construct general dynamics. The advantage of this construct is that in it one can ponder the scale of these gnomes and any "quanta" of action at the level at which they are discovered.

Thus, for example, we can accept half of Ostwald's energetics concept—and reject the other half which he well states in the following:

In the language of modern science I express these ideas by stating: *what we call matter is only a complex of energies which we find together in the same place.* We are still perfectly free, if we like, to suppose either that the energy fills the space homogeneously or in a periodic or grained way; the latter assumption would be a substitute for the atomic hypothesis. The decision between these possibilities is a purely experimental question. Evidently there exists a great number of facts—and I count the chemical facts among them—which can be completely described by a homogeneous or non-periodic distribution of energy in space. Whether there exists facts which cannot be described without the periodic assumption, I dare not decide for want of knowledge. Only I am bound to say that I know of none!

We reject the latter ideas as being incomplete. In our view, at each increasing scale of system, we find first a quantized graininess in space and time, then an apparent homogeneity independent of space and time but now with transmission and degradative processes, alternating from level to level. Beside such "permanent" process levels, we find identifiable transient phenomena at each level. It is by such a complex that we propose to try to make order out of nature. Furthermore, the full significance of inhomogeneity will not become clear until the end of the book.

SUMMARY

1. We are first concerned with the kinematic ideas that are common to all systems, physical, biological, and sociological. Kinematic descriptions deal with the motion of systems in space and time independent of the dynamic force structure which may be responsible.

2. We find two types of motion: the emergence of new paths of motion from the past states of the system, a "progressive" motion; and the indefinite repetition of what went on before, a "cyclic" motion.

3. For some perspective, in human societies the time constant for the emergence of new ideas is estimated to be about 200 years.

4. In systems, two classes of modes of propagation are found—diffusive and wavelike. "Noise" or disorganized input always exists as perturbations in the background of a system. If these can be entrained in a system, they can be formed into a coherent atomistic element that can persist in the system field. The precipitating entity in the field around which the larger atomistic element formed is an "elite" structure. An "elite" is an entity that can tie up more than its share of energy of the field.

5. Within the scope of an energetics doctrine, in which there are local degrees of freedom through which energy is transformed, there are two classes of structures that can be visualized—atomistic elements which are active, and collections of atomistic elements which are dense enough to act as a continuum. As a result of instability in the continuum, it breaks up into superatoms. Superatoms cluster in density to form continua on a sufficiently extended scale, etc. This makes up the hierarchical structure of systems. The atomistic elements are nonlinearly quantized from their continuum fields. They interact, in statistical mechanical fashion, so as to flood the phase space (of displacements and momenta) that is available to them.

QUESTIONS

1. Trace the emergence of some major idea in human thought, from the time or in the form that it might have been first enunciated, through the individuals who have contributed to its enunciation, up till the time it was fully enunciated by some master who put it in its known form (e.g., evolution). Note whether the nominal 200-year time constant proposed has any significance.

2. Discuss the transformations of some complex system (e.g., society in a particular epoch) from one historical state to another, as noise crept in through the borders in space and time and was responsible for a change of state.

3. Discuss the characteristics of elite structures in a number of systems from different hierarchical levels and note the different ways in which they help key their systems.

4. Examine Ostwald's claim that chemical dynamics may make the atomic hypothesis unnecessary.

BIBLIOGRAPHY

1. Mach, E.: "The Science of Mechanics," 5th English ed., The Open Court Publishing Company, LaSalle, Ill., 1942.
2. Iberall, A.: "Technical Forecasting, A Physicist's View," for Army Res. Off., AD #635–811, July, 1966.
3. Rogers, E.: "The Diffusion of Innovations," The Free Press, New York, 1962.
4. Miller, D.: Pierre Duhem, *Phys. Today*, vol. 19, p. 47, 1966.
5. Duhem, P.: "The Aim and Structure of Physical Theory," Princeton University Press, Princeton, N.J., 1959.
6. Ostwald, W.: Elements and Compounds, *J. Chem. Soc.* (*London*), vol. 85, p. 506, 1904.
7. Klein, M.: Thermodynamics and Quanta in Planck's Work, *Phys. Today*, vol. 19, p. 23, 1966.
8. Cline, B.: "Questioners, Physicists and the Quantum Theory," Crowell Publishing Company, New York, 1967.
9. Zemansky, M.: "Heat and Thermodynamics," McGraw-Hill Book Company, New York, 1957.

2

Some Preliminary Comments on Nonlinear Mechanics

It will not be possible to do very much more in nonlinear mechanics than to outline where we expect that it may lead.

A book that provides an extensive background into the working methodology of Newtonian mechanics is Whittaker (1). Its availability as a background will be assumed. Further, the content and background preparatory to Minorsky's book (2) on nonlinear oscillations will be assumed. It will also be assumed that the reader has some familiarity, real or potential, with the task of solving a nonlinear mechanical problem according to the following protocol:

1. Develop a set of differential equations that encompasses the phenomena to be found in a number of chains or loops of some physical system.
2. Set forth a statement of boundary conditions in space or time which the equation set must satisfy.
3. Investigate the singular states of motion of the equation set. When these

can be explicitly determined in a first solution step, the problem is infinitely easier than requiring their explicit determination, only after the dynamic states are found, by recursion.

4. Attempt to construct solutions that are relevant to each of the singular states or, more selectively, that are pertinent to the boundary conditions.
5. Satisfy the boundary conditions.
6. Seek to determine sufficient conditions, out of temporal and secular equations, that the solution is determinate in some sense.

Different from a linear equation set in which boundary conditions can clearly be set forth by which the solution is temporally and spatially unique, in a nonlinear equation set one may find that characteristic solutions are limit cycles whose specific performance "ultimately" (and practically, this often means only after a few cycles) do not depend on specific starting conditions. Thus—minimally—phase conditions are stochastic. In addition, in real interacting systems coherence is lost ultimately in time, except possibly in the sense of time-independent correlations. Even the amplitude conditions may be stochastic. Current mathematics may not be fully able to describe all of the phenomena that are observed. It can, in part, by involving time-dependent coefficients.

Broadly, nonlinear motion may be considered from the point of view of stability. A very brief outline [abstracted from Andronow and Chaikin (3)] follows:

1. The motion of a system may be considered, typically, to be a phase plane of displacement versus rate of change of displacement (velocity). More generally, the motion must be considered in a hyperspace of $2n$ dimensions corresponding to the n degrees of freedom of the system. We will discuss a 1-degree-of-freedom system and thus observe some necessary requirements for characterizing systems.

2. If we concern ourselves with autonomous systems,[1] then we can consider the singular points, those points at which all time derivatives except the lowest-order "displacement" variables vanish.[2] These singular points correspond to states of equilibrium of such systems. The question arises as to the kinds of equilibrium that can exist, and the kind of motions that can exist in the vicinity of these states.

3. A state of equilibrium is stable, in the Liapounoff sense, with regard to a singular equilibrium point if for each arbitrarily small neighboring

[1] Systems in whose description time does not appear as an explicit parameter; roughly, neither its structural nor functional parameters change with time, nor is it driven by time-dependent functions. Its behavior plays out by virtue of an initial start and interaction with boundaries.

[2] For example, in a hydrodynamic velocity field, the lowest order of "displacement" is the velocity. One typical singular state is the rest state at velocity equals zero.

region in a nested diminishing sequence of geometric regions in phase space around the singular point there exists a smaller region in which, if the system is started, it will always stay within the first region.

4. We wish now to classify the possible types of equilibrium, stable and unstable.

   a. System motion which consists of a nested family of closed shells, which can be of arbitrarily small size, may be referred to as motion around a *center*. This represents a stable state of equilibrium.

   b. System motion which forms a family of spirals concentric with and approaching arbitrarily close to the singular point may be referred to as motion relative to a *focus*. If all members of the family spiral down toward the focus, it is a stable focus. In an unstable focus, the system spirals away from the focus. The stability with regard to a focus is "stronger" than with regard to a center. Its stability is said to fulfill a condition stronger than stability in the Liapounoff sense. It is called asymptotic stability.

   c. If system paths exist which converge toward an equilibrium point, the point is known as a *node*. Convergence toward the point represents a stable node; away from the point, an unstable node.

   d. System paths which approach an equilibrium point only through an asymptotic family of curves (such as hyperboles) which never pass through the point form a *saddle-point* equilibrium. This is unstable.

5. Globally, motion in the whole phase plane may then exhibit some additional topological characteristics. Typically, there is concern with the stability of trajectories or of closed orbits.

   a. A *separatrix* is a path tending to a singular point—generally the asymptotes of a saddle point—so that the neighboring trajectories on both sides do not converge toward the point. These separate the space into regions of trajectories.

   b. There may be closed curves which are limiting sets for trajectories in the vicinity of the curve, as time approaches plus or minus infinity. These are known as limit cycles. The limit cycles represent closed paths independent of the initial starting conditions. If the trajectories spiral toward the limit cycle from both inside and out, the limit cycle is orbitally stable. If trajectories on one side wind toward and on the other side away, the system is orbitally semistable. If the spiraling on both sides diverges from the closed curve, the system motion is an orbitally unstable limit cycle. If the neighboring paths remain separate (as in concentric circles), the system has neutral stability and does not possess a limit cycle.

   As Minorsky states about the limit cycles of Poincaré, "Limit cycles, and in particular stable limit cycles, are fundamental in the theory of oscillations of nonlinear, nonconservative systems—the only kinds of systems in which they can arise. A stable limit cycle represents a stable stationary oscillation of a physical system in the same way that a stable singular point represents a stable equilibrium."

In GSS, we are concerned with the spectrum of sustained oscillations in nonlinear physical systems that are embedded in their open "isothermal" (more generally, "constant potential") environment. Thus we are fundamentally concerned with the stable near-stationary oscillations of such systems. These systems are lossy. Therefore, we are fundamentally concerned with limit-cycle performance at all times and at all hierarchical levels. The problem of conservative systems is an idealization that concerns us very little.

Stability, to us, will refer either to the static and linear stability represented by asymptotic stability in which the motion ultimately spirals down to a focus or to the dynamic and nonlinear open system performance in which the motion ultimately forms limit cycles. A third kind of motion, "open" trajectories that depart from a region of interest, will be considered to be unstable. (They may be entrained at some more remote region, which is of no concern to us in the specific instance.)

6. We may touch upon a few other pertinent concepts of nonlinear mechanics (2, 3).

a. The index of Poincaré. A necessary condition for the existence of closed trajectories, such as limit cycles, is desirable. Consider a simple smooth closed curve that does not pass through any singular points. Now consider the vector direction of all system trajectories passing through the curve. As the curve is traversed, say counter-clockwise, its tangent will rotate 2π radians per revolution. Please note the rotation of the vector direction of system trajectories that intersect the curve. The vector direction may rotate n times 2π radians. The number n is called the index of the closed curve. It depends, not on the curve, but on the singular points within the curve and is thus the index of the singular point or points enclosed by the curve.

The Poincaré indices of a center, a node, and a focus are all $+1$ [see 4a., b., c. for definitions]. The index of a saddle point is -1. With no enclosed singular point, the index is 0.

b. A closed trajectory surrounds at least one singular point.

c. There is no known general method for locating limit cycles. Some negative conditions are that limit cycles do not exist if there are no singular points; if there is only one singular point of index not equal to $+1$; if there are several singular points but no collection of them has a sum of indices equal to $+1$; if there are only simple[1] singular points but all those forming collections whose index sum is $+1$ are limit points of paths going to infinity.

7. A sufficient condition, limited by determining the existence of a certain domain, is given by the Poincaré-Bendixson theorem. It states that if the trajectory from some time on remains in a finite domain without approaching any singularities, then the trajectory is either a closed trajectory or approaches one. Typically, the domain may be sought as a ring between two closed curves that is free of singularities.

It is outside our scope to pursue the matter further. The reader is referred to (2–4) for a start to any deeper study.

Thus with a considerably lesser degree of determinism than for linear mechanics (the small-amplitude theory of mechanics near stable equilibria is considered to be a linear theory), the methods of nonlinear mechanics can be made use of in attempts to furnish a description of dynamic systems phenomena.

Intimations of difficulty can be found in the first problem in mechanics after a few elementary examples. If the mechanics of a simple linear mass-spring are considered and a more complex force such as a central inverse square force between two bodies (Newton's central problem for which he "invented" mechanics) is then considered, what happens with three attracting bodies? The general three-body problem with central force is already beyond solution in closed algebraic form. It may be traced in Whittaker (1) or Dziobeck (5).

In the three-body problem, there are nine degrees of freedom, the positions of the centers of gravity of the three bodies. There are 9 second-order differential equations or 18 first-order equations in coordinates and momenta. By elimination, it is said that the set can be replaced by a differential equation of the eighteenth order for any one of the variables. In principle, nine initial positions and velocities or momenta should fix the motion. Six integrals represent the conditions for conservation of motion for the center of gravity. These can be eliminated by considering the origin as a coordinate system, relatively at rest, at the center of gravity.

There are three areal integrals. They express the fact that the mass-weighted moments of velocity about each axis are respectively constant. (These moments of velocity represent the instantaneous rate of change of the projections of area swept out by the radius vector for each particle. These integrals express the conservation of the rotational momentum.)

---

[1] Consider a singular point in phase space x,y such that the singularity is located at the origin $x = y = 0$ (i.e., the coordinate system has been translated to the singular point). A simple singularity is one in which the singularity $\dot{x} = 0$, $\dot{y} = 0$ is approached as $\dot{x} = ax + by$, $\dot{y} = cx + dy$, where $ad = bc$. Namely it is approached as two intersecting lines.

Finally, there is the energy integral. This expresses the constancy of the sum of kinetic and potential energy. Since potential energy would become infinite upon a collision, there is an extra assumption concerning the system that no collisions occur during the time of observation.

These 10 integrals are the only ones discovered in general, and in fact, it has been shown that no other algebraic integrals exist. (There are two others known, associated with the conserved energy and angular momentum when treated as ignorable coordinates, but they are not algebraic.) Beyond this, as Dziobeck states, "Systems of differential equations are generally not submissive to integration; their very generality gives them an obscurity which is not likely to yield to any other than a complete general treatment." Thus the third and largest section of his book is concerned with a theory of perturbations.

However, what can be said about the stability of motion of these planetary systems? Consider this, even for the simplest interpretation of meaning no collisions. Dziobeck suggests that for the earth's planetary system—the sun plus lighter planets, small eccentricity, small inclination to an invariable plane—the solution arrived at by perturbation theory is stable, and there is no chance for collision. More generally, Whittaker (Sec. 189) refers to the work of Bisconcini (1905) for two relations that must be satisfied among the initial values of the variables in order that a collision may ultimately happen in the three-body problem.

Thus it might appear that "stable" orbits are conceivable (in the sense of being indefinitely persistent) for many orbital configurations involving three or more bodies. However, as these books indicate, the information is marginal. It is not possible to assure that the perturbation parameters cannot ultimately get out of hand.

In particular, consider the following case. Suppose a system with three or more particles had a persistent orbital motion. Suppose that two or more of the bodies have identical masses. The system is indeterminate in the sense, at least, that if the two particles could exchange their states the same solution would hold. These systems are considered locally determinate because, generally, there are energy barriers that separate the states. Classically, in a mechanical sense, there is no way for such exchange to take place except during collision. Thus in the case of n-body systems involving many similar particles, the most that results is a considerable degree of degeneracy of states that are hardly to be distinguished from each other. Nevertheless, as long as only this class of force system is considered, say unidirectional attraction toward a center, systems that are essentially persistent seem to emerge—albeit with an unclassifiable variety of open and closed trajectories. (While the planetary systems without losses are conservative and thus not capable of exhibiting limit cycles, we really are discussing systems which have some loss and some mechanisms for restoring these losses.)

However, the case that is more interesting for us is one in which there is coupling between two force systems so that additional energy exchange is possible. An example of such a coupling would contain one mechanical system (approximately as described) and one electromagnetic system (involving attraction and repulsion and field radiation). Further—most often neglected in theoretical discussion—there is complex interaction with bounding walls.[1] This is the case of the atomic particles. The nominal properties of the coupled mechanical and electromagnetic field were explored near the turn of the century. [See, for example, Abraham and Becker (6).] Experimentally, it is clear that a new kind of quantization emerges, the quantization of atomic orbits. This has been the subject of study by quantum mechanics. While not demonstrated by any known mathematical theory, it is not inconceivable that a nonlinear state might be imagined in which the e-m field radiation due to the mechanical acceleration of the orbiting electron was prevented for an epoch—say by interaction with the walls—then followed by mechanically discontinuous jumps (of the type proposed before, for which no mechanism existed) with a brief epoch of continuous radiation, etc. By some mechanism of this sort, a "classical" quantization might be conceived of. The difficulties, at this time, are the very careful sifting of coupled mechanical electromagnetic problems and the need to reconsider possible changes in the electromagnetic equations to bridge any remaining gap.

This is proposed conceptually as a substitute for (or hopefully an equivalent to) quantum mechanics. At this point, we cannot seriously defend this view as the solution for quantization. We are entitled to propose it as an idea of some plausibility to those concerned with some of the logic of science, or who could perhaps find grounds to reject it on direct mathematical grounds. However, we propose it as a different, plausible view on how "exchange" forces—which are generally considered to arise only from quantum theory —may come into existence. In fact, if quantum mechanics is viewed as a probabilistic theory governing the effects of interaction, rather than determining the path of motion dynamically, then some such nonlinear quantization perhaps may be considered as a description of such exchanges.

Such thoughts provide an appropriate transition to current controversy regarding the foundations of quantum mechanics. We can propose, for a brief

---

[1] One, speculatively, should perhaps recognize a mystique of the "wall," i.e., whatever bounds systems. Most often it is a source of loss. Typically, walls have a nonzero coefficient of restitution for mechanical bounces, a finite conductivity for electrical systems, etc. Yet the problem that self-sustaining systems face is how to get a "reflecting" system going within the cavity so that they persist without dying down. How they can feed from the mean-time-independent potentials that are available at the wall to maintain their time-dependent course of periodic variation is a nonlinear problem. It is much more transparent how to invoke wall considerations in limited boundary-value problems which involve field material of appreciable density within the enclosure than in very sparse systems with ill-defined walls (such as might be encountered in sociology or cosmology).

line of reading, (7) and (8) for introductory history to quantum mechanics and (9) and (10) for some of the current perplexity regarding the subject. Without overstressing its merit, we are in sympathy with Bohm's conclusion (9) that, "The current formulation of quantum mechanics must be regarded merely as a statistical algorithm which provides no conceptual structure in terms of which the movement of individual systems can be understood." The current foundation of quantum mechanics continues to be unconvincing to us in that it still appears to be an ad hoc prescription added to a classical mechanics for processes which are intrinsically electromagnetic in nature.[1] Each solution, whether Planck's, Bohr's, or Heisenberg and Schrödinger's, still seems to resemble a formal scheme to avoid tackling a very difficult mathematical quantization problem that exists in the particular coupled fields (namely, that in some way succeeds in characterizing the "eigenvalues" of a field that should represent nonlinear limit cycles).

It is quite appropriate to examine Heisenberg's more recent writings (10) in which he emphasizes the possible common insight to be gained from a particular line of physical problems that he is familiar with. We are quite thrilled because here is a master who refers to the same line with which we have been involved. "From simple problems to more complicated problems . . . the simplest problem can be considered to be the solution of a homogeneous linear equation; next simplest is the solution of an inhomogeneous linear equation; and finally, much more complicated is the solution of a nonlinear operation. . . ." He mentions the conceptual line from point mechanics to linearization to stability; from orbital stability; to field stability.[2] "One doesn't know how the solutions will behave after a very long time; I think that this may be a general feature of nonlinear problems." He thus proposes that one should consider ensembles of solutions, which leads to the statistical character of solutions. He then brings up the mechanics of continuous media, of hydrodynamic stability, as exhibiting the same range of nonlinear difficulties. Finally, he returns to the quantum theory—of atoms.

We have put forth some of our own preliminary thinking about the quantization problem in (11), pages 84–91.

We tried to show for complex systems and are now extending:

1. That the mechanics of every system of atomistic or "point" particles (the atoms are conceived as active, so that some lower energetic exchange structure is implied) involving center of gravity motion according to

---

[1] Quantum theory has also been useful, to some extent, for forces beyond e-m. It has been applied to the nucleus and to nuclear forces. It is definitely not satisfactory at present for gravitational forces.

[2] Citing the three-body problem, he states that the planetary motion can be followed by perturbational methods for a considerable length of time, for planets thousands of years. Yet when time increases, the perturbations may no longer be small, and ultimately it is not possible to state whether a planet will remain orbitally periodic or become aperiodic.

Newtonian or Einsteinian mechanics of the entire field of particles or of any local collection, may be sufficiently dense in space to treat asymptotically by differential equations in equivalent field form (a background of problems in which it was possible to proceed interchangeably from finite difference equations to continuous differential equations with acceptable error helped develop this point of view);

2. That other equations, describing the motion of the particles with regard to the center of gravity (cg) motion, relate to energy absorption into internal modes;

3. That in "long" time there are processes, governed by a hierarchy of "higher" frequency relaxations, by which the system jumps from one "stable" nonlinear state to others. The "stability" epochs between jumps mark domains in which the system's solutions may be described by convergent perturbations.

We will attempt to develop the character of this description in various systems. Suffice it to say that at present we will leave various of the elements poorly defined. Yet the theoretical structure we have proposed seems to us to deal with Bohm's and Heisenberg's requirements. To some extent, we can demonstrate the process in hydrodynamics. We can show that there exist solutions of the continuum equations of hydrodynamics which for some epoch should represent "stable" solutions for turbulence. We believe that these solutions represent—piecewise in time—convergent perturbations that are drawn from an ensemble of possible similar solutions that have different phasings and perhaps different symmetries, so that in some undefined sense they act stochasticlike without any certain or prior estimate of the statistics, but nevertheless they are probably quite close to "true" descriptions. We believe, from experimental data, that epochal breakdowns exist, but that they are not catastrophic (that is, they represent the flashes or intermittent nature of turbulence as was already well conceived by Stokes, Reynolds, and Kelvin). Nevertheless, the intermittent fluctuations continue to upset stable nonlinear solutions so as to produce near-stochastic but bounded performance. We believe that it is conceivable that the character of these epochal breakdowns may be made more explicable by embedding the problem in a field with a higher order of dimensions (specifically this could require coupling in elastic properties of the wall, dealing with thermal noise sources, etc.), but we do not have any demonstrable theory [although we have recently conceived of a relevant hypothesis (12)]. Alternately, the difficulty may lie in attempting to represent the higher-ordered field spaces of hydrodynamics within the confines of the geodesics of a three-dimensional geometric space. This is meant to imply that the perturbation solutions may be stable around the mean "streamlines" of the field but the intrinsic geometry of the streamlines may not be satisfactorily conformal with the geometric field. Less cryptically, this means we might view the space as being topologically open for epochs and then having its connection cut off by vortices from other epochs.

Thus, in our opinion, what we think we can bring to the nonlinear mechanics of systems is the following rough schema:

1. Attempt as full a classical description as possible of the field equations of the system (by fullness, we do not intend to stress or neglect its mathematical completeness, but we are more concerned with the completeness of the kinds of phenomena that can "phenomenologically" interact and be coupled at the time and space level under consideration);
2. Attempt to solve, or at least discuss, the solutions of such a set, to determine by perturbation methods pertinent solutions in various regions of stability;
3. By imagining the system operating in various boundary-value environments, determine whether there exists a larger scale for the phenomena which determine the salient epochs of the system.

While step 2 can represent, in our opinion, the hidden-variable description desired by Bohm as a substitute for a quantum mechanical theory, we have no complete theory for step 3, only suggestions. Yet, in step 3, we are concerned with such problems as the life of a system, the nonradiative atomic orbits between emission or absorption, the span of attentiveness of the human in his discrete "moments" of behavior, the shifting patterns of a turbulent field, and the "history" of an association of mankind. Although as yet unsatisfactory, this represents an improvement over our past discussion about flexible cycles (11), which cited Yule (1927) or Jeffries (1940),[1] and, more recently, contributions to econometric theory by Wald, on what sort of causality keeps changing the periods of a limit cycle, particularly in distributed fields. Yule's example was that of a group of boys with a small number of peashooters, shooting at a pendulum to spoil phasing. We now begin to see the emergence of epochs as more epigenetically bound to the system. In orbital mechanical terms, it is as if the perturbation parameters began to increase so considerably—in size—as to affect the timing and in fact even the stability.

The first difficult physical problem that must be faced is the coupled mechanical-electromagnetic (m/e-m) field. The traditional path has been to attempt to derive thermodynamics from classical mechanics (by statistical mechanics) and then correct the results to hold for quantum mechanics, too. However, a thermodynamic theory cannot be coupled to classical electromagnetic theory to produce a quantum theory of radiation. It was at this

[1] Jeffries was discussing the problem of the deterministic versus the stochastic nature of processes in the particular case of *The Variation of Latitude*.

point that the ad hoc nature of quantum assumptions was introduced (Planck, etc.). Thermodynamics is not really the resultant of the statistics of a mechanical system but of a coupled mechanical-electromagnetic (m/e-m) system. It is not clear whether nonlinear quantization of the coupled m/e-m atom can or cannot emerge solely from this basic problem of the coupled m/e-m field at the level more primitive than thermodynamics. It is clear that we are requesting a retracking of the Planck–Bohr–Schrödinger–Heisenberg–Pauli–Dirac foundations.

This is an unseemly demand for physicists not actively concerned with the foundation of theoretical physics, unless they have something quite positive to hang onto. What we have to hang onto, a reasonable share of it arising from original experimental discovery in system after system, are nonstationary orbits and cycles. These cannot be encompassed in satisfying linear theories. Thus, we made up our mind—the mystical step—that nonlinear mechanics was the more plausible foundation. We have wrestled with that thought for a long time, solving many nonlinear mechanical problems along the way, till we had one system in which we could be certain of the physics and the mathematical physics that posed the quantization problem and for which it was not possible to embed the quantization rules in any linear (theory of small vibration) mechanics. This was the field problem of hydrodynamics. We were reassured in our hope of being able to find a solution by the knowledge of G. I. Taylor's magnificent paper on spatial quantization of the flow field between rotating cylinders. The theory of turbulence is essentially one of the great problems of classical physics, and it really raises coupling problems as comparably complex as the m/e-m field. The availability of all the coupling terms makes it more accessible, perhaps deceptively so, to treatment by anyone with some technical competence. In fact, in time we were able to show how nonlinear fluctuating solutions may exist for the complete Navier-Stokes equations. This "missing link" probes at the power of demonstrating nonlinear solutions for complex field equations, and at the likely specialization of the solutions, in each boundary-value problem, by the boundary conditions under which quantization rules may tend to emerge. In the case of turbulent flow in a tube under a pressure gradient, we found the specialization related to propagative restrictions on elastic waves. This is quite controversially regarded since hydrodynamicists concluded that compressibility and elasticity had nothing to do with turbulence over a decade ago. Nevertheless, the mathematical demonstration is fairly straightforward, in a classical mathematical physical sense. For rotating cylinders, we are prepared to believe Taylor's mechanism for the spatial quantization; for thermal convective cells (Bénard cells), another mechanism, etc. We simply ask for detailed examination of the way we have built solutions for flow in the long tube or between parallel plates. Time will tell whether the methods will be accepted.

It would seem quite plausible to us, that if positional-momentum exchanges were possible between neighboring like particles—such as the electron—in a coupled m/e-m field, that the requirements for nonlinear stability itself might lead to the Pauli exclusion principle, by which the nonidentity of neighboring (i.e., within a particular space limitation) electrons can be assured. Thus, many of the requirements of a quantum theory are suggested by what we know about nonlinear mechanics. Why we are willing to parade our thoughts out now is that we now have enough background to show the common nonlinear features in a great number of fields and can point to enough expert criticism of the current quantum mechanics from points of view that do not damage a nonlinear theory. Whether this theory can be merely classical is, of course, moot.

Although this introduction would suggest that the detailed content of physics, a major tool of systems' science, could be introduced profitably, such detailed content would overburden us. For the interested reader, an elementary introduction enjoying current popularity is Feynman (13). A much more advanced level is Landau and Lifshitz (14). Instead we have proposed to limit ourselves to introductory remarks in the three divisions of physics that must concern us—nonlinear mechanics, quantum mechanics, and statistical mechanics. In this section we have introduced nonlinear mechanics.

SUMMARY

1. The nonlinear mechanics of systems deals with the singular states of their motion and the motion of the systems relative to these singular states. We imagine an equation set that describes the motion. Generally, we consider systems as autonomous, that is, only supplied with potential energy, or else driven by external cues.

2. Singular states are those at which the system might be viewed as instantaneously "resting"— all motional derivatives are imagined zero. The states presented by the lowest derivative (typically the zeroth derivative) are the singular states. Classically, these would have been viewed as stable, unstable, or neutral. Nonlinear mechanics classifies them somewhat more complexly.

3. Although there are degradative motions towards a rest state, explosive motions away from a rest state, and neutral motions around a rest state, a major concern in systems is with limit-cycle motions. These are states reached by a nonconservative system independent of initial conditions (within bounds), whether it starts from lesser or greater energy than in the limit-cycle state.

4. There are complex systems motions associated with conservative nonlinear systems. As yet, these are beyond the capability of modern mathematics to deal with in general (e.g., to classify the orbits of the three- or n-body problem). However, such systems are not capable of limit-cycle motions. Only nonconservative nonlinear systems can exhibit limit-cycle behavior.

5. Conservative systems can exhibit some specialized neutral orbits. We can make nearly conservative systems[1] exhibit more nearly distinguishable orbits by boxing the systems in. Most systems of interest to man (except the solar system) are boxed or play out on a substrate. The degree of certainty about orbits changes as one goes from the conservative unboxed system, to the boxed system with "perfectly" reflecting walls, to the boxed system with imperfectly reflecting walls. Limit cycles finally emerge when the system is nonconservative and conditions satisfied that supply energy at a phase that will continue the system through a cycle. Walls become potential contributors to these nonlinear processes in a number of ways.

6. The "mystical" characteristic (from an elementary point of view) that arises from such systems is a spatial or temporal quantization without the cued character of the conjugate (namely, spatial quantization without temporal cue or the opposite). Two means have been used to explain this character. One has been to embed the theory in nonlinear-nonconservative-system descriptions and seek limit cycles. The other, in the one exceptional case of nuclear particles (proton, neutron, electron, etc.), has been to seek a quantization by an ad hoc quantum theory. A potential

[1] Whereas a more general philosophic principle might like to consider planetary or stellar processes degradative, a solar system that persists unchanged for 5×10^9 years is difficult to view as highly nonconservative. Thus, its orbits are not strongly influenced by nonconservative theorems past start-up or prior to a terminal degradative phase.

conflict between these two ideas, in our opinion, is a major theoretical problem of this age. It cannot be settled by this book. However, among a great variety of possible resolutions, one is suggested here as a research for a classical quantization that may have been overlooked. Although the philosophic issues and physical issues are profound, they need not concern others than involved specialists.

QUESTIONS

1. Solve a nonlinear mechanical problem of some significance (e.g., a second-order mechanical system with losses coupled to at least a first-order system. There may be nonlinearities associated with almost any element. Make certain that you can isolate the singularities and identify the motions. Find a problem with at least one limit cycle. Some elementary but illustrative problems are: *a.* a rotameter connected to a high-vacuum source through a rubber tube, in which the flow is set by a valve at the vacuum source; *b.* a horizontal mass-spring dragging on a rough endless belt; *c.* a pressure relief valve tied to a large tank through an elastic hose to relieve overpressures in the tank. The relief valve has a valve of some mass, tied to a pressure-responsive surface, such as bellows or diaphragm, through a linkage which is tied to a dashpot for linear damping).

2. Discuss as far as you are able the kinds of motion exhibited by the bodies in the gravitational attraction of three bodies. First note how the motions for the two-body problem are easily and quickly classified.

3. Discuss, as far as you are able, the motion of the solar system in its nearly constant plane, with interacting losses due to plastic-elastic deformation or fluid atmospheric and tidal drags, from the time of formation 5–10 × 10^9 years ago to 80 × 10^9 years.

4. Discuss the problem of a series of small elastically colliding spheres in a cavity made up of vibrating elastic walls (both showing less than unity coefficient of restitution). In particular, classify the motions.

BIBLIOGRAPHY

1. Whittaker, E.: "Analytic Dynamics," Dover Publications, Inc., New York, 1945.
2. Minorsky, N.: "Nonlinear Oscillations," D. Van Nostrand Company, Inc., Princeton, N.J., 1962.
3. Andronow, A., and C. Chaikin: "Theory of Oscillations," Princeton University Press, Princeton, N.J., 1949.
4. Urabe, M.: "Nonlinear Autonomous Oscillations," Academic Press, Inc., New York, 1967.
5. Dziobeck, O.: "Mathematical Theories of Planetary Motions," Dover Publications, Inc., New York, 1962.
6. Abraham, M., and R. Becker: "The Classical Theory of Electricity and Magnetism," Blackie & Son, Ltd., Glasgow, 1932.
7. Klein, M.: Thermodynamics and Quanta in Planck's Work, *Phys. Today*, vol. 19, p. 23, 1966.
8. Cline, B.: "Questioners, Physicists and the Quantum Theory," Crowell Publishing Co., New York, 1967.
9. Bell, J., D. Bohm, and J. Bub: *Rev. Mod. Phys.*, vol. 38, pp. 447, 453, 470, 1966.
10. Heisenberg, W.: Nonlinear Problems in Physics, *Phys. Today*, vol. 20, p. 27, 1967.
11. Iberall, A.: "Advanced Technological Planning for Interdisciplinary Physical Research," for Army Res. Off., Final Report AD-467-051-L, June, 1965.
12. Iberall, A.: A Contribution to the Theory of Turbulent Flow Between Parallel Plates, *Seventh Symposium on Naval Hydrodynamics*, Rome, Italy, 1968 (in press).
13. Feynman, R., R. Leighton, and M. Sands: "The Feynman Lectures on Physics," vols. 1–3, Addison-Wesley Publishing Company, Inc., Reading, Mass., 1963–1965.
14. Landau, L., and E. Lifshitz: "A Course of Theoretical Physics," 9 vols., Addison-Wesley Publishing Company, Inc., Reading, Mass., 1958–1969 (vols. 4, 9 not yet out).

3

Some Preliminary Comments on Quantum Mechanics

We have touched on how a system of interacting particles leads up to the problem of the possible configurations of the motion, some stable in the sense of being indefinitely at rest (in a suitable coordinate system), some unstable in the sense of being indefinitely divergent, and some linearly unstable but nonlinearly stable in the sense that the system periodically or nearly periodically repeats its motion. We have seen the implication that the latter configurations may form the basis for quantized systems, in that as one element jumps from one possible equilibrium state to another, the system as a whole has to make jumps all compatible with some constraints. We now intend to move in the direction of the aggregation of the system toward continuum properties. The bridge towards such a description is statistical mechanics. However, we do not wish to be concerned with the vacuous problem of the statistical mechanics of a system of particles at rest but with a system composed of ac active atomistic elements,[1] a viable system. Thus, we must

[1] It should be noted that the ac active source for the system may arise from the particles themselves or from a bounding wall.

consider that the statistical mechanics of interest to us is based on some kind of quantum mechanical foundation.

Within a Newtonian framework, the transition to quantum mechanics is made as follows: One Newtonian theorem states that the action of a system of external forces on a system of particles results in the same motion at the center of gravity (cg) as that of a point mass at the cg of the system (whose mass is equal to the sum of the particle masses) acted upon by a parallel force system, plus some further motions with regard to the cg. The proof involves another Newtonian law that the internal forces (of interaction) cancel out in pairs as far as the cg motion is concerned. From the residual effects, it is possible, further, to extract the effect of external couples (force times normal distance to the cg) on the motion around the cg. The first theorem deals with changes in linear momentum of the system; the second theorem, with changes of angular momentum. Both changes are due to external forces. What remains now is the internally "unorganized" motion with respect to the cg.

Although the introduction to the three-body problem may not have fully explored the issue but only hinted at the fact, the number of states that may be assumed—particularly nonlinearly stable states—begins to grow tremendously as the number of interacting particles increases. However, it must be clear that this can only be pertinent if the interactions are lossy and the particles are themselves active (namely, possessing rechargeable internal energy which they can exchange) or if the walls of the enclosure are actively excited.

What is known from mechanics is that there exist integrals of isolated systems (we start our discussion with systems that are isolated from the rest of the universe in some sense, typically that other interactions may be weak or constant). Such an integral is the energy integral. It is postulated that such an energy integral was made of positive, definite forms involving displacements and momenta. A system whose energy was so retained was said to be conservative. The essential law of physics is that isolated systems will conserve energy. (For elementary particles the question of isolation becomes moot. In extensions beyond local Newtonian physics, as to cosmology, the concept of mass-energy conservation is more appropriate.) Thus, it is only possible for a system to retain configurational motion if it is fully conservative or if its losses to outside are compensated, for if there is energy loss (that is, by some means by which energy might be "radiated" away from the isolated system), it will lose such motion.

We are up to an ill-explored paradox in physics. Logically, a system may either be isolated or not. However, it is clear that gravitational effects penetrate every screen, and that while electrostatic fields may be attenuated by screens, it is impossible to prevent radiative equilibrium with any screen on both of its faces. The fact that unidirectional energy fluxes, i.e., the resistance to heat flux, can be severely attenuated by insulation does not gainsay this "fact."[1] Thus a so-called adiabatic envelope does not isolate a system. It only lags it for an epoch that may be designed to be moderately long.

Furthermore, as cosmic radiation and the possibility of pair-production indicate, it is not completely possible to shield against the penetration of material particles, nor is it possible to prevent the sudden appearance of material particles inside an enclosure if energy density becomes sufficiently high in very localized regions.

Thus the more plausible physical position is to view all systems as being open and nonisolated over periods of time sufficient to produce equilibrium, or isolated for relatively short periods of time, depending on the quality of insulation of the walls.

If we take an equilibrium, or at least quasi-equilibrium, point of view for the system of particles within the enclosure, then we can view the paradox in the following light.

First we assume that we are concerned with repetitive patterns of motion. We come to this conclusion because we have assumed an interest in a system of particles that remains within bounds—both as to their existence (they are not evanescent) and as to their physical localization. Then we also note that the particle interactions are of limited variation. The number of particles per unit volume is bounded and involves finite transit time or finite communications time between interactions. Thus these interactions are denumerable or, at most, infinitely denumerable. These conditions are sufficient to permit a mathematical description by harmonic functions, i.e., by a spectrum. Thus we can think about our systems as described by a prototype form for simple repetitive forms, typically as represented by the harmonic function.

The only exception to this status is a system of particles at rest. However, we are concerned with active systems. Thus, such stasis is excluded.

How can we achieve repetitive patterns using the harmonic description by way of illustration? One logically admissible process, and we fear the physical prototype that is most commonly assumed, is the linear description

$$E = \sum(\tfrac{1}{2}m\dot{x}_i^2 + \tfrac{1}{2}kx_i^2)$$

i.e., the linear mass-spring analogue, related to

$$m\ddot{x}_i + kx_i = 0$$

Now the basic reason that a system could not be isolated was because it could exchange energy with the walls, i.e., that there must be the possibility of gains or losses among particles and walls. Thus a second facet, the great tenet of classical physics, that as a result of natural processes ordered energy is lost, comes into view. The processes must be lossy. Thus, if one persists in using a linear form,

$$m\ddot{x}_i + c\dot{x}_i + kx_i = 0$$

is more appropriate. If in fact the losses are nonlinear—as for example, the extreme case of dry friction—the result is not that much different for small displacements. See, for example, den Hartog (1) for such a comparison of the response to linear damping and to dry-friction damping.

---

[1] The only way known for shielding against exchange is to compact the system within an enclosure to a sufficient density that relativistically the system is "out of this world."

However, another logically admissible way of getting repetitive patterns is from nonlinear systems. Furthermore such a procedure seems also to be logically complementary to the second law of thermodynamics. Of phenomena associated with nonlinear systems, the limit cycle is the only known sustained entity that emerges from the prescription of a lossy system.[1]

Of course, for there to be a sustained active system, the losses must be compensated for. Thus although the thermodynamic assumption requires that exchanges be lossy by "natural" processes, "artificial" processes (themselves often natural[2]) can make up the losses.

This brings us to the paradox. How can the nonlossy linear element, which the physicist uses so often for fundamental exposition, represent the sustained character of an active system. The physicist's answer seems to be, "Because it does. The ground state of an atom is not negotiable. At higher states these active systems are describable by a highly pure stationary spectroscopy, good enough to define time and space intervals to parts per billion and better." Further, he has buttressed the case by showing means for computing the spectrum as eigenvalues from suitable linear equations.

Yet in our opinion the logical weakness still remains. Periodic phenomena can only be associated with lossy processes when they are nonlinear limit cycles.

Is there any resolution of the paradox? Yes, we can think of one—which is easily shown in principle, and which we have also been able to show in the more complex physical field of hydrodynamics. Deferring discussion of the latter, we will illustrate the former.

Write the "linear" equation (illustrated for one degree of freedom) as

$$m\ddot{x} + c\dot{x} + kx = \epsilon$$

i.e., permit the loss $c\dot{x}$, whether linear or not, so as to achieve consistency with thermodynamics; permit the highly linear and isochronous mass-spring on the left hand; but add a nontime-dependent but small right-hand term. This is represented as an ϵ force (or corresponding energy) that dribbles in energy in phase. This is well known in the nonlinear theory of the clock. The left hand can create the linear eigenvalues that seem so desirable yet can satisfy the theoretical necessity for losses. The right hand, intermittently or steadily dribbling in energy, can overcome the losses and maintain the limit cycle.

Then, in our opinion as borne out by the success of quantum mechanics, the physicist may permit himself to write

$$m\ddot{x} + kx = \epsilon - c\dot{x} \approx 0$$

i.e., the right hand is made up of near-zero net gains and losses in which the losses are essentially compensated for and the system persists. Alternately, as a weaker

[1] It is worthwhile pondering the following elementary statements. Sustained oscillations can come from linear systems, but they must be nonlossy (i.e., conservative) and thus violate the second law of thermodynamics. Sustained oscillations can come from nonlinear systems, but they must be lossy and thus do not violate the second law of thermodynamics. Reader, would you prefer that sustained oscillations and thermodynamics be consistent, or would you prefer the sustained oscillations and the elegance of conservative mechanics?

[2] It is more the case that specialized processes rather than artificial processes can make up the losses. However, as we know it today, it has to be the drawing of energy for a "finite" time from a constant potential source.

form of this argument, we can imagine the result to hold only for time averages of both sides.

An interesting introduction to the quantization problem is contained in Klein (2) and the references he offers. His article illuminates the story of Planck's introduction of the concept of energy quanta and of his effort to bring a theory of blackbody radiation and thermodynamics into consistency.

Devoted to the study of thermodynamics as viewed by Clausius, Planck spent 15 years of his career on the exposition and application of the second law, particularly of the concept of irreversibility. In this area, he was more powerfully preceded by Gibbs. He fought vehemently with the "energeticists," Ostwald and Helm, believing that they tangled thermodynamics (Collected Papers 1, p. 459, 1896). He also started with strong objections to Boltzmann's thermodynamic formulation as a statistical law.

However, having tackled the thermodynamics of blackbody radiation, by 1900, only months before introducing the idea of the quantum, Planck concluded that the Wien formula (the high-frequency, low-temperature relation of spectral density, frequency, and temperature for blackbody radiation) is implied by the principle of entropy increase of the second law of thermodynamics. He was thus able to get around the inability of a statistical mechanics and energy equipartition to account for blackbody radiation (the position of Jeans, who had had to conclude that there was not true thermo-dynamic equilibrium in the radiation) and still did not have to accept the extreme view of a quantum of action (the view soon to be defended by Einstein).

However, the "further experimental tests" that Planck called for were soon made, and at longer wavelengths the Wien law broke down, so that Planck assumed that either the second law was not universally valid or there was an error in his arguments (Papers, 1, p. 668, 1900).

In 1900 Planck thus had to take a painful step, which was contained in his new, now familiar, black-box distribution law:

$$e_v = \frac{\alpha v^3}{\exp(\beta v/T) - 1}$$

where e_v = energy density per unit frequency
       α, β = constants
       v = frequency
       T = temperature

Planck's earlier analysis of the way that entropy increased with time had suggested this as the next simplest possibility after Wien's law, i.e., it was phenomenologically inferred. "The problem was to create a suitable theo-retical foundation for the new distribution law." This required a difficult and probably painful step. "The crux of the matter was still the energy-entropy relation for an oscillator." (The reader should note that it is at points like this, the introduction of some sort of quantum or discrete condition, that the central issues regarding nonlinear systems always tend to be found.) Planck had to have recourse to Boltzmann's ideas of fixing entropy in terms of the number of statistical complexions of a system. By assigning a quantized energy per oscillator and summing over a distribution without letting the energy go to zero, he arrived at the distribution law.

Writing about this in 1931, Planck describes it as "an act of desperation." As he states, the problem of the equilibrium of matter and energy was fundamental to physics and "a theoretical interpretation had to be found at any cost, no matter how high." He would sacrifice anything except the two laws of thermodynamics. He then considered the hypothesis of energy quantization as "a purely formal assumption."

An essential characteristic of this scheme was the introduction of two constants, the quantum of action h (Planck's constant), and Boltzmann's constant k [the ratio of the universal gas constant (per mole) to Avogadro's number—the number of atoms (per mole)]. Thus, two quantizing constants were implied, one a quantum energy per unit oscillator frequency, and the second a quantum equipartition of energy per degree of "atomistic" freedom. The fundamental atomistic nature of matter was thus again confirmed. The full mysteries of a quantum of action were still not in focus, although in 1906 Planck sought [see his lectures on *The Theory of Heat Radiation* (3)] some direct electrodynamic meaning. Both he and Einstein (in 1909) pointed out the dimensional equivalence of h and e^2/c (the square of the charge on the electron to the velocity of light). However, as Einstein put it, "Now one must remember that the elementary charge, e, is a stranger in the Maxwell–Lorentz electrodynamics. It seems to me to follow from the relationship $h = e^2/c$, that the same modification of the theory which contains the elementary charge as one of its consequences, will also contain the quantum structure of radiation."

> Thus one can point out that our thesis that quantization might be related to the coupling between mechanical and electromagnetic fields lies within the tradition of Einstein's thinking. The mystery here is further deepened as to what determined the quantization of the electric charge. Nevertheless, regardless of the strenuous efforts of Einstein, of Eddington, and others, 60 years later the two quantities h and e are not joined with any certainty.[1]

In January, 1910 (in his *Annalen der Physik* paper), Planck placed his views on radiation before the world of scientific opinion, accepting the reality

---

[1] For an illustration of the current numerology regarding these and others of the elementary constants of physics, see A. Unsöld, *The New Cosmos*, Springer-Verlag New York Inc., New York, 1969, p. 330.

of a quantum of action (to be reconciled with electrodynamics) but not holding a view as radical as Einstein's (who already saw a real quantum theory of radiation) or as conservative as Jeans' (who was trying to save the Hamilton equations and equipartition of energy).

In attempting to retain his radiation law without restricting oscillator energies[1] to quantized states, Planck tried to meet various objections (Lorentz, Einstein, 1910–1912). He finally settled on the following view. The oscillator absorbs energy continuously but emits only in multiples of hv. Emission had to be of full quanta. For any energy level nhv, a probability p of emission exists such that $(1 - p)/p$ is proportional to the intensity of incident radiation. The constant of proportionality is to be determined at the limits of high-intensity radiation by classical behavior (Klein points out that this is a first use of the correspondence principle, of seeking to evaluate parameters from limiting classical behavior). By this structure, Planck retained his radiation law, but it changed the average energy of an oscillator to

$$E = \frac{hv}{\exp{(hv/kT)} - 1} + \frac{hv}{2}$$

The last term indicated a zero point energy even at absolute zero of temperature. This was confirmed in Debye's work on thermal vibrations of crystals (the specific heat). However, it had lost its connection to Planck's second quantum theory.

In reviewing the then new Nernst heat theorem, a third law of thermodynamics that fixed the entropy of a pure substance in a condensed phase at zero for absolute zero temperature, Planck sought the atomic meaning of this new law at the level of Boltzmann's fundamental relationship between entropy and probability. He decided that it required that the size of the cells, in the phase space in which the statistical mechanical system was to be embedded, was not indifferently small but fixed at h as the phase cell for oscillators of any frequency. This, he thought, was the essential content of the hypothesis of quanta. For Planck, the quantum hypothesis came to roost in a positive definite value to the entropy of a state with a zero limit, rather than the minus infinity of classical thermodynamics.

One notes that the key questions have to do with large collections of atomistic oscillators, with electromagnetic radiation and absorption, with the relation to thermodynamics, all as precursors to or part of a quantum theory of the atomistic oscillators.

Reviewing the argument again, briefly but with a little more detail, Rayleigh (*Philosophical Magazine*, 1900) derived a classical mechanical result

---

[1] The internal "electrical" oscillators are the fundamental source of the continuous electromagnetic radiation energy being discussed as blackbody radiation.

that the (reversible) energy content per unit wavelength of blackbody radiation at large wavelength (low frequency) is given by

$$e_\lambda = \frac{8\pi R T}{\lambda^4}$$

where λ = wavelength
        R = ideal gas constant

Jeans (*Philosophical Magazine*, 1907) showed that this should hold classically at any wavelength. This result, however, is not valid at short wavelength (high frequency) or low temperature, where the energy is much smaller than given by the Rayleigh–Jeans law. In experimental fact, it is very nearly reduced in that region to

$$e_\lambda = \frac{8\pi R T}{\lambda^4} \frac{x}{\exp(x) - 1}$$

$$x = \frac{h\nu}{R T}$$

the Planck law. The substantive issue is to account for the large defect of energy at high frequency (or low temperature, i.e., in the regions for which the quantum theory had to be developed).

Before examining the Rayleigh–Jeans (R–J) derivation, it is useful to illustrate the elementary thermodynamics result, which had been known before the R–J law. It is the argument by which electromagnetic radiation is brought within the scope of thermodynamics (i.e., ultimately, of statistical mechanics).

Maxwell (1873) deduced from e-m theory that radiation normally incident to a surface, from a collection of oscillators, produces a pressure equal to the energy per unit volume of the radiation. If oblique, there is a $\cos^2\theta$ reduction in pressure, so that in a unit volume enclosure—in which the radiation arrives from all directions—the mean pressure is one-third of the energy per unit volume.

Boltzmann (*Wiedmanns Annalen*, vol. 21, 1884) applied thermodynamic reasoning to the pressure exerting e-m radiation in an isolated enclosure.

The Wien formula

$$e_\lambda = \frac{f(\lambda T)}{\lambda^5}$$

was derived (1893) from thermodynamic reasoning for radiation. It is clear that both the classical, mechanical R–J law and the Planck law are consistent with it. The difference is the form of f, and the difference is found essentially only at high frequency or low temperature. (This is quite fascinating for a reason that has not been pointed out. The high-frequency divergence would seem to be related to e-m properties associated with the constant of the propagation velocity of light. The low-temperature divergence would seem to be related to mechanical properties associated with the constant of the propagation velocity of sound. However, in both cases, the issue is a statistical mechanical one—a collection of e-m and mechanical oscillators are involved.)

It is interesting to note that the resolution that took place was for Einstein to accept Maxwell and Lorentz' e-m theory and to change mechanical kinematics; for Bohr to change the classical consequence of m/e-m coupling whereby an accelerating electrical charge (arising from its high-frequency revolution) no longer need radiate but is adjusted to "fit" Planck's quantum scheme; and for Heisenberg and Shrödinger to find means to formalize the quantum scheme, but thereby leaving Einstein dissatisfied and, within this scheme, leaving subsequent physics the very difficult task of dealing with the ensuing proliferation of fundamental e-m and mechanical particles. Very few, outside of professional modern particle physicists, can be impressed by the current logical state of foundational physics. (More detailed exposition of these issues for the nonspecialist may be found in many articles in *Scientific American*, during the 1960s in particular. Many such references are provided in a later chapter. A history may be found in (4). On the other hand, for any reader who wishes an exhaustive source of references for the foundations of quantum physics, (5) can be suggested.)

For a specific and quite authoritative statement of modern views, one might select Heisenberg's 1958 lecture (6) discussing the philosophic foundations of atomic physics and, in particular, the problem posed by Planck's action quantum.

Heisenberg points out that Planck's quality of discreteness in radiation phenomena, although related in a surprising manner to the existence of atoms, could not be explained simply by their existence. Also, it led, not to a local or material constant, but to a universal constant, a property of nature. Later, he hastens to state, Einstein's constancy of the speed of light also appeared as a constant of nature, and relativistic mechanics could be so understood. Heisenberg goes on:

> It was, however, much more difficult to understand the physical relationships connected with the existence of Planck's action quantum. It appeared probable from a paper of Einstein in the year 1918 that the laws of the quantum theory in some way or other involved statistical relationships. But the first attempt to thoroughly study the statistical nature of the laws of the quantum theory was made by Bohr, Kramers and Slater in 1924. The relationship between electromagnetic fields, which had been considered to be the propagators of light in classical physics since Maxwell, and the discontinuous, i.e., quantumwise,

absorption and emission of atoms as postulated by Planck was interpreted in the following manner : The field of electromagnetic waves, to which the phenomena of interference and diffraction are manifestly due, determines only the probability that an atom will absorb or emit light energy by quanta in the space under consideration. The magnetic field was thus no longer considered as a field of force that acts on the electric charge of the atom and causes movement. Its action takes place more indirectly: the field determines only the probability that emission or absorption takes place.

Later this interpretation was shown to be not quite exact. The actual relationships were still somewhat uncertain ; somewhat later they were correctly formulated by Born. Nevertheless, the work of Bohr, Kramers and Slater contained the decisive concept that the laws of nature determine not the occurrence of an event, but the probability that an event will take place and that the probability must be related to a wave field that obeys a mathematically formulable wave equation.

This was a decisive step away from classical physics; basically, a concept that played an important part in Aristotle's philosophy was used. The probability waves of Bohr, Kramers and Slater can be interpreted as a quantitative formulation of the concept of $\delta \acute{v} \nu \alpha \mu \iota \varsigma$, possibility, or in the later Latin version, *potentia*, in Aristotle's philosophy. The concept that events are not determined in a peremptory manner, but that the possibility or 'tendency' for an event to take place has a kind of reality—a certain intermediate layer of reality, halfway between the massive reality of matter and the intellectual reality of the idea or the image—this concept plays a decisive role in Aristotle's philosophy. In modern quantum theory, this concept takes on a new form ; it is formulated quantitatively as probability and subjected to mathematically expressible laws of nature. The laws of nature formulated in mathematical terms no longer determine the phenomena themselves, but the possibility of happening, the probability that something will happen.

This introduction of probability corresponded at first quite closely to the situation found in experiments with atomic phenomena. If the physicist determines the intensity of a radioactive radiation by counting how often this radiation activates the tube in a given time, he admits implicitly that the intensity of radioactive radiation regulates the probability of the counter's responding. The exact interval of time between impulses does not interest the physicist—he says they are statistically distributed. What matters is only the average frequency of the impulses.

The fact that this statistical interpretation reproduces exactly the experimental situation has been proved in many investigations.

Quantum mechanics has also obtained exact confirmation in experiments that permit quantitative evidence, as for example, on the wavelength of spectrum lines or the binding energy of molecules. There could be no doubt of the correctness of the theory.

The problem of the compatibility of this statistical interpretation with the large store of experience collected in so-called classical physics was, however, more difficult. All experiments depend on an unequivocal relation between the observation and the physical phenomena on which it is based. If, for example, we measure a spectrum line of a definite frequency with a diffracting grating, we take it for granted that the atoms of the radiating substance must have emitted light of that frequency. Or, if a photographic plate is blackened, we suppose that it has been struck at that point by rays or particles of matter. Physics, in collecting experimental data, thus utilizes the unequivocal determinateness of events and thus apparently finds itself somewhat opposed to the experimental situation in the atomic field and to the quantum theory. It is precisely here that this unequivocal determinateness of events is questioned.

This apparent inner contradiction is eliminated in modern physics by establishing that the determinateness of phenomena exists only insofar as they are described with the concepts of classical physics. The application of these concepts is on the other hand limited by the so-called uncertainty relationships; these contain quantitative data on the limits that are placed on the application of classical concepts. The physicist thus knows in which cases he may consider events as determined and in which cases he may not; he can consequently use a method devoid of intrinsic contradictions for the observation and its physical interpretation. Of course, the question arises why it is still necessary to use the concepts of classical physics, why it is not possible to transform the whole physical description to a new system of concepts based on the quantum theory.

Here it is first of all necessary to stress, as von Weizsäcker has done, that the concepts of classical physics plays a role in the interpretation of the quantum theory similar to that of the a priori forms of perception in the philosophy of Kant. Just as Kant explains the concepts space and time or causality a prioristically, because they already formed the premises of all experiences and could, therefore, not be considered as the result of experience, so also the concepts of classical physics form an a priori basis for experiments in quantum theory, because we can conduct experiments in the atomic field only by using these concepts of classical physics.

In the meantime, in the last twenty years the development of atomic physics has led us even further away from the fundamental concepts of materialistic philosophy in the ancient sense. Experiments have shown that the bodies that we must undoubtedly regard as the smallest particles of matter, the so-called elementary particles, are not eternal and unalterable, as was supposed by Democritus, but can be transmuted into one another. Here, of course, we must first state our grounds for describing these elementary particles as the smallest particles of matter. Otherwise it would be possible to believe that these particles are composed of other smaller bodies, which in their turn would be eternal and unalterable. How can the physicist exclude the possibility that the elementary particles themselves are composed of smaller particles, which escaped our observation for one reason or another?

I wish to explain in detail the reply given by modern physics to this question, because it gives prominence to the nonintuitive character of modern atomic physics. To ascertain experimentally if an elementary particle is simple or complex, it is evidently necessary to try to break it up with the strongest means at our disposal. Naturally, as there are no knives or tools with which we might attack the elementary particles, the only remaining possibility is to make the particles collide with each other with great energy to see whether they break each other apart.

The large accelerators, which are today in operation in many parts of the world or are still under construction, serve this very purpose. One of the largest machines of this kind is, as you know, being constructed by the European organization CERN here in Geneva. With such machines, it is possible to accelerate elementary particles to very high velocities (in most cases protons are used) and to make them collide with elementary particles of any other material being used as recipient. The results of such collisions are then studied case by case. Although much experimental material on the results of such collisions must still be collected before we can hope to be completely clear about this branch of physics, it is nevertheless even now possible to say qualitatively what happens in such collision processes.

It has been found that scission can undoubtedly take place. Sometimes a great many particles originate in such a collision, and surprisingly and paradoxically the particles that originate in the collision are no smaller than the elementary particles that were being broken up. They are themselves again elementary particles. This paradox is explained by the fact that, according to the theory of relativity, energy can be converted into mass. The elementary particles to which the accelerators have given a large amount of kinetic energy, with the help of this energy, which can be converted into mass, can generate new elementary particles. Therefore, the elementary particles are really the last units of matter, that is, those units into which matter breaks up when maximum forces are used.

We can also express this phenomenon in the following manner: All elementary particles are composed of the same substance, that is, energy. They are the various forms that energy must assume in order to become matter. Here the pair of concepts, "content and form," or "substance and form" from Aristotle's philosophy, reappears. Energy is not only the force that keeps the "all" in continuous motion, it is also— like fire in the philosophy of Heraclitus—the fundamental substance of which the world is made. Matter originates when the substance energy is converted into the form of an elementary particle. According to our knowledge today, there are many such forms. We now know about 25 types of elementary particles, and we have good reason to believe that these forms are all manifestations of certain fundamental structures, that is, consequences of a mathematically expressible fundamental law of which the elementary particles are a solution, in the same manner as the various energy states of the hydrogen atom represent the solution of Schrödinger's differential equation. The elementary particles are, therefore, the fundamental forms that the substance energy must take in order to become matter, and these basic forms must in some way be determined by a fundamental law expressible in mathematical terms.

This fundamental law sought by present-day physics must satisfy two conditions, both of which follow immediately from experimental knowledge. In the researches on elementary particles, for example, in those performed with large accelerators, so-called rules of selection have been obtained for the transformations that take place following collisions or following radioactive disintegration of particles. These rules, which can be formulated mathematically by means of suitable quantum numbers, are the direct expression of the symmetrical properties inherent in the fundamental equation of matter or its solutions. The fundamental law must, therefore, contain in some form these observed symmetries or, as we say, represent them mathematically.

Second, if it is conceded that there is such a simple formulation, the fundamental equation of matter must contain, together with the two constants, velocity of light and Planck's action quantum, of which we have already spoken, at least one further similar constant of measure; since the masses of the elementary particles can for purely dimensional reasons follow from the fundamental equation only when, apart from the known constants of measure of which I have already spoken, we introduce at least one more. Observations on atomic nuclei and elementary particles suggest that this third constant of measure should be represented as a universal length whose order of magnitude should be about 10^{-13} cm.

In the fundamental natural law that determines the form of matter and thus the elementary particles, there must be three fundamental constants. The numerical value of these three constants of measure no longer really contains any physical expression...(6).

It is hard to escape the feeling that the issues warrant further examination by a nonlinear theory for the elementary form of matter.

> In our view, similar to Bohm's criticisms, a probabilistic quantum theory seems to be a formal theory that accounts for net system transactions among elementary particles. We are concerned with the detailed motions of systems, those that also occur between transactions. We wonder whether the quantization may not arise from the nonlinear interaction among large collections of entities, perhaps by more classical laws of motion aptly applied. We cannot perform this application to nuclear and atomic particles at this time, certainly not by any easy exposition. However, we find the spectral response of nonlinearly quantized fields to be a common experimental thread among practically all systems. Thus we raise the question, may we not proceed as if there were a formal theory—whether ad hoc quantization in the sense of Bohr or Schrödinger and Heisenberg, or a nonlinear limit-cycle theory in the sense of a classical or some new as-yet-unstated physics? Some further physical exposition may be helpful to bring forth some of the issues.

The issues that existed, pre-quantum theory, may be well noted in Jeans' writing [see, for example, (7)].

We wish to consider a dynamic system of particles possessing a large number of degrees of freedom, which is self-contained, conservative, and not dissipative. Its energy can be described as a homogeneous quadratic function of its generalized displacement and momentum degrees of freedom, i.e., in some suitable coordinate system as

$$E = q_1{}^2 + q_2{}^2 + \cdots + q_s{}^2 + p_1{}^2 + p_2{}^2 + \cdots + p_s{}^2$$

where E = a particular energy level near which the system energy lies

q_i, p_i = both displacement and momentum degrees of freedom (here suitably diagonalized)

$2s$ = number of degrees of freedom

and it can be represented, more restrictively, in general statistical mechanics and thermodynamic terms by comparison with a system providing a thermodynamic scale of temperature as

$$E = (2s)\frac{kT}{2}$$

that is, by essentially $kT/2$ for each of the degrees of freedom concerned in the quadratic energy function.

In a vacuum (for Jeans, an "ether" capable of supporting two independent transverse waves), e-m waves propagate according to

$$\frac{\partial^2 \varphi}{\partial t^2} = c^2 \nabla^2 \varphi$$

where φ = a potential

c = velocity of propagation (of light)

For a restricted volume for which the e-m field vanishes outside, the number of free vibrations of natural frequency v in an interval dv is given by

$$\frac{v^2\,dv}{\pi^2 c^3} = \frac{8\pi\,d\lambda}{\lambda^4}$$

since $v = 2\pi f$; $f = c/\lambda$.

For a collection of simple harmonic oscillators, for which

$$E_i = \tfrac{1}{2}a_i\dot{q}_i^2 + \tfrac{1}{2}b_i q_i^2$$

where q_i = displacement

$\dot{q}_i$ = "velocity"

the energy distribution becomes

$$\frac{8\pi RT}{\lambda^4}\,d\lambda$$

for a structureless medium (i.e., one without graininess).

However, integrating over a range of wavelengths, one obtains $1/\lambda^3$ terms. This would go to infinity at small wavelength unless $T = 0$. One must surmise that structureless media are thus excluded, except possibly at $T = 0$, else the system has infinite energy.

Basically, Jeans attempts to show that the same exchange would take place for any assumption of resonators based on classical dynamics, not only as in the Rayleigh derivation for long wavelengths.

Continuing on to the thermodynamics of radiation, he visualizes an adiabatic enclosed volume (if by no other means, we can do this by using mirror image enclosures) which is filled with a radiation field (the same e-m field considered before) by bringing it to equilibrium with some heated matter, so that a thermometer in the enclosure indicates a mean temperature T.

Change the volume V reversibly

$$T\,dS = d(e_V V) + p\,dV$$

where e_V = energy of radiation per unit volume

S = entropy

p = pressure

Maxwell (1873) deduced from e-m theory that the e-m radiation incident on a surface produces a pressure.

By the cosine law

$$e_V = 3p$$

$$dS = \frac{d(e_V V)}{T} + \frac{e_V}{3T}\,dV$$

$$= \frac{4}{3}\frac{e_V}{T}\,dV + \frac{V}{T}\frac{de_V}{dT}\,dT$$

The assumption is made that the total energy per unit volume is just a function of temperature. Then, since dS is a perfect differential

$$\frac{\partial^2 S}{\partial T \, \partial V} = \frac{\partial^2 S}{\partial V \, \partial T}$$

$$\frac{1}{T}\frac{de_V}{dT} = 4\frac{e_V}{T^2}$$

$$e = aT^4$$

Stefan's law (1879; theoretically by Boltzmann in 1884).

For an adiabatic change in volume, Wien (1893) showed that

$$TV^{1/3} = C_1$$

[Since $d(e_V V) = -(e_V/3) \, dV$, the Wien relation easily follows.]

Thus a change in volume must change the temperature. From a Doppler argument, it can be shown that

$$\lambda = C_2 V^{1/3}$$

or Wien's law

$$\lambda T = C_3$$

holds for each constituent of radiation through changes in volume and temperature.

Thus, thermodynamically, the energy partition by Wien's law should be

$$e_\lambda \, d\lambda = f(T)\varphi(\lambda T) \, d\lambda$$

$$E = \int_0^\infty e_\lambda \, d\lambda$$

$$= \frac{f(T)}{T}\int_0^\infty \varphi(\lambda T) \, d(\lambda T)$$

$$= C_3\frac{f(T)}{T}$$

$$= aT^4$$

Therefore

$$e_\lambda \, d\lambda = \frac{f(\lambda T)RT}{\lambda^4} \, d\lambda$$

Wien's displacement law; or

$$e_\lambda \, d\lambda = \frac{8\pi RT}{\lambda^4}f\left(\frac{hc}{RT\lambda}\right) \, d\lambda$$

The classical mechanical result was $f = 1$. The experimental result is the Planck result.

Jeans proposes that the simplest way of arriving at Planck's formula is possibly as follows:

He considers a linear oscillator system

$$E = \frac{a}{2}\dot{q}^2 + \frac{b}{2}q^2$$

The equilibrium distribution, according to statistical mechanics, will be

$$C_1 \exp\left(-\frac{E}{kT}\right) dq\, d\dot{q}$$

for a system with one integral (here the energy integral).

If $q = A \cos(\omega t + \eta)$

$$dq\, d\dot{q} = \frac{dE\, d\theta}{\sqrt{ab}}$$

Integrating, with respect to all phases, θ, the energy distribution becomes

$$\frac{2\pi C_1}{\sqrt{ab}} \exp\left(-\frac{E}{kT}\right) dE$$

If, instead of all continuum values of energy being possible, the energies are segregated into discreet ranges $E = 0, E = \epsilon, E = 2\epsilon$, then by the distribution law, the number inside these ranges will be in the ratio

$$1 : \exp\left(-\frac{\epsilon}{kT}\right) : \exp\left(-\frac{2\epsilon}{kT}\right) : \exp\left(-\frac{3\epsilon}{kT}\right) : \cdots$$

so that if N have zero energy, the number having energies $\epsilon, 2\epsilon \ldots$, will be

$$N \exp\left(-\frac{\epsilon}{kT}\right), N \exp\left(-\frac{2\epsilon}{kT}\right), \cdots$$

The total number of systems whose energy lies within the small ranges will be

$$N\left[1 + \exp\left(-\frac{\epsilon}{kT}\right) + \exp\left(-\frac{2\epsilon}{kT}\right) + \cdots\right] = \frac{N}{1 - \exp(-\epsilon/kT)}$$

or the total energy will be

$$N\left[\epsilon \exp\left(-\frac{\epsilon}{kT}\right) + 2\epsilon \exp\left(-\frac{2\epsilon}{kT}\right) + 3\epsilon \exp\left(-\frac{3\epsilon}{kT}\right) + \cdots\right]$$

$$= \frac{N\epsilon}{\exp(\epsilon/kT)[1 - \exp(-\epsilon/kT)]^2}$$

The mean energy of all the vibrations will be

$$\frac{\epsilon}{\exp{(\epsilon/kT)} - 1} = kT\frac{x}{\exp{(x)} - 1}$$

$$x = \frac{\epsilon}{kT}$$

The fundamental "quantum" has the value

$$\epsilon = h\nu$$

i.e., the vibrational energy can occur only in integral quanta. It is by such arguments that the first round of quantum theory was introduced. The second round was Bohr's (1913) theory of the atom, which really amounted to a very similar thing in that it asked that the orbitally accelerating electron not radiate but instead emit in quantum jumps, e.g., the Wilson-Sommerfeld quantization rule (1915)

$$\oint p_i \, dq_i = nh$$

or the Bohr frequency postulate (1913)

$$\Delta E = h\nu$$

plus Bohr's correspondence principle, which related transition probabilities between quantized states to the radiation expected by classical e-m theory. The third round was the Heisenberg–Schrödinger quantum mechanics–wave mechanics equivalence, which provided a formal structure related, by rules, to classical dynamics for obtaining quantized fields.[1]

An interesting review of three recent popular books on Bohr and quantum theory is given by V. Weisskopf (9). It is clear that Cline's book (10) is the best one. Although she is not a physicist, she has been correctly steered by experts so that she mirrors their logical position quite well. It is worth the reader's while to scan her book.

Writing with great clarity, Weisskopf points out that Bohr's 1913 paper proposed to lock up the atom by the concept of the quantum state. In the 10 years that followed, an electrodynamics of atoms was hewed out.

All of this rested on Bohr's assumption that the atomic orbits were quantized, i.e., that only certain specified patterns of orbits were admitted within the atom, at that time still a provisional hypothesis. Bohr's contemporaries, however, took this assumption quite literally, although Bohr warned them both in his papers and at meetings that his could not be the final explanation, that something fundamental remained to be discovered before what was going on in the quantization of the atom could be properly understood.

[1] It is quite fascinating to compare the grudging and uncertain acceptance by Jeans (1925) of the Bohr quantum dynamics (7) with the rise of the first new wave thinkers [see (8)].

Although we are not able to argue out a full case, simply because we don't have one, we would like to state our position. Fifty years later we still cannot find the arguments of quantum mechanics completely compelling[1] for the following reasons. Consider just the classical argument. (Note that quantum mechanics is supposed to be built on top of a classical mechanics.) A system of linear, isochronous e-m oscillators (according to the theory of small vibrations) are required to be consistent with mechanics, thermodynamics, and e-m theory. In our frame of reference, we do not see this consistency. One of the equivalent statements of the second law of thermodynamics [Zemansky (11) for example] is "As a result of natural processes, energy is becoming unavailable for work," or "... the disorder of the universe is increasing." Yet an isolated system of oscillators within the walls of a blackbody enclosure (which can be one of a group of a large cluster of cells to make each of them very nearly adiabatic) may be characterized by a thermodynamic temperature T and will maintain a sustained mechanical order, the Maxwellian distribution, without degradation. The atoms will vibrate indefinitely. This is not the degradative system promised by thermodynamics. Furthermore, as spectroscopy can demonstrate, there is a stationary and continuing interconversion of mechanical energy and e-m energy. The space is filled by a detectable energy density in flux. Thus mechanical, thermal, and e-m energies are involved. These have classical theories. However, the classical exposition is immediately contradictory to a classical thermodynamics. Thus one may even raise doubts whether a self-consistent, applicable classical physics has been written. It appears that the dilemma exists and ought to be treated from the beginning before setting up a quantum mechanics. Classical microscopic theory has been treated solely as a contracted macroscopic mechanics. When one switches levels so drastically, there may be new facets to bring into focus.

The elementary classical paradox, stated again, is that, since the spectrum is stationary and isochronous, the oscillators are linear, of the form

$$E = \frac{a}{2}q^2 + \frac{b}{2}\dot{q}^2$$

The only "exact" oscillator that fulfills this condition is the linear one, i.e.,

$$m\ddot{x} + kx = 0$$

[1] Nevertheless, quantum mechanics has a well-developed, widely accepted corpus among physicists as a few references, (12–16) among many others, well testify. In the main, its professional adherents are fairly well satisfied with its power, capability, and scope. One may examine, for example, the difference in views of Jeans regarding light-quanta [Sec. 493, (7)] with the Feynman rationale (17). Whereas Jeans was troubled by the "indivisible" quantum of light still diffracting and interfering through a double slit with extremely reduced illumination, the modern view (17) is that diffraction is a statistical property of a single photon (or electron) and that the concept of classic causality regarding the determinacy of particle motions must be modified or abandoned.

Yet the second law of thermodynamics really implies that if a system degenerates toward small-amplitude linearity, it must be of the form

$$m\ddot{x} + c\dot{x} + kx = 0$$

i.e., it must be dissipative. It is only possible for the system to be globally persistent if it is nonlinear and if there exists a limit cycle of Poincaré's. This requires that the system be locally dissipative. As Minorsky points out, the limit cycles of Poincaré are the stable states of nonlinear mechanics, just as the linear oscillator or rest state is the stable state of linear mechanics. Thus the logical question remains whether the quantum state of a group of atomic oscillators is not to be described as a possible limit-cycle state of nonlinear mechanics, rather than requiring an ad hoc change in classical mechanics. In other words, since the low-frequency R-J energy law was perfectly valid and it is only the high-frequency law that is bad, is it possible that one can change high-frequency e-m[1] rather than small-size mechanics? Is quantum mechanics only a disguise for instability of a small number of regions in classical, or slightly modified classical, physics, in particular, electrodynamics? (For the high-frequency, spatially small fields of atomic or nuclear physics in particular. On such a score, the author can only point with hesitation to O'Rahilly's *Electromagnetics* as an example that there are those who would carry on discussion critical of classical electrodynamics. To any other than specialists, the reference is a red herring.)

Einstein changed kinematics to make mechanics fit with e-m. Heisenberg and Schrödinger changed mechanics to get around electrodynamics. Is it not also logically possible to change e-m or nuclear dynamics? (The fact that efforts by such great scientists as Bohr or Born may not have succeeded does not change the issue. We would like again to point to the possible example of a many-bodied system which can jump from a nonlinearly stable configuration with energy E_1 to a nearby stable configuration of very nearly the same energy and in which the jumps may then be represented by such "radiative" quantized jumps as might be furnished by a coupled e-m field. Although the mechanical field by itself may pose high-energy barriers that make the process unlikely, it is not precluded that the coupled e-m field cannot lower the barrier.)

After this introduction to the logic of quantum mechanics, it is desirable next to review the statistical mechanics that bring one up to the details of quantum mechanics mentioned in the preceding paragraph.

SUMMARY

1. We introduce the subject of quantum mechanics and quantum electrodynamics within its "modern" framework of the classical Hamiltonian of conservative systems. By first introducing

---

[1] "High frequency" here is to be defined as beginning at $h\nu = kT$. This is not really high frequency by many standards, but the frequency at which the quantum issues arise.

the dialectics of classical thermodynamics versus the blackbody radiation problem of Planck's, we point to its reconciliation by the introduction of an ad hoc quantization of electromagnetic energy. Second, by alluding to the dialectics of a catastrophically radiating "solar system" of electrified particles interacting with electromagnetic radiation in a cavity, we point to its reconciliation by the ad hoc introduction of a probability wave function to replace the particle, which provided quantized states by a linear eigenvalue problem.

2. It is clear that this type of program has been accepted in modern physics, although reservations have currently begun to appear on the horizon. Its philosophy is quite remote from nonlinear limit-cycle theory. If it is accepted in toto, nuclear, atomic, and radiation systems tend to have the same character as other systems (namely, long-term viability) but a seeming theoretical foundation of different nature from other systems (namely, a linear probabilistic theory with discrete, quantized dual nature at its foundation). If not accepted in toto, it might be possible to reform the theory to make it agree with that for "all" other systems.

3. For those who have caught a whiff of the dissent but want the question clarified, the burning issue is the following: We are concerned with conditions under which a system involving sustaining processes might exist. The linear model of a harmonic process

$$m\ddot{x} + kx = 0$$

or its equivalent

$$x \propto \exp\left(\pm j\sqrt{\frac{k}{m}}\,t\right) = \exp(\pm j\omega_0 t)$$

will not do because a fundamental physical assumption is that current cosmological processes run down. They exhibit losses. Thus even the linear model must be

$$m\ddot{x} + c\dot{x} + kx = 0$$

or

$$x \propto \exp(-\alpha \pm j\omega_0)t$$

However, there is another "stable" process, the nonlinear limit cycle. This can only come into existence if the system is degradative (if it fits a major prescription for real physical processes). We have no general procedure for finding limit cycles. One approximate procedure[1] is referred to currently as Kochenberger's describing function technique. It involves seeking out among all the possible processes, degradative and divergent, those that might persist. The necessary condition is that sustained harmonic components can be demonstrated to be solutions of the dynamic equations of motion, i.e., that solutions of the form $\exp(\pm \lambda t)$ (whether a lumped or discrete system or a distributed system) may exist, where λ is purely imaginary.

Then the system (or its mathematical description) may be looked at as a nonlinear filter, which, of all possible modes of oscillation that the collection permits (that is, for all possible modes, which in general are complex), may draw from an available dc energy source sufficient power to support a particular harmonic spectrum. All other permissible spectral segments (lossy and divergent processes) will, however, only be transient.

This, it is submitted, is the fundamental source of nonlinear quantization, logically the only general process at present discernible by which sustained ac processes can be maintained.

---

[1] It is essentially Poincaré's concept of characteristic exponents. As Whittaker's *Mechanics* states in discussing stability of types of motion of dynamical systems, "Hence a necessary condition for stability of the periodic orbit is that all the characteristic exponents must be purely imaginary."

The evanescent lossy and divergent processes may stir the field into whatever stochastic components it may exhibit (for example, spectral line broadening), but the mean spectral processes will have already been selected as eigenvalues in this quasi-linear decomposition [illustrated in (18)].

We feel entitled to wonder whether some such nonlinear scheme may not be at the root of the dynamic character of nuclear, atomic, gravitational and radiation fields and particles.

QUESTIONS

1. Discuss various mathematical means for obtaining a stationary spectral line or lines in which the line remains stationary to moderate excitation perturbations, or at most only transiently disturbed. Distinguish between linear and nonlinear cases.

2. Examine the foundations of electromagnetic theory for any clue as to how a nonlinear quantization might emerge in a cavity.

BIBLIOGRAPHY

1. Den Hartog, J.: "Mechanical Vibrations," McGraw-Hill Book Company, New York, 1947.
2. Klein, M.: Thermodynamics and Quanta in Planck's Work, *Phys. Today*, vol. 19, p. 23, 1966.
3. Planck, M.: "The Theory of Heat Radiation," 1913 ed., Dover Publications, Inc., New York, 1959.
4. Jammer, A.: "The Conceptual Development of Quantum Mechanics," McGraw-Hill Book Company, New York, 1966.
5. Kuhn, T., J. Heilbron, P. Forman, and L. Allen: "Sources for History of Quantum Physics," American Philosophical Society, Philadelphia, 1967.
6. Heisenberg, W., M. Born, E. Schrödinger, and P. Auger: "On Modern Physics," Clarkson N. Potter, Inc., New York, 1961.
7. Jeans, J.: "The Dynamical Theory of Gases," 1925 ed., Dover Publications, Inc., New York, 1954.
8. Van Der Waerden, B. (ed.): "Sources of Quantum Mechanics," Dover Publications, Inc., New York, 1968.
9. Weisskopf, V.: *N.Y. Rev.*, April 20, 1967.
10. Cline, B.: "Questioners, Physicists and the Quantum Theory," Crowell Publishing Co., New York, 1967.
11. Zemansky, M. W.: "Heat and Thermodynamics," McGraw-Hill Book Company, New York, 1957.
12. Pauling, L., and E. Wilson: "Introduction to Quantum Mechanics," McGraw-Hill Book Company, New York, 1935.
13. Schiff, L.: "Quantum Mechanics," McGraw-Hill Book Company, New York, 1968.
14. Heitler, W.: "The Quantum Theory of Radiation," Oxford University Press, New York, 1954.
15. Landau, L., and E. Lifshitz: "Quantum Mechanics—Non-relativistic Theory," Addison-Wesley Publishing Company, Inc., Reading, Mass., 1958.
16. Lamb, W.: An Operational Interpretation of Nonrelativistic Quantum Mechanics, *Phys. Today*, vol. 22, p. 23, 1969.
17. Feynman, R., R. Leighton, and M. Sands: "The Feynman Lectures on Physics," vol. III, Addison-Wesley Publishing Company, Inc., Reading, Mass., 1965.
18. Iberall, A.: A Contribution to the Theory of Turbulent Flow Between Parallel Plates, *Seventh Symposium on Naval Hydrodynamics*, Rome, Italy, 1968 (in press).

4

Some Preliminary Comments on Statistical Mechanics

Statistical mechanics deals with systems of like ac active interacting entities. The implicit assumption is that so many interactions take place that they cannot be followed in individual detail. A more basic assumption is that the interactions lead to chains that really are more nearly like limit cycles so that any particular sequence cannot lead to a chain far removed from a limit cycle. Yet the individual process step is described within a conservative Hamiltonian framework. Thus in our opinion it is necessary to either accept an ad hoc quantization among the interacting entities or to assume some nonlinear process that compensates for degradative losses and permits a sustained interaction.

In providing some background for statistical mechanics, this issue must be kept in mind. The reader may accept quantum mechanics or nonlinear mechanics as the physics for the actions of his entities.

A number of excellent books (1–10) that introduce statistical mechanics or that introduce quantum ideas into statistical mechanics were reviewed in preparation for writing this chapter. They were not found to be completely satisfying with regard to atomic quantization, the assumed foundation upon which statistical mechanics is built.

The simplest path taken by these books is to state Newton's laws, apply them to a conservative closed system, and derive or show that Lagrange's equations, Hamilton's principle, and the existence of an energy integral are all consistent. (Obviously some idea in this group must be brought in ad hoc.) They then proceed in one of two directions, either to derive Liouville's theorem for a system of particles and then the theorem of the virial, equilibria states, Boltzmann's equation, and the equations of change of continuous media; or to assume, ad hoc, the quantum apparatus of Heisenberg or Schrödinger. Planck's radiation law, the beginning of the quantum controversy, is left for later or omitted.

Let us reexamine the radiation problem in a classical context to see if we can rediscover the unsettling problem of quantization. We wish to show first that the quantization problem cannot be deferred classically because of the very nature of the coupling between the electrical and the mechanical nature of "atomic" particles. Yet it is quite clear that classical or well-known textbooks in electromagnetic theory or electrodynamics, e.g., (11), (12), or (13), do not treat the problem adequately. We wish to introduce the problem at least in some primitive fashion.

Thus, imagine a very thin, spherical shell made of some "atomic" element which exhibits sufficient interatomic binding "forces" to form a self-articulated structure. Thermodynamics tells us (that is, through Newtonian force laws) that a metric of "temperature" can be associated with equipartitioned internal motions associated with various degrees of freedom of the atomic system. Since this initial problem of thermal motion itself will cause us so many difficulties, we prefer to avoid contact of the shell with any other material externally. Thus, let the sphere be highly evacuated. We are first going to be concerned with radiation inside the sphere in equilibrium with it. Since there may be losses in maintaining the radiation field, we prefer to isolate the sphere as much as possible. We can surround the sphere by a concentric sphere that is uniformly heated and evacuated so as to maintain the internal sphere at some desired temperature.

It is clear that the test sphere has no "temperature" unless it is brought up to one by surrounding material. It is not sufficient to mount the internal sphere "adiabatically" in an isolating space. The temperature, under these conditions, would drop more nearly to zero, albeit with a considerable time delay. Thus the "equilibrium" temperature has to be maintained from outside. It is clear, in such a guard-sphere array, that at equilibrium there is little or no transfer of energy required to maintain the internal test sphere at constant temperature.

Yet, consider the problem from the point of view of classical electrodynamics. Suppose that the sphere material had infinite conductivity. It would then be possible to insert in the cavity a denumerable sequence of electromagnetic modes that would continue to oscillate indefinitely. But, actually, it is not possible to name a temperature arbitrarily that will permit the sphere to exhibit infinite conductivity.[1] Therefore, consider instead some high-quality conductor—a metal such as copper, silver, gold, or aluminum. It has the further property that it can be made into a physically achievable thin shell.

Now according to classical electrodynamics, there will be a skin-effect loss of power because of the finite resistivity. One must note that coupling between the electromagnetic field and the mechanical field (the oscillatory mechanical nature of the atomic motion associated with the thermal characteristics of the shell) arises immediately in order to maintain the potential blackbody radiation spectrum in the cavity. We may proceed further with the argument.

In order that we understand what it means for the sphere to possess a large but finite conductivity, we must consider the requirements and consequences of a theory of conductivity, in this case the so-called Drude–Lorentz free-electron theory of conduction.

If the "atomic" array were completely regular, e.g., a perfect crystalline array, it is theorized that its conductance would be infinite; that is, perfect regularity is associated with zero resistivity.[2] Thus, normal resistivity in metals is associated with dislocations. The elementary notion is that there are unbound orbital electrons which can drift through the metal with mean-free-path motions (linear increase in velocity, collision and loss in velocity, linear increase, etc.). The mean free paths are of the order of dislocation distances. However, in its most elementary view, at such dislocations it appears that there is not a single collision but a lockup—a pinwheel of n revolutions around the dislocation before the subsequent release. [Discussed so by Frenkel (14).] One must ask whether such description can be made consistent with the physically coupled fields that must be involved.

Note one common way this is done is to derive a theory of thermal conductivity on the basis that the same free electrons (in good conductors) are the carriers of the heat energy. From this parallel derivation one obtains the Wiedemann–Franz law, which relates the ratio of thermal to electrical conductivity to a temperature function. The obvious question is how did any

[1] The question of superconductivity and superfluidity, both considered at present purely low-temperature quantum mechanical phenomena, must be deferred. We have to avoid the near-zero regime at present. Besides, this is not the essential problem in radiation equilibrium. Thus at present, we will say that "classically" we do not expect superconductivity to hold over any arbitrarily wide temperature range.

[2] Some such regularity is likely the basis for superconductivity. Classically, such a state—consider for example Nernst's theory, the so-called third law of thermodynamics—was viewed as only feasible at $T = 0$. It has turned out that the limit has been raised at present to $T = 20°K$. Room-temperature superconductivity in specialized materials is now expected on the agenda of history. [See (15), (16), and (17) for the current status of superconductivity.]

complete electrical-field problem ever get solved, without the assumption of a sort of pseudo-linear field through the metal which urged the free electron. [A typical first-order theory, based on Fermi–Dirac statistics for the electrons, may be found in (10) in Secs. 9.5, 9.6.]

The stumbling block to an adequate classical theory always seems to originate from the same problem, quantization of many interacting electrified atomistic particles. Suppose one assigned some potential geometric structure to such a three-dimensional array. Phenomenologically, it appears that the regular array cannot be maintained indefinitely, except at low temperature. It breaks up into crystallite domains. One is tempted by this observation alone to believe that there is an instability associated with the center-of-gravity (cg) vibrational motion of the atomic nucleus itself as its amplitude (or energy) exceeds some level. This fits in with a thesis of ordering forces and disordering vibrations.

However, one finds it hard to believe that there are not out-of-phase lossy modes of vibration to associate with the array. Thus one must start a problem that geometrically looks like Fig. 4-1. A regular array of atomic lattices is depicted that persist out to some dislocations. Such elementary crystallites are then arrayed, somewhat randomly, to create polycrystalline aggregates. It is to some such configurations of polycrystalline material that a rigorous electrodynamics has to be applied.

It is some such attack that we would consider necessary for a re-examination of the classical quantization of the electromagnetic field in a cavity and a determination of its spectral density. It is obviously a nonlinear conversion by which the "constant" potential of temperature both determines the permissible e-m modes in the cavity and is itself supported by mechanical modes in the wall.

As a second problem, an extension of the first, we can imagine the inner sphere filled with a low- to moderate-density gas like hydrogen. The problem now is to determine the state of this hydrogen gas, i.e., to review the theory of the hydrogen atom, *ab initio*, by a classical mechanical and electromagnetic

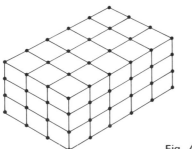

Fig. 4-1 A crystallite.

(m/e-m) theory of the field. This means that the problem is to be solved without bringing in any a priori thermodynamics and with the reality in which atoms and molecules exist and possess real-state properties such as finite conductivity, lattice imperfections, etc. From phenomenological temperature equilibrium and energy exchanges between inside and outside the shell, the difficulty consists of solving the e-m problem of interaction of radiating hydrogen-atom systems with wall-atom systems and solving the mechanical problem of characterizing their respective states.

It seems to us that real solutions of such problems are inordinately difficult. It would seem that one must start from the assumption of a possible, steady, nonlinear limit-cycle state, not necessarily the "real" one that actually exists and which in fact must likely be the quantum mechanical one, but a classical one which must also exist. The test of the validity of classical mechanics—or an appropriate modification—is whether the limit cycle of a nonradiative-except-by-quantum-jumps theory, the Bohr hydrogen atom, can emerge. The reason for asserting the likelihood of some classical non-linear quantization is that if the shell temperature T is appreciably above the critical temperature of hydrogen one could not expect a state in which all the hydrogen condensed on the shell. The thermal processes, if nothing else, would boil off the hydrogen. Thus minimally, a process of "condensation" and "emission" would take place. A limit cycle to make up for "dissipative" forces would have to exist. Boltzmann statistics would guarantee rather rapid approach of this ensemble to equilibrium because of the large numbers. It is not clear, from inspecting classical e-m or mechanical literature, what the results might be. Thus a few classical problems of some interest still exist, without adequate exposition.

Following such position statements, it is timely to sketch the current foundation for these problems—of coupled mechanics and e-m—on which thermodynamics is built and which involves quantization at some point, by starting from statistical mechanics.

An elementary derivation of statistical mechanical theory will follow Tolman [1927, (4)].

4-1 PRELIMINARY MECHANICS

1. Assume a system of particles whose condition is determined by a set of generalized coordinates, $q_1, \ldots, q_n$ (such as positions, angles, etc.), and their corresponding generalized velocities, $\dot{q}_1, \ldots, \dot{q}_n$. (We will also use, subsequently, their momenta, $p_i = m_i \dot{q}_i$.) The number n must be sufficient to completely specify the system configuration. In principle, n can be reduced to the minimum number. However, these coordinates may be subject to constraints. In a nonholonomic system, all the constraints are not integrable

functions of the coordinates. In a holonomic system, conversely, the minimum number of coordinates are all capable of independent variation, and they represent the number of degrees of freedom of the system. In a nonholonomic system, the number of degrees of freedom are fewer by the number of holonomic constraints. (Although at this point it may appear insignificant, the concept of nonholonomic constraint will become more important in our later discussion. To loosely identify the concept, it is tied up with the problem of particles contacting or penetrating boundaries.)

2. For a conservative system of particles, i.e., one which neither gains nor loses energy to a region outside the one under consideration except to or from reservoirs of potential energy, Hamilton's principle is postulated as

$$\delta \int_{t_1}^{t_2} L(q_1, \ldots, q_n, \dot{q}_1, \ldots, \dot{q}_n) \, dt = 0$$

It states that the integral of the Lagrangian function is a minimum for the system's motion from t_1 to t_2. For low velocity, nonrelativistic systems

$$L = T - V$$

where T = kinetic energy
V = potential energy

3. One may derive Lagrange's equations of motion.

$$\delta \int L \, dt = \int \sum_1^n \left(\frac{\partial L}{\partial q_i} \delta q_i + \frac{\partial L}{\partial \dot{q}_i} \delta \dot{q}_i \right) dt = 0$$

$$\delta \dot{q}_i = \frac{d}{dt} (\delta q_i)$$

Integrating by parts between the fixed times then leads to

$$\int_{t_1}^{t_2} \sum \left(\frac{d}{dt} \frac{\partial L}{\partial \dot{q}_i} - \frac{\partial L}{\partial q_i} \right) \delta q_i \, dt = 0$$

For a holonomic system, the variations δq_i are each arbitrary. Thus

$$\frac{d}{dt} \frac{\partial L}{\partial \dot{q}_i} - \frac{\partial L}{\partial q_i} = 0$$

Lagrange's equations for a conservative holonomic system.

(For a nonconservative system

$$\frac{d}{dt} \frac{\partial L}{\partial \dot{q}_i} - \frac{\partial L}{\partial q_i} = Q_i'$$

where Q_i' = forces which are not derivable from a potential; these must be otherwise identified.)

4. Proceeding toward a canonical form, momenta are derived from

$$p_i = \frac{\partial L}{\partial \dot{q}_i}$$

Let

$$H = p_1 \dot{q}_1 + \cdots + p_n \dot{q}_n - L$$

where H = Hamilton function.

Differentiating

$$dH = -\frac{\partial L}{\partial q_1} dq_1 - \cdots - \frac{\partial L}{\partial q_n} dq_n + \dot{q}_i \, dp_i + \cdots + \dot{q}_n \, dp_n$$

$$= \frac{\partial H}{\partial q_1} dq_1 + \cdots + \frac{\partial H}{\partial q_n} dq_n + \frac{\partial H}{\partial p_1} dp_1 + \cdots + \frac{\partial H}{\partial p_n} dp_n$$

Thus

$$\frac{\partial H}{\partial q_i} = -\dot{p}_i \qquad \frac{\partial H}{\partial p_i} = \dot{q}_i$$

These are the equations of motion in canonical form, namely $2n$ first-order equations, rather than the n second-order equations of Lagrange.

5. From

$$\frac{dH}{dt} = \sum \frac{\partial H}{\partial q_i} \dot{q}_i + \sum \frac{\partial H}{\partial p_i} \dot{p}_i$$

$$= -\sum \dot{p}_i \dot{q}_i + \sum \dot{q}_i \dot{p}_i = 0$$

$$H = \text{const}$$

From the theory of small vibrations, in which the kinetic energy becomes a positive definite form, $T = \sum a_i \dot{q}_i^2 / 2$, where each a_i is constant.

$$L = \sum \tfrac{1}{2} a_i \dot{q}_i^2 - V$$

$$\frac{\partial L}{\partial \dot{q}_i} = a_i \dot{q}_i = p_i$$

$$H = \sum a_i \dot{q}_i^2 - \sum \tfrac{1}{2} a_i q_i^2 + V = T + V - E$$

$$E = \text{the conserved energy}$$

Thus this relation is true at both large and small displacement and velocity levels (if the system field can be kept conservative).

4-2 SOME RUDIMENTARY STATISTICAL MECHANICS

1. In a single conservative system of particles, identification is required of n generalized coordinates and their associated momenta. Consider a phase

space of $2n$ such dimensions (called the μ space)[1]. The instantaneous "phase" (see Gibbs) of the system is a point in that space. In time, the system will describe a trajectory in that μ space.

Now, instead of the single system, consider a collection or ensemble of essentially identical systems. The intent is to determine the probable behavior of any single system from the properties of this ensemble (whether we are discussing 10^{23} molecules or 10^7 current-era American teenagers, theory should have similar applicability). In the phase space, the ensemble will be represented by a streaming distribution of points. When the entire ensemble is depicted, the phase space is generally called the γ space.[2]

In γ space, we can watch the "streamline" motion of an individual system. At any point in that space, there will be a density ρ of points per unit volume.

2. Gibbs theoretical structure (3) really begins by assigning a priori characteristics to the ensemble in phase space and using these to derive information about the probable behavior of the individual system. He chose two:

    a. The microcanonical ensemble

$$\rho(E) = \text{const} \qquad E_0 - \tfrac{1}{2}\delta E \leqslant E \leqslant E_0 + \tfrac{1}{2}\delta E$$
$$\rho(E) = 0 \qquad\qquad \text{outside the interval}$$

This distribution is representative of an isolated system with total energy E_0 which is constant in time. However, by providing a uniform distribution in a small interval, a finite region of phase space is accessible with a priori equal probabilities for all those systems in equal volumes. It is an appropriate distribution for a system of known energy in a macroscopic steady state.

> Note that at this point we have already been asked to accept some sort of quantization, for it is not quite comprehensible to assume particles that are not in the long run degradative in a classical sense. That these systems of particles ("atoms") do not decay, on the simplest interpretation, would mean that the energy losses are restored. Thus although we prescribed an isolated system, it is doubtful that we really meant it. More probably it seems that we permit just enough transfer to establish an equilibrium. However, this transfer can only come from or through walls, i.e., from an external universe. Classically, the statement might be added that there are walls, and they are to be made adiabatic. However, this will not do. Regardless of what is done, there is either radiation loss to the rest of the universe or equalizing change with the walls. Thus we are intrinsically faced with electromagnetic exchange with "walls," however remote. If, by whatever "electromagnetic" model is necessary, the wall is held "isothermal" to represent no degradation of its state then it—the wall—represents some sort of potential source for the motion of the internally confined system of particles.

[1] Implying the molecule space.

[2] Implying the gas space.

Shall we regard the walls as made up of particles, too, or shall we seek to model them as an equivalent continuous source and barrier? The philosophic question is whether we can mix two hierarchical levels. The "practical" answer is usually to do some problems to find out if you may or may not. However, it is to the same very difficult problem that we continue to return. When the particles (wall and system's) are atomic or molecular, a quantization takes place. The molecular-atomic systems do not radiate; they show quantum jumps. Is this a function of these atoms, their exchanges, the size and time domain? The principle of Occam's razor seems our only recourse at this moment. Find the simplest quantizing scheme. One consistent way has been quantizing mechanics. However, whether it can be done better or simpler or at all by a modification of electromagnetic theory seems to remain an alternate possibility.

For any reader who may regard the insistent theme of the wall as being somewhat mystical or mystifying, we must expand our logic. When we speak of systems, we consider some sense in which the system—even cosmological—can be identified or isolated by imagining it contained within a geometric boundary. In general, we surmise that there must be some kind of interelement forces that bind the system, else the elements would diffuse away. One kind of binding is provided by physical walls which literally prevent any loss of elements by their binding and impenetrability. A second kind is like surface tension in which the barrier is a little less than literal. The "bounding" results from internal binding forces that form a cluster of the elements of the system. Yet likely in all cases, it is important to note that there are surface elements that make up the barrier, and they involve interaction. There is a third kind of binding, as in a solar system, in which a few elements stay "indefinitely" or a very long time in orbit with no contact, yet with interelement forces. This latter kind of system does not interest us too much.[1]

At boundaries we must be concerned with the question of energy fluxes as well as mass fluxes. Classically, we generally require that barrier conditions between the extremes of isothermal and adiabatic be postulated. Although there certainly are processes which permit little energy flux, in general it appears to be an illusion to accept the point of view of the existence of an impermeable barrier

---

[1] It may demonstrate nonlinear orbits but not the significant interaction by nonholonomic constraints. Thus Tolman (4) can validly point out that in 5 billion years planetary orbits off the ecliptic in our solar system have not occurred. However, this system has a certain kind of wall, namely an expanding universe, resulting in insufficient internal collisions to involve the interactions that interest us. Actually, as subsequent discussion will show, cosmological space is occupied by galaxies. These result from matter-radiation interaction. They form with definite structure throughout the "bounded" cosmos. And the stars and solar systems are entities that form within the galaxy (the alternate would still be a continuous or uniform density of particles— it is hardly an accident that the galactic "particles" that do form are not nuclear but stellar particles distributed along a definite sequence exhibiting historical formation). It is more plausible to consider that there is a relatively deterministic process (perhaps rare, perhaps not) by which solar systems "explode" onto the scene. That an epoch emerges with little interaction may very well be. It happens to be long compared to our brief span, but it is hardly valid to conclude that the motion of the system is then isolated and independent of its universe. The time scale defining the "equilibrium" process, if any, or, alternately, possibly only the transient process is just very long.

as an absolute abstraction. Instead, it is perhaps better to postulate different degrees of resistive "insulation" as a phenomenological characteristic of barriers.[1]

However, what seems to arise characteristically, when there is sufficient confinement of mass and energy fluxes, is a quantization out of complex interactions between active internal elements and the active elements that are involved in the wall barrier.

We can avoid the issue of seeking a unique philosophic foundation for this hypothesis, until we have gained more experience with it, by saying that, for whatever reason, on a lower hierarchical level[2] an orbital quantization takes place among the atomistic elements. Else, philosophically, one could imagine that these atomistic entities would coalesce, merge, and dissolve their own bounds. There are a limited number of energetic constellations that can be formed among these elements. There are transfer or exchange mechanisms by which a transit is made from one constellation to another. In a gross linear sense, such a system of particles acts as if it were nondissipative; that is, losses in collisions or exchanges are somehow made up from the walls. How? By radiative absorption and emission between themselves and any confining walls.

Practically, therefore, this is approximately representable by positive, definite forms, such as in a harmonic oscillator, i.e., by

$$H = \sum[\tfrac{1}{2}a_i\dot{q}_i{}^2 + \tfrac{1}{2}b_iq_i{}^2]$$

if the displacement amplitude is not too large. Our general philosophy is the non-linear mechanical concept contained in Poincaré's characteristic exponents. Very crudely stated, it is that limit-cycle oscillators tend to resemble in some primitive fashion ordinary linear mass-spring oscillators with no damping.[3] Thus it is not really the theory of small vibrations which is being appealed to but some equivalent linearization process at the actual operating points. Will this work satisfactorily? No, it will not discover or permit classifying the domain of stability, such as frozen-out degrees of freedom, regions isolated by thresholds, and internally unconnected systems. It will have difficulty in discovering that such atomistic molecules can arrive at a state such that living properties can emerge from the organization. It is this emergent evolution which is another part of general systems science that must come into focus.

This evolution of new organized forms is represented, on one hand, in man's discovery of various governing processes (quasars, pulsars, the DNA molecule, more coherent resonance in masers and lasers than he suspected before, capitalism, socialism, communism, black power), and, on the other hand, in the evolutionary process in biological systems or stellar systems. We will discuss the process in some more detail in describing three logic levels in the brain.

For those who may suggest that system aggregation (or gain in ordering) has no basis, we can point to the following fact: In any region of space, there is a background noise level—call it cosmic radiation, or pair production, or call it Eddington's description of interconverting mass-energy density at astrophysical levels. The noise cannot be eliminated indefinitely by shielding. At "practical"

---

[1] Systems that are classified as "open," may be open with regard to mass and energy fluxes. All writers are not clear on their precise specification. We shall regard a system as open when either mass or energy fluxes are permitted access to the bounded system. Our concern is essentially with open systems.

[2] Namely, whether it happens in molecules or atoms for one "internal" reason and in humans for another reason. In humans, we will later discuss the "walls" and degrading sources.

[3] Namely, that it is a necessary condition for the stability of periodic orbits that purely imaginary characteristic exponents must exist.

local levels, there are, of course, many other disturbances that may produce aggregation. However, as long as a background of sustained energetic noise exists, the possibility of entraining formal processes exists. It is from this background that form and function of new dynamic aggregates can come about. (A current race is on to show origins to life from aggregation of a chemical chain. Pertinent work will be referenced later.)

Thus with regard to systems which are bounded by walls, the effective continuum walls of the container are not indefinitely smooth. The very processes we are modeling require accommodation coefficients, etc., at the walls. Therefore and thereby, fluctuations exist. Thus, there are statistically fluctuating processes below the grosser ones we are interested in describing. Without their adequate characterization, one cannot guarantee that these fluctuations cannot change the existing stability. A common means of avoiding the question of coupling from outside is to suggest that they are far removed in space or time (that they are so far removed in space that they do not have high expectation of occurrence). However, this really does not jibe with our growing knowledge of cosmic radiation. Scientists are much closer to being able to demonstrate directly that particles, life, and stars are born out of the content of cosmic radiation. Thus our present descriptions must be viewed as having a kind of crude gross local character. However, it is to the credit of Gibbs, writing as a theoretical chemist—just as did Faraday and Ostwald —that he could see a way to describe gross phenomena by such a marvelously simple program.

b. The canonical ensemble

$$\rho = A \exp \left(\frac{E_0 - E}{\theta} \right)$$

where E_0 = const (energy)
        E = energy
        θ = const (ultimately to be related to temperature)

It will turn out that this is convenient for an ensemble of systems each in thermal contact with a temperature source (e.g., our walls).

The canonical ensemble is useful for prediction because a system chosen at random from the ensemble tends to be a fair sample of a system left to itself, both giving rise to somewhat normal fluctuations. The microcanonical ensemble provides fair samples for a single system left to itself with a near-constant energy content.

3. Having specified an initial ensemble distribution, the next question (as Gibbs discussed it) is the change of density-in-phase with time.

First, it is unlikely that an ensemble of systems of constant or near-constant energy, whether "isolated" or maintained by thermal equilibrium, will change its density in time (if the ensemble systems are maintained in thermal equilibrium, one does not expect the macroscopic properties of the system to change in time).

Thus one expects

$$\frac{\partial \rho}{\partial t} = 0$$

This means that the numbers of particles in any of the systems of the ensemble between displacement and momentum limits do not change in time (or practically do not change much). Such ensembles in which the density of distribution is constant with time (conservation of density in time) are said to be in statistical equilibrium. The canonical and microcanonical ensembles are examples.

4. Now examine the total derivative $d\rho/dt$. The equation of continuity of an ensemble moving in space is

$$\frac{\partial \rho}{\partial t} + \nabla \cdot (\rho \overline{V}) = 0$$

where ρ = density
$\overline{V}$ = vector velocity (average for the ensemble)

We require the generalization for a multidimensional space. It is

$$\frac{\partial \rho}{\partial t} + \sum \frac{\partial}{\partial x_i} \rho V_i = 0$$

where V_i = generalized "velocity" in the phase space.

Thus, since

$$\frac{\partial \rho}{\partial t} = 0$$

$$\sum \frac{\partial}{\partial x_i} \rho V_i = 0$$

Then

$$\sum \left(\frac{\partial}{\partial q_i} \rho \dot{q}_i + \frac{\partial}{\partial p_i} \rho \dot{p}_i \right) = 0$$

since the flux is in both coordinate and momentum space. Similar relations could have been derived in other spaces. However, the utility of coordinate-momentum space and the Hamilton formulation will become more evident.

From

$$\frac{\partial H}{\partial p_i} = \dot{q}_i \qquad \frac{\partial H}{\partial q_i} = -\dot{p}_i$$

$$\frac{\partial \dot{q}_i}{\partial q_i} = -\frac{\partial \dot{p}_i}{\partial p_i}$$

Thus

$$\frac{d\rho}{dt} = \sum \left(\frac{\partial \rho}{\partial q_i} \dot{q}_i + \frac{\partial \rho}{\partial p_i} \dot{p}_i \right) = 0$$

This is Liouville's theorem, the principle of the conservation of density in phase (according to Gibbs). Its derivation was made simple by Hamilton's equations of motion in canonical form.

5. With closed systems, i.e., ones that do not interact with other bodies, such theorems would be valid indefinitely. However, if we are only dealing with quasi-closed systems, these results would only be valid for intervals over which the systems behaved with sufficient accuracy like closed systems. For those physical systems whose macroscopically observed physical parameters show little detectable deviations (are highly insulated), the closure can be for long intervals.

We note that in our contained system, we proposed that exchanges with the wall were essentially perfectly elastic although the walls were not perfectly smooth (that is, we may disregard any lossy mechanical exchange with the wall and accept the radiative equilibrium). Thus the system appeared to be nominally closed in that reflection by the wall becomes simply a device to keep folding the geometric field back on itself. Typically, we can get rid of cg motion by a choice of coordinate systems so that we do not have to carry certain irrelevancies along. However, there are two classes of systems whose behavior fits the prescription. In one, as with sand particles, the motion dies out. In the other, molecules and other systems in which a nonlinear transfer can be fixed up, as in the saltation example in rivers given earlier, there is a sustained, motional steady state. In both, the system settles down to little deviation from a steady state. Statistical mechanics, however, becomes interesting only for the second class of motions. The first class is perhaps better analyzed by estimating some sort of loss mechanism. For example, even if each process were a linear or nonlinear relaxational step (in plasticity-elasticity, there is concentration on linear relaxational steps as in the Voigt and Maxwell models for internal "friction," erroneously in our opinion, whereas the true steps are nonlinear), the gross process, consisting of many steps, acts pseudo-linearly, i.e., described by

$$\exp\left(-\alpha + j\beta\right)t$$

Thus, forms like

$$f(x, y, z)\exp\left(-\alpha + j\beta\right)t$$

with estimates for the effective time constant $1/\alpha$ and some theory by which the spatial function f can be estimated, are suitable for relaxation systems.

Thus, what we are talking about here are nondecaying motions, which act "quasi-indefinitely" (for periods that are very long compared to the individual relaxation times) over spatial domains that are very large compared to the spatial domain of the individual relaxation process. Such motions "ultimately" do not depend on the starting conditions of initial coordinates and momenta.

One must note the very intricate state of logic that is involved at this point. For a given hierarchical level, conservative systems cannot lead to persistent motions of the limit-cycle type that are independent of starting conditions. Only lossy systems can do that. Yet by mixing hierarchical levels, we assume a conservativelike bookkeeping scheme for the lower level and assume that its "atomistic" motions will appear, in summation, at the higher level as limit cycles.

The literature, typically Landau and Lifshitz (1), regards sustained distributions as emerging probabilistically rather than deterministically. We propose to differ with this, in a way that can reconcile the issues. We will have the same issue in sustained turbulence in the hydrodynamic field, which is also regarded as probabilistic rather than deterministic.

If we were asked what set of abstract mathematical ideas seems best suited, operationally and "practically," for the physical phenomena we note here (not of the decaying-type motions which classical physics or mathematical physics seems reasonably capable of handling, but of persistent motion which has given rise in almost every class of phenomena in which it is observed to ad hoc quantization), we would say the nonlinear-continuous-function mathematics that can lead to Poincaré limit cycles. We realize that conceptually some readers may consider this proposed technique for developing a discrete character from continuous functions as not that much different from the ad hoc assumption of a new class of phenomena. Yet the issue we are hinting at is the relation between the eigenvalue problem and the appearance of distinct domains that seem parametrically controlled in nonlinear fashion.

When marked by such nonlinear characteristics, what emerges, it seems to us, is a spectrum of deterministic amplitudes and indeterminate and nearly statistically distributed phase that looks nearly harmonic—it may be discrete, continuous, or mixed. Since the characteristic equation sets may have more than one singular state of motion, there may appear trapped and frozen-out regions or degrees of freedom. Different near-constant environments can excite different states, and there may be more than one state associated with any dc boundary. Where we would agree with the statistical views is that, as a result, the amount of information that characterizes the system is much much less than the summation of information of the state of each individual particle. Typically, a result which is characteristically derived from Liouville's theorem is that the density ρ, the distribution function, can depend only on constants of integration of the equation of motion such as the energy integral, the momenta, or angular momenta integrals. However, more generally, it is the mean value status and some overall characterization of the spectrum which are eked out. For example, for the hydrogen atom what is required is the mean quantization, i.e., confinement to sustained orbits and discrete line spectra, e.g., the Balmer, Lyman, Brackett series. In the hydrodynamic field, say flow in a long tube under a constant pressure gradient, what is required is the time-averaged mean velocity field across the section, the relation between mean pressure gradient and mean velocity, and the spectrum of turbulence.

6. With open systems of a distributed field nature, even with nominally "isothermal" boundaries (that is, boundaries which are fairly regulated in their mean state but nevertheless allow noise to pass), there seems to be a class of phenomena for which we are not certain that we can account fully. This is the appearance in spatial domains and temporal epochs of an intermittency or wobbling of the variables of the field that blurs it into a statistical smear. (Typical phenomena are aging,[1] the ultimate lack of coherence of independent frequency sources, and the statistical nature of the turbulent field. In the first illustration, two or more systems with similar boundaries and interiors show a statistical distribution of aging characteristics. In the second illustration, two or more "oscillator" systems, built of the same internal components and exposed to similar environments, show a statistical distribution of frequency and phasing. In the third instance, with similar internal and external conditions, the frequency states exhibited by the field are statistically distributed.) We believe that nonlinear limit-cycle theory may be able to account for characteristic quantization of fields, generally by perturbation theory. However, at most this is "typical" or averaged performance, not performance for all time. The actual motion seems to be a wobbling among all such possible sets. It is as if the perturbational parameters could vary either monotonically toward some "ultimate" state or variably from epoch to epoch. For this, we have no completely satisfactory theory. It will be necessary to bring this in ad hoc on top of any nonlinear mechanics or statistical mechanics.

> An "explanation," tentatively, for such phenomena—intermittency in the turbulent field is a specific example (18)—may be the following: In distributed fields the nonlinear excitation which may occur at many points in the field is not necessarily unique. Thus briefly, all sorts of modes—decaying, sustaining, or building up—may come into existence. It does not appear to be the case that all of the field energy necessarily goes to the sustaining reactions. Thus, some kind of local directed transformation emerges (in the turbulent field, typically, a breakdown of large eddies toward smaller eddies). However, this transformation does not necessarily lead to a "winner take all" resultant (as it does, for example, in the G. I. Taylor cells that form between rotating cylinders). An epoch of excitation, aggregation, and degradation emerges. This may travel stochastically from point to point in some cases, or it may form a more stationary spectral field in other cases. By such means, warbling or flexible limit cycles may emerge, nonlinearly excited epoch by epoch.

7. A corollary to Liouville's theorem is Gibbs' principle of the conservation of extension-in-phase. Let

$$N = \rho(\delta p_1, \ldots, \delta p_n \delta q_1, \ldots, \delta q_n) = \rho(\delta \omega)$$

---

[1] In aging, for example, the following phenomena may appear indeterminately—embrittlement, fracture, precipitation, and replacement, loss or appearance of salient trace "impurities."

where N = number of representative points in the hypervolume $\delta\omega$ (or the probability)

$\delta p_1, \ldots, \delta q_n$ = the hypervolume (of momenta and displacements)

$$\frac{dN}{dt} = \frac{d\rho}{dt}(\delta\omega) + \rho\frac{d}{dt}(\delta\omega) = 0$$

since the number of systems is conserved.

Thus

$$\frac{d}{dt}(\delta\omega) = 0$$

This states that a boundary drawn around points in a given region in phase space may change in shape, as the system points move, but not in volume.

We may now try to discover how the statistical (that is, mean or stationary) properties of the particle system may be obtained.

8. The fundamental hypotheses that is generally proposed at this point is the ergodic hypothesis or its substitute, a priori probability in phase space. In this regard, one may read and contrast Tolman, 1927 [(4), pp. 37–42] with Tolman, 1938 [(7), pp. 59–70]. Eyring et al. (10), for example, opt for Tolman, 1938 "The modern approach ... is to regard without proof, as a fundamental postulate, that all accessible regions at phase space have equal a priori probabilities for equal volumes" (pp. 83–86).

The ergodic hypothesis, developed through the work of Boltzmann, Maxwell, Gibbs, and others, did not seem to be provable by classical mechanics and in its original form seemed to be wrong. In the quasi-ergodic form, it essentially states that the trajectory of an individual system in time would pass quite close to any point in phase space whose energy was compatible and whose region was accessible to it. Thus, one may expect time averages for an individual system to be near ensemble averages over accessible phase space. Criticisms are twofold. There may be closed paths, such as astronomical orbits, all with the same energy content. In fact, Tolman (1927) states, "Indeed no single example of a dynamical system of more than one degree of freedom has ever been discovered in which the motion carries the system through all possible phases consonant with its energy content." Secondly, as in the nonoccurrence in 5 billion years of planetary orbits off the ecliptic with the same possible energy levels, it may take an inordinately long time.

Thus the more elementary and modern view may be adopted. Simply stated, the representative point of a system is equally likely to be in any region of phase space it visits. However, we will regard this as being essentially equivalent to a quasi-ergodic form.

This requires discussion. The following remarks, while speculative, seem to us essential. In a simple system with highly idealized characteristics, such as perfectly elastic and geometric spheres that do not interact with walls or interact with perfect walls, the coherence may be quite extensive in time. However, such a system fashioned with any kind of reality would not persist. It would likely decay in some simple fashion. (The cumulation of perturbations would ultimately take it toward some catastrophic end.) In order to persist, some kind of graininess and internal structure and some kind of interaction with the walls must exist. Questions of the nature of nonholonomic constraints become very important. The philosophy of Hertz with regard to mechanics is most worthy of study for these questions. More specifically, it appears that the conditions for persistence are quantizable particles within the system, quantizable particles within the wall, transfer mechanisms, transport mechanisms with losses, and a lack of sharpness—namely, a possession of fuzziness—at the boundary of each particle. The interactions are not perfectly coherent, but as a result there is a diffusive propagation throughout the system's phase space. This diffusive quality of geometrically imperfect and lossy but quantizable elements is what brings the quasi-ergodic hypothesis into being. For example, we have thereby regarded the topography of a large geographic unit—the United States—as being quasi-ergodic in geological time. The landscape is worn down, but earth-disruptive forces, volcanic action, keep building the land up. Thus, the sequence of such topographic states makes up an ensemble. The fuzziness of statistical results is much greater in this field than for molecules, but the underlying ideas remain the same. The difficult problem, at every level, is to account first for the quantization and second for the smearing of coherence into epochs.

To clarify this, let us ask, from our point of view, why an ergodic hypothesis is necessary. In our view, the interactional quantizations—particles or radiation and wall—arise from some nonlinear process and form limit cycles. The motions bring very similar particles near each other. Jumps or interchanges between one limit-cycle stable pattern to neighboring ones are easy. "Phenomenologically" they only seem to involve energy steps of the order of the small quantity hv. Such changes, commensurate, as fluctuations, with the nominal energy available in toto and commensurate with the time scale that is provided by collisions or collisionlike interactions, quickly spread the system out through space-filling phase space tracks. In the idealized case mentioned before, such steps led to divergence. Here, in interacting nonlinear jumps, they stumble upon each other and lead to a fixed statistical spread. It is the assurance that epochs exist for nonlinear reasons, but which so far must be brought in ad hoc, that leads us to accept the same point of view as ergodic theory. If the "vital principle" of the system leaked off, it could not remain ergodic. Physically, we can only see that this means the "vital principle" of energy—because the nonlinear interactions all involve energy states.

9. We can proceed by considering a microcanonical ensemble. There will be a uniform density in phase in the hypershell bounded by $E_0 + \delta E/2$,

$E_0 - \delta E/2$ for all time. Any system at random can come from any one of the $\delta\omega$ equal differential hypervolumes into which we may divide this space. We propose in this section to examine a few properties of this equilibrium system.

For a system of many similar particles, say n, each with s degrees of freedom, each particle will have a representative point in the $2s$ dimension μ space. Divide μ space into a large number of equal small cells (in quantum mechanical formulations there may be limits to the subdivision)

$$(\delta\omega)_i = (\delta q_1, \ldots, \delta q_s \delta p_1, \ldots, \delta p_s)_i$$

where $(\delta\omega)_i =$ the i divisions of equal magnitude cells.

The γ space may be taken as the $2ns$ dimension-space that comprises all of the particles. The elementary volume is

$$(\delta\omega)_\gamma = [(\delta q_1, \ldots, \delta p_s)_1 \ldots (\delta q_1, \ldots, \delta p_s)_n]_\gamma$$

We are interested in the number of distributions of particles per unit cell independent of identification of the individual. The interchangeability of particles in μ space lessens many of the distinct states of the system. The number (i.e., the a priori probability) is

$$N = \frac{n!}{(n_1!) \ldots (n_i!)}$$

This enumeration is based on putting n_1 of n objects in a cell, n_2 of $n - n_1$ objects in a second cell, etc. Thus, i is the number of cells, since at that point all the objects have to be exhausted.

We wish to find the most probable distribution. Thus, we have to extremize the number of states. However, we have some constraints.

$$\sum n_j = n$$
$$\sum n_j \epsilon_j = E$$

Now to maximize

$$\ln N = \ln n! - \sum \ln n_j!$$

The Stirling approximation to $n!$ (for large n) is

$$n! = \sqrt{2\pi n} \left(\frac{n}{e}\right)^n$$

$$\ln n! = n \ln n - n + \tfrac{1}{2} \ln n + \tfrac{1}{2} \ln 2\pi$$

$$\approx n \ln n - n$$

Taking the variation of each equation

$$\delta n = \sum \delta n_j = 0$$
$$\delta E = \sum \epsilon_j \delta n_j = 0$$
$$\delta \ln N = -\sum (1 + \ln n_j) \delta n_j = 0$$

by the use of Lagrange's multipliers, we have

$$\sum(-1 - \ln n_j + \alpha - \beta\epsilon_j)\delta n_j = 0$$

Since δn is arbitrary

$$1 - \ln n_j + \alpha - \beta\epsilon_j = 0$$

$$n_j = C \exp(-\beta\epsilon_j)$$

Similarly, if instead there were a number of different particle species,

$$a_i = C_1 \exp(-\beta\epsilon_i)$$

$$b_j = C_2 \exp(-\beta\epsilon_j)$$

This law, the Maxwell–Boltzmann distribution law, expresses the distribution of a system of ac active particles (that are likely "involved" in limit cycles) which are at or near equilibrium and which have a uniform density in phase (or satisfy the quasi-ergodic hypothesis in some constructive sense).

What is viewed as having very deep significance in statistical mechanics is that, in order to obtain a valid invariance as a result of collisions, it is the logarithm of the distribution functions that should possess summational invariance; that is, that the logarithm of any distribution function can only be a linear combination of the scalar invariants of the particular species. This path also leads to the Maxwell velocity—really kinetic energy—distribution. However, we would like the reader to note that it is the invariance after collisions—as with a wall, as well as with other members—that leads to the most probable distribution.

The evaluation of the constants can be performed as follows: Since the number of particles is large, we shall replace sums by integrals. Since the equal space elements are $dq_1, \ldots, dp_n$ and n_i is the distribution of a particular species in this μ space, we may expect

$$dn = nC \exp(-\beta\epsilon)\, dq_1, \ldots, dp_n$$

$$dn = nC \int \cdots \int \exp(-\beta\epsilon)\, dq_1, \ldots, dp_n$$

$$C = \frac{1}{\int \cdots \int \exp(-\beta\epsilon)\, dq_1, \ldots, dp_n}$$

Thus

$$dn = \frac{n \exp(-\beta\epsilon)\, dq_1, \ldots, dp_n}{\int \cdots \int \exp(-\beta\epsilon)\, dq_1, \ldots, dp_n}$$

To evaluate the constant β, the essential idea is that because of the summational invariance of the logarithm of the distribution function one may expect β to remain constant as other systems in equilibrium with the given system are added. (It must be recognized that somehow the constancy depends on achieving the equilibrium through the walls, i.e., in a phenomenological sense through "isothermal" walls, not "adiabatic" in any isolational sense.)

An easy way is to put the system in contact with an ideal gas, say, a volume V carrying n molecules of a dilute monotonic gas, such that

$$pV = nkT$$

where p = absolute pressure
k = Boltzmann's constant (R/N)
N = number of molecules in one mole of gas.
m = mass of the gas

For that system

$$dn = nC_1 \exp(-\beta\epsilon) \, dx \, dy \, dz \, d\dot{x} \, d\dot{y} \, d\dot{z}$$

Integrating over the volume

$$dn = nC_2 \exp(-\beta\epsilon) \, d\dot{x} \, d\dot{y} \, d\dot{z}$$

The rebound of particles over the wall represents the change of momentum and thus the force, such as the force per unit area (or pressure) on the wall.

$$p \, d\sigma = 2\left(\frac{\dot{x} \, d\sigma}{V}\right)\left[\int_0^\infty \int_0^\infty \int_0^\infty m\dot{x}^2 nC_2 \exp(-\beta\epsilon) \, d\dot{x} \, d\dot{y} \, d\dot{z}\right]$$

where $d\sigma$ = cross-sectional area of a column extending from the wall

$\dot{x} \, d\sigma$ = volume whose particles will reach the wall in a second

$\dfrac{\dot{x} \, d\sigma}{V}$ = fraction of all molecules per second

$\left(\dfrac{\dot{x} \, d\sigma}{V}\right)\left(\displaystyle\iiint_{\dot{x}}^{\dot{x}+d\dot{x}}\right)$ = actual number that will reach per second (call these terms T_1 and T_2)

$2m\dot{x}(T_1)(T_2)$ = the momentum exchange

The last expression finally requires integration for all $\dot{x}$ from 0 to ∞

$$pV = \int_{-\infty}^\infty \int_{-\infty}^\infty \int_{-\infty}^\infty nC_2 m\dot{x}^2 \exp(-\beta\epsilon) \, d\dot{x} \, d\dot{y} \, d\dot{z}$$

But

$$n = \int_{-\infty}^\infty \int_{-\infty}^\infty \int_{-\infty}^\infty nC_2 \exp\left[-\beta\frac{m}{2}(\dot{x}^2 + \dot{y}^2 + \dot{z}^2)\right] d\dot{x} \, d\dot{y} \, d\dot{z}$$

$$= \int_{-\infty}^\infty \int_{-\infty}^\infty \int_{-\infty}^\infty nC_2 \exp\left[-\beta\frac{m}{2}(\dot{x}^2 + \dot{y}^2 + \dot{z}^2)\right]\beta m\dot{x}^2 \, d\dot{x} \, d\dot{y} \, d\dot{z}$$

the last result being the remnant after integrating by parts.

Thus

$$pV = \frac{n}{\beta}$$

from which

$$\beta = \frac{1}{kT}$$

Thus we find that β is a measure of the ideal gas temperature, which can ultimately be shown to be the thermodynamic temperature. One must note that the property is transmitted intensively from system to wall to system.

This derivation from mechanistic energetics concepts to a system invariant, the temperature, provides an excellent tie between mechanics and thermodynamics. For example, (9) and (10) develop the formalism quite thoroughly. Thermodynamics can be well structured on the basis of statistical-mechanical foundations. It has not (by the possible path of radiation equilibrium with walls) provided an adequate tie from mechanics to electromagnetic theory. It is in such a tie that the quantization issues, and perhaps the forcing mechanism that keeps the system going and which cannot be shielded against, arise.

Our distribution law can thus be expressed as

$$dn = \frac{n \exp(-\epsilon/kT) \, dq_1, \ldots, dp_n}{\int \cdots \int \exp(-\epsilon/kT) \, dq_1, \ldots, dp_n}$$

10. Averages for any property F of the system may now be found from

$$\bar{F} = \frac{1}{n} \int F \, dn = C \int \cdots \int \exp\left(-\frac{\epsilon}{kT}\right) F \, dq_1, \ldots, dp_n$$

$$C = \frac{1}{\int \cdots \int \exp(-\epsilon/kT) \, dq_1, \ldots, dp_n}$$

As a few examples:

a. What is the temperature coefficient of C?

$$\frac{dC}{dt} = -\frac{1}{(\int \cdots \int)^2} \int \cdots \int \exp\left(-\frac{\epsilon}{kT}\right) \frac{\epsilon}{kT^2} \, dq_1, \ldots, dp_n$$

$$\frac{1}{C} \frac{dC}{dt} = -\int \cdots \int \exp\left(-\frac{\epsilon}{kT}\right) \frac{\epsilon}{kT^2} \, dq_1, \ldots, dp_n$$

$$= \frac{\bar{\epsilon}}{kT^2}$$

that is, it is determined by the average value of the energy.

b. What is the average molecular velocity for simple mobile molecules?
Suppose that the energy of the molecule can be decomposed as

$$\epsilon = \epsilon(x, y, z) + \tfrac{1}{2}m(\dot{x}^2 + \dot{y}^2 + \dot{z}^2) + \epsilon_i$$

a potential   kinetic   energy associated with
energy of   energy   other coordinates
position   and momenta

$\qquad\qquad\qquad\qquad\qquad\qquad\qquad\qquad\qquad\qquad (q_4, \ldots, q_n, p_4, \ldots, p_n)$

If this is possible (i.e., weak coupling), then

$$dn = nAV \exp\left[-\frac{1}{2}m\frac{(\dot{x}^2 + \dot{y}^2 + \dot{z}^2)}{kT}\right] d\dot{x}\,d\dot{y}\,d\dot{z}$$

having integrated over the other variables.
Let

$$\dot{x}^2 + \dot{y}^2 + \dot{z}^2 = c^2$$

$$d\dot{x}\,d\dot{y}\,d\dot{z} = c^2 \sin\alpha\,d\alpha\,d\beta\,dc$$

where c = total velocity magnitude

$\qquad\quad \alpha, \beta$ = polar coordinates

$$dn = nVA \exp\left(-\frac{mc^2}{2kT}\right) c^2 \sin\alpha\,d\alpha\,d\beta\,dc$$

Integrate α from 0 to π, and β from 0 to 2π

$$dn = 4\pi nVA \exp\left(-\frac{mc^2}{2kT}\right) c^2\,dc$$

Integrate c from 0 to ∞, then

$$n = 4\pi nVA \frac{1}{4}\sqrt{\frac{8\pi k^3 T^3}{m^3}}$$

$$VA = \left(\frac{m}{2\pi kT}\right)^{3/2}$$

$$dn = 4\pi n\left(\frac{m}{2\pi kT}\right)^{3/2} \exp\left(-\frac{mc^2}{2kT}\right) c^2\,dc$$

$$\bar{c} = \sqrt{\frac{8kT}{\pi m}} \left(= \sqrt{\frac{8RT}{\pi M}} = 14{,}500 \sqrt{\frac{T}{M}}\,\frac{\text{cm}}{\text{sec}}\right)$$

where $\bar{c}$ = average velocity
$\qquad\quad M$ = molecular weight

If, instead, the square of the velocity is sought

$$\left(\sqrt{\overline{c^2}} = \sqrt{\frac{3kT}{m}} = \sqrt{\frac{3RT}{M}}\right)$$

(Note: the velocity of sound is given by $c = \sqrt{\gamma RT/M}$. Thus these molecular velocities all possess near-sonic-velocity magnitudes.)

c. Derive the theorem of the virial and the pressure of a gas. Consider a system of mobile molecules. Newton's law for the cg motion will be $M\ddot{x} = X$, etc., where X = force, x = displacement. Write

$$\frac{d}{dt}(mx\dot{x}) = m\dot{x}^2 + mx\ddot{x}$$

so that

$$(m\dot{x}^2 + xX)\,dt = d(mx\dot{x})$$

Integrate over a long time interval, so that

$$\overline{m\dot{x}^2} + \overline{xX} = \frac{(mx\dot{x})_{t_2} - (mx\dot{x})_{t_1}}{t_2 - t_1}$$

If the parameters are bounded, then the right hand approaches zero as the interval grows. Thus

$$\overline{m\dot{x}^2} + \overline{xX} = 0 \qquad \text{etc.}$$

or

$$\sum \overline{{\scriptstyle\frac{1}{2}}mc^2} = -\sum {\scriptstyle\frac{1}{2}}(\overline{xX} + \overline{yY} + \overline{zZ})$$

This is the theorem of the virial. It relates the total kinetic energy of all the molecules to the right hand, Clausius' virial. It can be developed further to give the pressure of a gas (i.e., its equation of state).

The stationary walls of a container exert a (countering) pressure (force per unit area) to the molecules of gas. This is assumed to be the same throughout the vessel. The pressure exerted by any area element $d\sigma$ of the wall is $-p(l\,d\sigma)$, $-p(m\,d\sigma)$, $-p(n\,d\sigma)$, where l, m, n are direction cosines of the normal to the surface. Thus the contributions to ΣxX by pressure (an external force) is

$$-\iint lxp\,d\sigma = -p\iint lx\,d\sigma$$

or the total contribution to the virial is

$$\frac{1}{2} p \iint (lx + my + nz)\, d\sigma = \frac{1}{2} p \iiint \left(\frac{\partial x}{\partial x} + \frac{\partial y}{\partial y} + \frac{\partial z}{\partial z} \right) dx\, dy\, dz$$

$$= \frac{3}{2} pV$$

(using Green's theorem).

Now for the internal (intermolecular) forces. If the intermolecular forces are central forces depending only on the separation φ, and a pair of molecules have coordinates $x, y, z; x', y', z'$; with the component forces X, Y, Z; and X', Y', Z' acting on them, then:

$$X = \varphi(r)\frac{x - x'}{r} \qquad X' = \varphi(r)\frac{x' - x}{r} \qquad \text{etc.}$$

The contributions to ΣxX by these forces are

$$xX + x'X' = \frac{\varphi(r)}{r}(x - x')^2$$

and the total contribution to $\frac{1}{2}\Sigma(\overline{xX} + \overline{yY} + \overline{zZ})$ is

$$\frac{1}{2}\frac{\varphi(r)}{r}[(x - x')^2 + (y - y')^2 + (z - z')^2] = \frac{r\varphi(r)}{2}$$

Thus in total

$$\tfrac{1}{2}\sum mc^2 = \tfrac{3}{2}pV - \tfrac{1}{2}\sum\sum r\varphi(r)$$
$$pV = \tfrac{1}{3}\sum mc^2 + \tfrac{1}{3}\sum\sum r\varphi(r)$$

Jeans (2) points out an important observation of Maxwell's. Suppose we were to consider that pressure arose only from intermolecular force. By Boyle's law, $pV = $ const. $\Sigma\Sigma r\phi(r)$ would have to be constant, or $\phi(r) = 1/r$. Molecules would have to repel with a force inverse to distance; i.e., remote molecules would predominate in effect over the nearby ones. Adding vessels would change external effects drastically. Since this is so remote from the inverse square law, which retains a geometric similarity, such a state of affairs seems quite inconceivable. Thus pressure cannot be explained solely from repulsive forces between molecules but must have a considerable component depending on the sustained motion of the molecules and thus on the interaction with the wall.

When we come to interaction between entities which do not have a cg Newtonian law of motion—such as humans—and which at most have only inter-entity binding forces, we will detect a considerable change in the nature of the equation of state, since momentum conservation or exchange will no longer be an independent equation of motion. The coalescence of such entities thus tends to be much more liquidlike.

However, we are again compelled to inquire whether the sustained motion does not arise from the wall or substrate, for if we gradually slow down the molecular motion at the walls (cool the walls), the molecular motion slows down.

Thus in case after case, if we overlook the question of the detailed transfer, we logically seem to be driven to assuming excitation from and equilibrium (in a non-linear sense) with the wall or substrate.

Without seeking any mystical or metaphysical character to the problem, we are forced to the following crude visualization. Any flux, continuous or discrete, when it meets a wall discovers that there is particle quantization, i.e., the atomic theory of matter. From a m/e-m interaction minimally, in a classic sense, or currently from probabilistic rules determined by quantum theory, energy quantization takes place. The transmitted flux thereby is also quantized by the wall. Insofar as many fluxes originate from bounded systems, except for the possibility of intermittent or continuous particle pair production which thus emerge already quantized, there tends to be a continual transfer from quantizing system to quantizing system by means of walls. The extracosmological nature of pair production, of course, remains a mystery, although it seems to require a particle as a "wall" or substrate on, against which, or in whose neighborhood the pair can form.

In a classical sense, one would expect degradation in systems. However, the issue is how much. From a period of belief not so long ago, the cosmological processes were pushed back from an age of 5,000 years to tens of millions of years to the present "firmer" estimate of 10^{10} years (or infinity—depending on one's current theoretical views). In any case, radiation in "space" does not degrade rapidly. Our earth processes are largely maintained by the sun; i.e., if the solar source were removed our temperature would notably diminish. Thus, the vaunted continuation of atomic processes—here on the earth at least—has a very proximal "cause." A space voyage with little material capacitance, involving the vicissitudes of helium-liquefaction temperatures and metal-boiling temperatures in its track among the stars, would quickly provide a different picture of the time constancy of processes we tend to take for granted in our sheltered earth home. It is only the ubiquitous but weak and long-range gravity force that is then left to provide the remaining tie that binds. Much more locally, of course, there remain nuclear forces.

"Practically" the virial relation is not immediately too useful in that the central problem remains how to compute the force fields. However, it stresses the fact that deviations from ideal-gas laws depend on the intramolecular forces and is useful, following Boltzmann, for the derivation of such second-order terms as

$$pV = nRT\left[1 + \frac{b}{V} + \frac{5}{8}\left(\frac{b}{V}\right)^2 + \cdots\right]$$

or to begin more general virial expansions for real gases in the reciprocal of volume

$$pV = nRT\left[1 + \frac{b_1}{V} + \frac{b_2}{V^2} + \cdots\right]$$

or in pressure

$$pV = nRT(1 + a_1p + a_2p^2 + \cdots)$$

d. Equipartition of energy. Starting from

$$\int \cdots \int C \exp\left(-\frac{\epsilon}{kT}\right) dq_1, \ldots, dp_n = 1$$

partially integrate, say, with regard to any q_i or p_i—illustratively q_1

$$\int \cdots \int C \exp\left(-\frac{\epsilon}{kT}\right) q_1 \, dq_2, \ldots, dp_n \Big|_{(q_1)_1}^{(q_1)_2}$$

$$+ \int \cdots \int C \exp\left(-\frac{\epsilon}{kT}\right) \frac{q_1}{kT} \frac{\partial \epsilon}{\partial q_1} dq_1, \ldots, dp_n = 1$$

If ϵ is independent of q_1, the second term vanishes. The result is not interesting. We are concerned with ϵ depending on q_1. In particular, we are concerned with the case when the dependence on q_1 (or q_i or p_i) is such that for remote limits $(q_1)_1$ and $(q_1)_2$ (or similar for other q_i or p_i), $q_1/\exp(\epsilon/kT)$ approaches zero. This is true for such things as springlike potentials in forces associated with displacement or with the quadratic nature of momenta. (Generally it is the quadratic character of the positive, definite form of the energy which is being appealed to.) If true, then

$$\int \cdots \int C \exp\left(-\frac{\epsilon}{kT}\right)\left(q_1 \frac{\partial \epsilon}{\partial q_1}\right) dq_1, \ldots, dp_n = kT$$

or

$$\overline{\left(q_i \frac{\partial \epsilon}{\partial q_i}\right)} = kT \qquad \overline{\left(p_i \frac{\partial \epsilon}{\partial p_i}\right)} = kT$$

for other such similar variables.

In particular, if

$$\epsilon_i = \tfrac{1}{2} a_i q_i^2$$

$$\epsilon_i = \tfrac{1}{2} b_i p_i^2$$

then

$$\bar{\epsilon}_i = \tfrac{1}{2} kT$$

the law of equipartition of energy for such coordinates or momenta. One must note that temperature is a measure of the average kinetic or potential energy associated with the fluctuating positional and momenta variables.

This is as far as it pays to take statistical mechanics from the elementary point of view of Tolman (1927).

The topics that still require discussion are:

Thermodynamics
The Boltzmann equations
Equations of change for near-equilibrium processes
Radiative and quantum mechanical processes
Some attempt at a provisional summary for the continuum and the
     quantizing process

4-3   SOME RUDIMENTARY STATISTICAL THERMODYNAMICS

We have provided some of the rudiments of the equilibrium statistical
mechanical treatment for active entities. This—to first approximation—
can serve as the foundation for thermodynamics. Thus, sketchy application
of equilibrium statistical mechanics to thermodynamics is a desirable next
step. Fowler and Guggenheim [(9), pp. 55–65] provide an excellent short
introduction.

As they point out, the ideas of thermodynamics are foreign to the
foundations of statistical mechanics, which are dynamic. The proper course
is to prove that the laws of thermodynamics are true for the assemblies of
statistical mechanics, if suitable analogies are used for interpretation. Gibbs
is offered as a persuasive source for this concept, and Caratheodory's
formulation in simplified form is proposed.

Postulate 1—the "zeroth" law of thermodynamics   If two assemblies
are each in thermal equilibrium with a third assembly, they are in thermal
equilibrium with each other.

From this, it follows that there is equality of a single-valued function
of the "thermodynamic state" of the ensembles which may be called temperature t. Any ensemble (e.g., the near-ideal gas or ideal gas) may be used as
a "thermometer," reading some t on a scale. Thus, the existence of temperature
may be known as the zeroth law.

As Tolman (1938) points out, in considering appropriate ensembles for
representing a system in thermodynamic equilibrium, the systems are never in
perfect isolation but purposely placed in thermal contact with each other. He uses
this to justify a canonical ensemble as an ensemble whose systems do not possess
identical energies but closely distributed energies.

Our basic stumbling block always remains the walls. For the ensembles we
have required that they be enclosed systems, which can be obtained, without
mixing, by permitting both "thermal" contact and flexible diaphragms or pistons
at the walls. However, to get around the radiation problem, we have to insulate all
walls (i.e., it appears we must make the walls both "isothermal" and "adiabatic").
This can be done if both $T = $ const and the gradient $\nabla T = 0$ (namely, all temperature derivatives are zero. "Practically" this can be done by guard-ring techniques.
One may enclose a composite system of ensembles in a "concentric" enclosure

whose walls are maintained at a desired temperature with reference to its surrounds by heaters or refrigerators. Alternately, instead of a "concentric" enclosure, a collection of neighboring enclosures can be used to surround and guard the fluxes.) The transfer of energy thus has to take place by exchange with the wall, namely, by electromagnetic, not mechanical, exchanges. It is an assumption that now lifting a partition between two assemblies to permit exchange of their material particles, if the particles are identical, will not lead to any significant difference. We can accept this. However, we continue to wonder. How did the particles come to their nonlinear, active motional mode?

This may pose the particular problem for the case of the mechanistic system of actual molecules. However, we have long speculated on thermodynamiclike concepts in most active oscillator systems. Thus, more generally, we are concerned with the "closure" of all such systems and the "identical" nature of the particles in each system. "Thermal-mechanical" contact allows transfer of many fluxes of communications, energy, and mass nature, but not of the segregated individuals. Excitation of the bounding walls is necessary for the transfer that does take place. The property, it seems to us, which emerges most commonly is a number of quantizations that take place within each enclosure, generally not as sharply quantized as atomic phenomena but nevertheless quantized. The autonomous biochemical oscillator is an illustration.

The thought formally expressed by Landau and Lifshitz [(1), p. 39] for mobile systems in thermodynamic equilibria is that in thermal equilibrium such additive quantities as energy and entropy do not depend on the shape of a body, only its volume. There is a hint in this that the quantization does not depend upon shape or volume, only its temperature. Since temperature is to be associated with energy, particularly kinetic energy, one is faced with the possible problem of always relating kinetic energy in wall components with electromagnetic radiation.

This first postulate follows from statistical mechanics because, as one notes, we already used the intensive property of comparing our system with a thermometric substance, say an ideal gas, $pv = nRT$, to derive the magnitude $\beta = 1/kT$. We would simply repeat the derivation.

Postulate 2—the first law of thermodynamics If a thermally insulated path can be taken from state 1 to state 2 by alternate paths, the work done on the assembly has the same value for each such (adiabatic) path.

From this, it follows that there exists a single-valued extensive function U of the state of an assembly, called its internal energy, which increases with the work done on the assembly. If mechanically isolated as well,

$\Delta U = \Delta W$ (adiabatic)

$\Delta U = 0$ (isolated and adiabatic)

Heat Δq absorbed by an assembly is now defined by

$\Delta q = \Delta U - \Delta W$ (for all transfer processes)

i.e., this is the first law of thermodynamics.

In our view, the system enclosure is maintained at both isothermal and adiabatic equilibrium by guard volumes. Thus work can be done through communicating pistons or diaphragms. Loosely, this theorem states that the energy added through work is immediately partitioned both spatially in displacements and temporally through momenta, that is, in the phase space.

However, the energy exchanges must pass through the walls, and the transmission through the walls must always involve electromagnetic fluxes. Further, to keep the walls "adiabatic," the surrounding systems must also be raised in "temperature" if the processes are to be kept fully at equilibrium. Thus the process of adding energy is not a simple one.

This postulate states that an equipartition of energy in phase space will take place and that the energy found within the system only depends on how much energy is added to the system. It proposes that the only way known to add energy in the first instance (that is, by mechanical means) is to do work on the system (i.e., work and energy are equated. In fact, there is the essential implication that the energy of a system can only be changed mechanically by doing work on the system.). However, it then suggests that there may be other (equivalent) energy transfers. This transfer is known as "heat" transfer.

The postulate does not make the idea explicit, but it is transfer, as through walls. This is a transfer by a second essential means, namely electromagnetic energy, not mechanical. As prototype: in conduction, there is transfer from system 1 to the walls by e-m, and from the walls to system 2 by e-m; in convection there is a combination of mixing of systems in phase and "conductive" transfer; in "radiation," as with "transparent" gases, there is e-m transfer. As background to all such equalizations, there is the theory of transfer which is identified as Einstein's emission and absorption. See Einstein (1917) in (19), or see the Dirac theory in Heitler (20).

It appears that one may also infer that the internal energy is a function of temperature and pressure. The assumption was not made explicit but it seems also assured that such "heat" transfer will take place from regions of one temperature —the higher—to regions of a second temperature. However, it is our contention that this assumption is essentially identical to the one of e-m transfer from a region of one mechanical-energy level to that of a second mechanical-energy level.

Phenomenologically, we are able to proceed with this concept because we can find a thermometric substance, fix a temperature scale, measure differences of temperature, fix a calorimetric substance as a receiver (such as a water-substance system), and, through the equivalence of work and heat transfer, change the same amount of internal energy (as detected by temperature and pressure, namely, as originally done by Rumford or then refined by Joule).

Postulate 3—the second law of thermodynamics There exist two single-valued functions of state—*T*, called the absolute temperature, and *S*, called the entropy—such that *T* is a function of *t* (the thermometric temperature only), the entropy is an extensive property (namely, the entropy of any assembly is equal to the sum of the entropies of its parts), and for any infinitesimal change in any completely homogeneous assembly

$$dq \leqslant T\,dS$$

The equality holds for quasi-static processes, the inequality for natural, rapid, and thus irreversible processes. If the assembly is not completely homogeneous, the relation holds for the homogeneous parts.

We have the "absolute" temperature of the ideal gas, $pV = nRT$, as a likely candidate for temperature. As a matter of fact, it is the one we used for $\beta = 1/kT$. Once again by the use of such substance, with very slow steps, we can derive $\Delta S = \int dq/T$.

These postulates are a sufficient basis for thermodynamics. We have yet to show that the second law is derivable from or equivalent to a statistical mechanical law. We can reach this status as follows:

For a system in "complete" equilibrium, $dq = T\,dS$, so that

$$T\,dS = dU - dW$$

For quasi-static changes that are adiabatic, and mechanically isolated,

$$dW = 0 \qquad dS = 0$$

Therefore,

$$dU = 0$$

Illustrative of the process is a two-compartment container in which a frictionless gate is opened so as to permit gaseous exchanges from one filled compartment to the second, empty compartment. This is a so-called free expansion. For an ideal gas, the ultimate adiabatic exchange is $p_1 V_1 = p_2 V_2$ for the pressure and equal temperature.

This is not the same as a quasi-static adiabatic exchange in which the gas does work against the outside. Then $dS = 0, dW \neq 0$, and $dU \neq 0$. In fact for ideal gases, $p_1 V_1{}^\gamma = p_2 V_2{}^\gamma$, with a related temperature change.

For example, Zemansky's *Heat and Thermodynamics* (1937) is confusing at this point. He points out validly (p. 96) that for a thermally isolated, mechanically isolated gas process, a free expansion, "no internal energy change takes place...." On the other hand, for a quasi-static adiabatic process (p. 101–102), $pV^\gamma = $ const for all the equilibrium states that the ideal gas passes through in this process. However, he then states, "It is important to understand that a free expansion is an adiabatic process but not quasi-static," and that it is therefore fallacious to attempt to apply $pV^\gamma = $ const to the states traversed by an ideal gas during a free expansion. This is misleading. The free-expansion result would obtain whether the process was quasi-static or fast. The difference is in the mechanical isolation. In the first case $dW = 0$, in the second $dW \neq 0$.

1. Free adiabatic expansion (whether quasi-static or not)

$$T\,dS = dU + p\,dV = C_v\,dT + p\,dV$$
$$dS = 0 \qquad p\,dV = 0$$
$$dU = 0 \qquad dT = 0$$

2. Adiabatic expansion doing work (quasi-static)

$$T\,dS = C_v\,dT + p\,dV$$

$$dS = 0 \qquad p\,dV \neq 0$$

$$pV = RT = (C_p - C_v)T$$

$$dT = \frac{p\,dV + V\,dp}{C_p - C_v}$$

$$\frac{C_v}{C_p - C_v}(p\,dV + V\,dp) + p\,dV = 0$$

$$C_p\,p\,dV + C_v\,V\,dp = 0$$

$$\gamma\frac{dv}{v} + \frac{dp}{p} = 0$$

$$pV^\gamma = \text{const}$$

In our view, the work that the gas has to do against the outside is what "abstracts" energy from the system and "cools" the ideal gas down. It is not the speed with which it happens. Later we will discuss at greater length the meaning of equilibrium or near-equilibrium.

We bring the illustration up, not to do battle with a straw man, particularly since we hold an unbounded admiration for Professor Zemansky, but as a token that elementary thermodynamics books leave too myopic a view of conditions for thermodynamic or near-thermodynamic equilibrium.

However, these two cases do not cast adequate light yet on the meaning of the entropy. In both cases, the walls were involved. In the first case, the walls of both containers could have been kept at both constant temperature and no temperature difference. The distribution of momentum flux changed at the walls. Yet the internal energy and thereby the temperature—for ideal gas systems—did not change. The full e-m interactions at the walls are unclear.

For quasi-static changes that are mechanically isolated, in a temperature-regulated system

$$dW = 0 \qquad dT = 0 \qquad d(U - TS) = 0$$

The quantity $F = U - TS$ is the Gibbs free-energy function. It may be used as a substitute for entropy.

In any assembly if not completely homogeneous, each homogeneous part—called a phase—can be identified with an energy, entropy, and temperature. U and S are extensive quantities. At equilibrium, T is the same in all phases. A description of the state of the entire assembly involves describing the phases.

For each equilibrium phase, its variables are $N_A, N_B, \ldots$, the number of particle types $A, B, \ldots$, it contains; its geometric parameters x, such as volume; and, say, its entropy. Thus

$$dU = \left(\frac{\partial U}{\partial S}\right)_{x,N} dS + \sum \frac{\partial U}{\partial x}\,dx + \sum \frac{\partial U}{\partial N}\,dN$$

At constant configuration (such as volume, etc.) and composition, no work can be done on the phase. Thus, from the first law

$$dq = T\,dS = dU$$

Thus

$$\left(\frac{\partial U}{\partial S}\right)_{x,N} = T$$

if in mechanical equilibrium; $-(\partial U/\partial x)$ is simply the generalized force exerted by the phase in attempting to increase x. Call this X. Gibbs introduced $\mu = \partial U/\partial N$, partial potentials, to allow for concentration changes. Thus

$$dU = T\,dS - \sum X\,dx + \sum \mu\,dN$$

U is a function of these variables. It is the associated thermodynamic potential. There is one such formula for each phase. (Gibbs used such fundamental formulas to determine equilibrium properties.) It contains implications that heat flows from higher to lower temperature; that geometric boundaries are mechanically urged from higher pressure to lower pressure; that concentration changes, such as from chemical change, always proceed so that certain linear combinations of the partial potentials decrease. Since, by the first law, thermal equilibrium implies equal temperature, we shall postulate equal temperature throughout the assembly.

Most elementary thermodynamics books, though they adopt a number of caveats about equilibrium, do not make clear what are the "facts" in the real world. As will appear from examination of irreversible but near-equilibrium thermodynamics, most systems are quite nearly in local mechanical and chemical equilibrium, so that real processes can be nearly and correctly computed from thermodynamic parameters if, additionally, some irreversible production of entropy linearly associated with transport processes and the added parameters they introduce are allowed. These processes (although representing nonequilibrium states) change the thermodynamics very little. Thus dynamic exercises based on irreversible thermodynamics may be carried out by scientists and engineers with considerable purity, with modifications subsequently superimposed. This will gradually become clearer.

It is equally convenient (anticipating chemical problems) to use F instead of S. Thus

$$F = U - TS$$

$$dF = -S\,dT - \sum X\,dx + \sum \mu\,dN$$

Since

$$\left(\frac{\partial F}{\partial T}\right)_{x,N} = -S$$

$$d\left(\frac{F}{T}\right) = -\frac{U}{T^2}\,dT - \frac{1}{T}\sum X\,dx + \frac{1}{T}\sum \mu\,dN$$

is an alternate version of the second law.

For complete equilibrium (i.e., all changes are quasi-static),

$$dT = 0 \qquad dF = dW$$

For chemical equilibrium, keeping the geometry unchanged,

$$dT = 0 \qquad dx = 0 \qquad dF = 0$$

The issue we are up to in this subsection is how to derive the second law from statistical mechanics. In examining the various books on statistical mechanics, one notes a number of derivations. Likely Gibbs (3) or Tolman (1927) interpreting Gibbs offer the most transparent argument (the issue of validity is entirely another matter).

Tolman states that of the three thermodynamic quantities—S, dq, T— S is the only one requiring elaborate study. The heat dq absorbed by a single system can be identified "with the energy which the system receives by molecular impacts or other elementary disordered processes which defy ordinary observational methods."

We call attention again to the point that "heat" is transferred by some otherwise unidentified action-at-a-distance forces, which we have little choice but to identify as e-m. Hirschfelder et al. (21), for example, take the course of defining the internal energy—that is, the average for a large collection of systems—by

$$U = \frac{\sum E_j \exp(-\beta E_j)}{\sum \exp(-\beta E_j)} = \sum_j \bar{a}_j E_j$$

whence

$$dU = \sum E_j\,d\bar{a}_j + \sum \bar{a}_j\,dE_j$$

They then formally identify $dq = \Sigma E_j\,d\bar{a}_j$ as the change in energy due to a redistribution of the total energy over the various "quantized" states of the system. We put "quantized" in quotations because—whereas a quantization is necessary in some nonlinear sense—it is not proven fact it must be quantum mechanical. Thus, a possible m/e-m quantization remains open.

Similarly, they put $dW = -\Sigma \bar{a}_j\,dE_j$ as a shift in energy state by alternation of "volume" or, really, of any other external variable. This then is how they interpret the first law by statistical mechanics.

With regard to temperature, as we showed previously, Tolman (1927) proposes that the measure of temperature T comes from the contact at equilibrium with a thermometric substance.

Thus the problem within the second law is entropy. Phenomenologically (that is, from phenomenological thermodynamics in which $dU = T\,dS - \Sigma X\,dx + \Sigma \mu\,dN$) for an equilibrium system which is isolated, any heat fluxes which were added quasi-statically before isolation give

$$dq = T\,dS$$

an increase in entropy. (We will regard isolation as effective only when a system is immersed among a collection of systems at the same temperature, i.e., temperature regulated and "adiabatically" isolated by whatever means is necessary to prevent a flux of heat in. Since real walls are both conductive and "radiative," it requires both thermostating and prevention of gradients.) If isolated parts of the same homogeneous phase had different temperatures, the same kind of optimalization (that is, flow of local heat) would result in increases in local entropy unless it was extensively additive.

Now the probability w of the state of a system is proportional to its γ weight. It would seem natural that an active isolated atomistic system would change from lesser probability to greater probability (namely, toward what has been expressed as its equipartition of energy). Thus, entropy may be functionally related to probability.

$$S = f(w)$$

However, entropy is additive, whereas probability—for combined states—is multiplicative. Thus, one might expect

$$S = a - k \ln w$$

This is Boltzmann's result, to bridge thermodynamics and statistical mechanics. A scale factor k is taken as Boltzmann's constant.

However, we have for the γ weight of a state

$$\ln \gamma = C - \sum_i N_i \ln N_i$$

Thus

$$S = -k \sum N_i \ln N_i + C$$

In Gibbs' first analogue for identifying the thermodynamic parameter, he starts from the canonical ensemble distributed in γ space with density

$$\rho = N \exp\left(\frac{\psi - E}{\theta}\right) = N \exp\left(\frac{F - E}{kT}\right)$$

where ψ has been tentatively identified as the free energy.

$$w_n = N \exp\left(\frac{F - E_n}{kT}\right)$$

will be the probability for an individual

$$\sum_n w_n = \exp\left(\frac{F}{kT}\right) \sum_n \exp\left(\frac{E_n}{kT}\right) = 1$$

$$F = -kT \ln \sum \exp\left(-\frac{E_n}{kT}\right)$$

is the basis for thermodynamic applications. In principle, the thermodynamic function of the system can be computed from its energy spectrum [Landau and Lifshitz (1)]. (One should always remember this as the bridge from experimental quantization to continuum thermodynamics.)

$$S = a - k \ln w_n$$

$$= (a - k \ln N) - \frac{F}{T} + \frac{U}{T}$$

Since

$$\ln w_n = \ln N + \frac{F - E_n}{kT}$$

$$\overline{\ln w_n} = \ln N + \frac{F}{kT} - \frac{\bar{E}}{kT} = \ln N + \frac{F - U}{kT}$$

Thus, if

$$a = k \ln N \qquad \text{then} \qquad F = U - TS$$

which is the valid definition of F.

(What we have shown is that

$$F = -kT \ln \sum \exp\left(-\frac{E_n}{kT}\right)$$

is a valid statistical mechanical definition of F, consistent with its thermo-dynamic definition.)

The quantity Z

$$Z(T) = \sum_n \exp\left(-\frac{E_n}{kT}\right) = \exp\left(-\frac{F}{kT}\right)$$

is called the partition function.

Fowler and Guggenheim (9) prefer discarding Boltzmann's hypothesis relating entropy to probability as a basis for showing that F and T have all the correct statistical mechanical and thermodynamic interrelations. Their argument cannot really be simplified. Interested readers should read their pages 1–13.

However, their ultimate results, for classical statistics (actually for large number of states), are that if

$$F = \sum F_A$$

then

$$F_A = -N_A kT[\ln Z(T) - \ln N_A + 1]$$

$$\mu_A = kT \ln \frac{N_A}{Z(T)}$$

$$Z(T) = \sum \exp\left(-\frac{E_n}{kT}\right)$$

expresses the sum of noninteracting free energies; their dependence on the partition function; and the partial chemical potentials in terms of the partition function. Although very sketchy (the reader may examine the references for greater detail), this section must stand as all the introduction to statistical thermodynamics that this study can afford.

4-4   INTRODUCING THE STATISTICAL MECHANICS FOR RADIATION

1. It is now appropriate to put forth some introductory material for the statistical mechanics of radiative processes. We can draw from Tolman (1927, 1938) for guidance.

We do not wish to adopt an entire quantum mechanics in ad hoc fashion. Instead we shall try to confine the assumptions.

Basically, we are constrained to the observation that an enclosure held in thermal equilibrium by its "thermostating" surroundings (so that neither temperature in the enclosure changes nor gradients near the enclosure exist) will show molecular-atomic processes of sustained oscillations in a single material phase (i.e., any atomic material within the enclosure will form possible molecular complexes. These atoms or molecules will exhibit extensive spectra, characteristic of the species and of the temperature.).

To simplify our problem, we will regard the material phase to be a near-ideal gas (i.e., one well above the critical temperature, with a moderate magnitude for the ratio of its pressure to critical pressure). The "kinematic" phenomenology is a spectrum of oscillator modes that show nearly iso-chronous lines with narrow (high Q) stationary line widths. The spectrum is demonstrated by passing weak electromagnetic waves, e.g., light, through the medium and noting their absorption and dispersion.

One may note the following implications and inferences: the key observation of isochronism of the oscillators with changes of e-m wave amplitude or with changes in pressure; the linear nature of the electrodynamic equations and the implication of superposition properties of the radiation field (i.e., the spectrum does not depend on the exciting amplitude or on the particle density).

In the fairly recent source, Landau and Lifshitz [(1), p. 171] say, "Black-body radiation may be considered as a 'gas' made up of photons. The linear nature of the electrodynamic equations indicates that photons do not interact with one another (the principle of superposition of the electromagnetic field), so that a 'photon gas' is an ideal one... this gas obeys Bose statistics.

"If the radiation is not in a vacuum but in a material medium, then for the photon gas to be ideal, the interaction of the radiation with the medium must also be small. This condition is fulfilled for gases....

"It must be borne in mind that the presence of some matter, however little, is in general necessary to enable thermal equilibrium to be set up in the radiation, since one may assume there is no interaction between the photons (apart from negligible interactions due to the possible existence of virtual electron positron pairs). The mechanism which brings about the establish-ment of equilibrium is, in this case, the absorption and emission of photons by the matter... leads to a very important property peculiar to the photon gas.... [The] number of particles in the... gas is variable [and] must be determined from the conditions of thermal equilibrium."

These thoughts all suggest near-linear, that is, nearly pure, harmonic oscillators, without losses. Thus, whether taken from Planck; from Bohr, as expressed in the Wilson–Sommerfeld rules; or as expressed by Heisenberg and Schrödinger, the tack taken in physics has been the quantization by an ad hoc linear quantum theory of mechanics. The reference to its ad hoc nature is contained in the fact that quantum mechanics still has formal connections with specific steps of classical derivation.

Yet there appears to be an uneasiness contained in (1) or [(9), page 5,] "We have neglected radiation... processes... in the formulation of Schrö-dinger's equation... have assumed... all the interactions... with the external world—can be represented by conservative forces. . . . [I]t is impossible for a strictly conservative Schrödinger's equation to be set up. . . . [T]ime-dependent disturbances . . . inevitably causing transitions, . . . it will never be possible for any assembly . . . to be regarded as existing permanently in a true stationary state. . . ." The radiation coupling with thermodynamic, namely, statistical mechanical, phenomena requires one to deal with a dual statistical m/e-m field. In particular, the enclosure is not brought up to its equilibrium state (remembering that it is at temperature appreciably different from zero) by any linear process, mechanical or e-m. Thus, its singular state (of motion), in which it stays put for long periods of time and acts as a source and sink for the small radiative conductive exchanges with its interior and exterior contents, is a very complex one requiring more sophisticated characterization.

However, the enclosure walls then act as a unifying source for both e-m and mechanical transfer to the interior. By filling the enclosure with a simple substance, a near-ideal gas, we can visualize the processes a little better. The wall source, by a highly nonlinear process (the gradients are not small), takes a pooled and condensed collection of molecules—viewed as the very pure "ice" of the simple substance—and melts and evaporates the solid or liquid state by adding energy to the molecules. The exchanges are m/e-m. Finally, an equilibrium statistical mechanical state is reached and then maintained. It would seem perfectly apparent that the interactional status in time and any final "quantum" maintained state depended on the nonlinear processes that took place and not on ad hoc quantization at the operating point. (Later, in viewing emission and absorption, we will try to make this clear.)

That the processes are degradative can simply be shown by removing the external thermostat. There is a decay in the process (the system loses "heat") and the collection recondenses to its pool. This is hardly indicative of a nonlossy process.

Thus what we consider missing is a study, without thermodynamics (which tends to act as a crutch), of the e-m and mechanical coupling of enclosure and content to further seek the "fly in the ointment."

> In our opinion, we see, much too casually proposed at this time, the following possibilities other than the quantization of mechanics: nonlinearization of e-m; perhaps no need for a nonlinear e-m, but just a working-out of the "quantization" of dislocations in the states of regular matter; "noise" sources at the wall such as the vibrational or rotational character of systems in the wall, as well as the translational and trapped character of migrant electrical particles such as free electrons in the wall; "noise" sources such as pairs, cosmic rays, or others in the enclosure space.
>
> Although far from accepted yet, this type of nonlinear process is what we have been trying to prove in hydrodynamics: namely, that a factor commonly neglected, compressibility, whether in gases or liquids, could provide a coupling in some problems such as flow in a tube to permit quantized modes; that when operated at a second singular state of motion—not the zero velocity around which

the creeping motion of laminar flow develops, but the high-mean-Reynolds-number velocity field around which turbulence develops—any kind of normal "noise," say from statistical mechanical sources, would be sufficient to drive this basically unstable system toward its stable nonlinearity of turbulence. Thus, we are willing to comment that pair production or cosmic radiation illustrates such potential noise in an e-m field even though neither its magnitude nor the process will appear in any way in the actual quantized state. However, it is our belief that it is precisely such characteristics that illustrate the nonlinear nature of the quantization.

Currently, our hydrodynamic solutions do not reveal what determines the amplitude of the limit cycle or what determines the statistically wobbling or intermittent nature of the solution. These logical difficulties are common to most nonlinear quantization problems.

Thus, we can continue the problem by bringing in the minimum ad hoc quantization. (The premise of good science fiction has always been that the reader may be asked to suspend belief on only one factor.) It defers the problem of what physical theory, in mechanics and e-m, was responsible for the quantization and, hopefully, separates the hierarchical problem, which is our self-consistent need in general systems science.

Thus let

$$\epsilon_j = (j + \tfrac{1}{2})hv$$

where ϵ = energy
   h = a quantizing number (Planck's constant)
   v = frequency
   j = integers

represents the limit-cycle response of one of our particles, viewed as responding as a highly pure harmonic oscillator (it is not assumed to be a highly pure harmonic oscillator but one whose quantization measures are the same as a harmonic oscillator).

Note we do not raise the issue whether this is an e-m or a mechanical quantization. Yet both fields are now interrelated. Some unexplained nonlinear laws prevent electronic radiation (whether by a wave guide or quantum mechanics, etc.), and the involved elements are a particle and its surroundings. The particles tend to show linear superposition for low density but some coupling at higher density.

The Maxwell–Boltzmann distribution is given by

$$n_j = (\Delta j) \exp\left(-\alpha - \frac{(j + \tfrac{1}{2})hv}{kT}\right)$$

$$\Delta \epsilon_j = hv\, \Delta j$$

representing the number of equilibrium oscillators which would have values of the quantum number j falling in the range Δj, indicated as the number of eigenstates falling in the energy range. Then

$$\bar{\epsilon} = \frac{\displaystyle\sum_{j=0}^{\infty} \exp\{-\alpha - [(j + \tfrac{1}{2})hv/kT]\}(j + \tfrac{1}{2})hv\, \Delta j}{\displaystyle\sum_{j=0}^{\infty} \exp\{-\alpha - [(j + \tfrac{1}{2})hv/kT]\}\, \Delta j}$$

Choosing a fixed interval Δj, and let $x = (hv/kT)$

$$\bar{\epsilon} = hv \left\{ \frac{\sum j \exp\left[-j(hv/kT)\right]}{\sum \exp\left[-j(hv/kT)\right]} + \frac{1}{2} \right\}$$

$$= hv \left[\frac{\exp(-1x) + 2\exp(-2x) + 3\exp(-3x) + \cdots}{1 + \exp(-1x) + \exp(-2x) + \cdots} + \frac{1}{2} \right]$$

$$= hv \left[\frac{1}{\exp(hv/kT) - 1} + \frac{1}{2} \right]$$

In a given enclosure, the number of modes quantized on the wall in a conservative system [Jeans (2), Sec. 459–465], per unit volume, is

$$\frac{\omega^2\, d\omega}{2\pi^2 c^3} \qquad \text{or} \qquad \frac{\omega^2\, d\omega}{\pi^2 c^3}$$

where c = propagation velocity

for a single mode of vibration (such as compressional waves) or for two independent modes (such as two transverse waves) respectively. The latter is appropriate to e-m in a cavity. In a volume, the number dz in a frequency range dv is

$$dz = \frac{8\pi V v^2\, dv}{c^3}$$

If we want to compute the radiation law for the density of energy, which here is to represent the nonmaterial (i.e., photon) energy in oscillating modes, we must realize that we have no atomistic particle on which to hang our statistical mechanics. However, we might recognize that the quantization must be similar or related but not involve the null-point energy $\tfrac{1}{2}hv$. Thus

$$\epsilon_j = jhv$$

is more reasonable for each oscillating equivalent mode (supported somehow by the oscillator particles). Running through the same derivation leads to

$$\bar{\epsilon} = \frac{hv}{\exp(hv/kT) - 1}$$

(i.e., Bose statistics for photons).

The radiation law for the density of energy is then

$$du = \frac{\bar{\epsilon}\, dz}{V} = \frac{8\pi h v^3}{c^3} \frac{dv}{\exp(hv/kT) - 1}$$

the density of radiant energy in a given frequency range in equilibrium with the wall at temperature T. [This derivation is from Tolman (1938).] This is Planck's distribution law. Integrated,

$$u = \int_0^\infty du = \frac{8\pi^5 k^4}{15 h^3 c^3} T^4$$

the Stefan–Boltzmann law.

2. We should now proceed toward the questions of nonequilibrium. However, it will be sufficient to just introduce it at this time with a paragraph.

First, in passing, one should note Boltzmann's H theorem which is concerned with the tendency of a system, not in equilibrium, to return to equilibrium and with the rate at which it does so. Eyring et al. (10), for example, point out the lengthiness and detail of such a discussion and recommend (rightly so) Tolman (1938). The essential problems [see Tolman (1938), pp. 98–179] are to consider mechanical reversibility and reflectivity after collisions among the particles of a conservative system; to classify collisions and their cycles of occurrence; to define a suitable function of the distribution, the H function, which gives the Maxwell–Boltzmann distribution as an extreme and whose rate of change in time changes with collisions to approach the equilibrium.

If the Boltzmann process is then applied to real systems such as molecules, and suitable methods of integration found, the equations of change—for different approximate levels of nonequilibrium processes—can be formed. However, this is the subject of the continuum description of molecular physics. Sources such as (21) or (22) are quite suitable.

3. We have neglected the theory of absorption and emission of energy, from Einstein through Dirac, etc. Although it is not suitable to include in this very sketchy presentation, yet its content is basic to complete the quantum-sustaining process. A more satisfactory source that relates it to quantum theory is Heiter (20).

4-5 SOME CONCLUDING REMARKS

Thus, in some sketchy fashion, the dialectics of particle mechanics, quantum mechanics, and statistical mechanics have been begun (but not carried on with sufficient completeness for our needs). The nonphysicist reader may draw whatever conclusions he wishes about pertinence. Nevertheless, our problem will be to highlight the case that it is the same kind of issues that have received fairly formal consideration in atomistic physics which beset

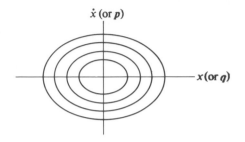

$\dot{x}$ (or p)

x (or q)

Fig. 4-2 Illustrating a "center."

the scientific synthesizer at every level in GSS. We have not stinted in offering good sources to follow the past masters of exposition of these topics in the physical sciences.

It is useful to restate the issue that faces us, now with the rudiments of nonlinear, quantum, and statistical mechanics behind us, as follows. The linear conservative system exhibits the orbital stability of a center; namely, there are nested families of noninteracting closed trajectories. Particularly in the form of normal modes, which "diagonalizes" the matrix into a set of independent system oscillators,

$$H = \sum(\tfrac{1}{2}ax_i^2 + \tfrac{1}{2}b\dot{x}_i^2)$$

the Hamilton description of the field has considerable elegance. In addition, there is reason to believe that, in the local domain around any stable singularity, motion might become consistent with such a linear description (the theory of small vibrations), except for the question of losses. Unfortunately, the entire apparatus is really quite analogous to Ptolemy's epicycles. As long as nature can be well described by them, the system is excellent and elegant.

Now the first class of problems beyond linear conservative systems, nonlinear conservative systems, no longer permits an elementary classification of orbits. What emerges are systems of motion that no longer can be characterized by normal coordinates. Yet a typical line of defense has been that it does not matter. "Practical" systems such as our solar system seem to be indefinitely stable. Why be concerned about the more general question?

Part of the difficulty may be understood when it is reduced to the result that one cannot say or can only say with marginal certainty whether any orbit visualized, such as shown in Fig. 4-3, might or might not be achieved. (These are no longer normal coordinates, so that this simply illustrates two dimensions in some complex $2n$-dimensional space.) This still represents chaos, with the possibility of a number of chaotic collisions.

By boxing the system, the spatial motion is confined. It becomes more "chaotic." To some degree [witness Tolman (1938)] cycles of collision can be

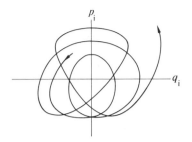

Fig. 4-3   Illustrating a wild orbit.

discussed. Are these the "real" cycles? No. The concept is that if these are added to molecular chaos an ergodic character can emerge, of cycles of orbits and phase-space–filling character.

Is this the real motion of systems? We believe not. We believe that the description of a particular system under consideration will have similar statistical measures, of various summational invariants, but will markedly differ in character. As a descriptive illustration, we would consider the problem similar to today's status of computer-composed music or poetry. Various Markov chain processes can characterize cycles of connection. However, the product does not have the coherence of human-produced "masterpieces." The reason may appear in discussing the next descriptive problem.

The second class of problems is nonlinear nonconservative systems. The expectation of a linear nonconservative character is that practically all systems should run down "eventually." (In their terminal states, by the theory of small vibrations, they become linear.) This is so distasteful that it is avoided. Thus, we quickly turn toward the nonlinear ones. Now among such systems, there is hope for systems to persist. At present such description cannot be presented with the elegance of the Hamiltonian theory, but then it has not had the advantage of the great minds working on it that the other description had.

In nonlinear nonconservative systems, we have the following possibility. Regardless of where a system is orbiting, if it has losses, which many might consider an apt description of natural process (this does not preclude, in any cyclic theory of cosmology, that there are not also aggregative processes, but these may not be at the same place, same scale, or same time), then the trajectory may be viewed as closing down in phase space. At some region there may be a limit cycle. The orbit is then trapped.

In the case of human experience, one may view the process by the following experience drawn from theatergoing: You, the onlooker, walk into the theater. The playwright has made the effort to structure a number of centering foci. First, he tries to trap your attention to his focal theme. When he has your attention (love—he perhaps has declaimed, he takes you around

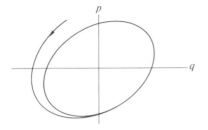

Fig. 4-4 Illustrating a limit cycle.

in orbits, choosing as he may those that may trap and entrain you. If he is fortunate and skillful, he may have you by the end of the first act.), then he may very well switch to other focal points, etc. This is what is characteristic of human patterning as a nonlinear nonconservative process. The present generation of computer-composed material does not reach you in that way. The short, nominally "recognizable" segments do not "grip" and "send" you.

However, this wrapping up or warping of the system toward limit-cycle domains that are marginally entered into, for particular epochs, is what seems to emerge characteristically in nonlinear nonconservative systems; at least this is one of our major surmises.

Although a much more difficult description of "real" systems, we believe that it lies closer to the truth of systems. However, if some exceptions have to be made, such as the indefinite orbital stability of a nonlinear conservative system like the gravitationally attracting solar system, the indefinite orbital stability of a linear "conservative" (i.e., since characterized as being based on linear harmonic oscillators) quantum-mechanical description of radiation, nuclear particles, atoms, ions, and molecules, we will accept it. We have declared that we do not wish the logic of any level to be indefinitely beholden to the logic of other levels. However, we will reexamine the question when we discuss cosmological problems.

SUMMARY

1. The foundation of the transition from an atomistic level to a continuum level is statistical mechanics; namely, the application of the dynamic laws describing the element-to-element interaction between atomistic particles, suitably "integrated" over the ensemble of particles to obtain laws of the ensemble, which under suitable conditions of spatial and temporal density behave the same as a phenomenological description of a continuum.

For systems science to be meaningful, the elements must be active energy transformers— likely nonlinear "limit-cycle" or "engine-cycle" elementary systems. Else, degradative states would be the common end of such ensembles.

2. The difficulty of getting an active atomistic element into the play is illustrated by the atomic oscillator. A macroscopic continuum shell made of atomic material shows degradative performance in its two coupled modes of thermal conductivity and electrical resistivity. An energy flux—radiation—is required to maintain its state of excitation. Atoms internal to the shell—the active elements of concern to us—require the thermally fed external wall to maintain their average energy excitation—again by radiation equilibrium—whether "linear" (by quantum mechanics) or not (by some other nonlinear description, perhaps classical, perhaps not). This is a minimal schematic description of how this system of active elements comes into being. Statistical mechanics then begins within this field.

3. The most elementary description of statistical mechanics commonly starts with a conservative system of particles. Reader, ponder. Is the atomic system conservative? If it is, its motions cannot belong to the class of limit cycles, which we would like to leave as an open question. (Recall that the atom has the character that a disturbance either to orbits outside or inside the "equilibrium" state will be followed by returns to the "equilibrium" orbit, an essential limit-cycle characteristic.) Yet the Hamiltonian of the atomic particle is commonly written as if it were a conservative harmonic oscillator equivalent (e.g., description by potential wells. The distortions of the force "wells" are not essential questions here).

4. Starting with the Lagrangian and Hamiltonian formalism (as an equivalent to Newton's laws) and applying it to a system of indistinguishable particles (atoms) in a multidimensional phase and momentum space in which all hypervolumes of that space are equally accessible (except for prohibited regions), we can investigate conditions for "statistical" equilibrium of the ensemble. By imagining an ensemble of such ensembles and a phase space in which this ensemble moves (i.e., it represents all such systems) with certain constants, such as all systems having the same energy, it is possible to derive the equilibrium distribution of such particles, in the present instance the Maxwell–Boltzmann probability distribution of particle species in phase-space variables.

5. Methods for deriving average properties of one ensemble are developed. The theorem of the virial, how a derivation for the equation of state[1] of the ensemble is begun, is presented. The dependence on interparticulate forces is exhibited.

6. The rudiments of thermodynamics are sketchily derived, that is, statistical mechanics and thermodynamics are bridged. The extensive quantities, entropy and free energy, and the concepts of work and heat are introduced. How the radiation field may be quantized and brought within the scope of statistical mechanics is also shown.

7. It is not essential that a system be conservative or that its constraints be holonomic for a statistical mechanics of a system of particles to be developed (i.e., it is clear, phenomenologically, that systems of particles can be maintained in sustained motion). However, there is probably no known general procedure for deriving the properties. It is likely that many dissipative systems would decay. Thus, achieving a rich occupancy of phase space (both displacement and momentum) is not a priori guaranteed. The assumption of a Hamiltonian, in particular any Hamiltonian which would lead to an intrinsic frequency (i.e., sustained cycle), is one way to avoid difficulties. The system does not have to be linear or nondissipative as long as an active element can arise. Then its "effective" conservative or linear Hamiltonian is an adequate descriptor. The systems, thus treated, are really multihierarchical (the quantization come from some other level).

---

[1] The equation of state relates momentum exchange at the bounding walls and an as yet undefined extensive property, "the temperature of an ideal gas," which is in equilibrium with and, thus, shares the property of the system. The flux associated with this parameter controls energy exchange between systems outside the wall and those inside the wall.

Thus, the crudest conjecture that can be made is that any process, linear or nonlinear (we really believe only nonlinear, but we must leave the issue open), which can form a sustained elementary cycle (and which may encompass more than one cycle, whether the amplitudes are superimpossible or not) then may be brought within a statistical mechanics. However, it does not necessarily lead to Boltzmann statistics. Correspondingly, a statistical thermodynamics can be derived. The sharpness of results depends on there being a sufficiently large number of such interacting particles.

QUESTIONS

1. Review carefully the issue of quantization for an isolated particle or photon. Can its quantization be tested experimentally?

2. Attempt to show that blackbody radiation from an evacuated sphere maintained at a constant temperature T_0 to an internal spherical metallic shell filled with hydrogen is consistent with the m/e-m modes in the shell wall and the molecular quantization of the gas by any consistent theoretical construct you care to use. The shell has finite conductivity, and the absorption molecules are capable of emission and of e-m energy.

3. Examine the thesis in (18) that a lossy nonlinear field with steady-state boundary conditions may be decomposed into a mean state and a time-dependent state; that the limit cycles of the field may be isolated by examining the first-order harmonic stationary components; and that its stochastic character—if any—emerges from a residue of nonstationary time-dependent components which disappear in time by being swept into decay modes. If you cannot accept the thesis, propose another more valid decomposition for the field phenomena.

BIBLIOGRAPHY

1. Landau, L., and E. Lifshitz: "Statistical Physics," Pergamon Press, Ltd., Oxford, England, 1958.
2. Jeans, J.: "The Dynamical Theory of Gases," 1925 ed., Dover Publications, Inc., New York, 1954.
3. Gibbs, J.: "Elementary Principles in Statistical Mechanics," 1902 ed., Dover Publications, Inc., New York, 1960.
4. Tolman, R.: "Statistical Mechanics with Applications to Physics and Chemistry," Chemical Catalogue Company, Inc., New York, 1927.
5. Fowler, R.: "Statistical Mechanics," Cambridge University Press, New York, 1936.
6. Kennard, E.: "Kinetic Theory of Gases," McGraw-Hill Book Company, New York, 1938.
7. Tolman, R.: "The Principles of Statistical Mechanics," Oxford University Press, New York, 1938.
8. Mayer, J., and M. Mayer: "Statistical Mechanics," John Wiley & Sons, Inc., New York, 1940.
9. Fowler, R., and E. Guggenheim: "Statistical Thermodynamics," Cambridge University Press, New York, 1952.
10. Eyring, H., D. Henderson, B. Stover, and E. Eyring: "Statistical Mechanics and Dynamics," John Wiley & Sons, Inc., New York, 1964.
11. Abraham, M., and R. Becker: "The Classical Theory of Electricity and Magnetism," Blackie & Son, Ltd., Glasgow, 1932.
12. Stratton, J.: "Electromagnetic Theory," McGraw-Hill Book Company, New York, 1941.

13. Bleaney, B., and B. Bleaney: "Electricity and Magnetism," Oxford University Press, London, 1965.
14. Frenkel, J.: "Wave Mechanics," 2 volumes, 1934–1936 ed., Dover Publications, Inc., New York, 1950.
15. Little, W.: Superconductivity at Room Temperature, *Sci. Amer.*, vol. 212, no. 2, p. 21, 1965.
16. Mendelssohn, K.: Frontiers of Physics: States of Aggregation, *Phys. Today*, vol. 22, no. 4, p. 46, 1969.
17. Firth, I.: Superconduction without Supercold?, *New Scientist*, vol. 42, no. 654, p. 644, 1969.
18. Iberall, A.: A Contribution to the Theory of Turbulent Flow Between Parallel Plates, *Seventh Symposium on Naval Hydrodynamics*, Rome, Italy, 1968 (in press).
19. Van der Waerden, B. (ed.): "Sources of Quantum Mechanics," Dover Publications, Inc., New York, 1968.
20. Heitler, W.: "The Quantum Theory of Radiation," Oxford University Press, New York, 1954.
21. Hirschfelder, J., C. Curtiss, and B. Bird: "Molecular Theory of Gases and Liquids," John Wiley & Sons, Inc., New York, 1964.
22. Rossini, F.: "Thermodynamics and Physics of Matter," Princeton University Press, Princeton, N.J., 1955.

part two

Introduction to Viable Systems

5

The Many-body Problem Exhibited by Atoms and Molecules

An apt starting point for GSS is the level of atoms and molecules. Hierarchically, this is an atomistic level. Undefined at this point, there exists "continuous" material possessing electrical, inertial, magnetic, and gravitational properties. This material is quantized into a series of particulate entities which have some permanence of form. Our major concern, because of their persistent temporal properties, is with the electron, proton, and neutron. Another quantized entity that concerns us is the photon. The issue is whether with these, and with some sort of science, we can build up the properties of other permanent systems that make up the world around us. We barely can identify the continuum level that lies beneath these particles. Thus, speculation about sub-atomistic levels beyond is irrelevant in this section.

We must remember under what condition particles of this hierarchical level have permanence in form for us. Typically, we have an enclosure, or at least a bounding wall or substrate, in whose vicinity a reasonably uniform temperature exists. The temperature is low enough that some condensation can take place. This likely restricts us to the range 0–10,000°K. (Stellar processes—in energetic level, spatial domain, or temporal domain—are not our present concern.) It is likely that some sort of weak force, such as gravity, has been responsible for bringing enough material together on an astrophysical scale that the local "continuous" aggregation persists. A potential segregation into three states of matter is found. These are the gas, liquid, and solid phases.

> We have thereby named those ingredients that are necessary for every level of GSS. From above, we require a box or stage that will hold the system in space and time; from below, a content that will provide the players with substance in space and time. Our present concern is with the properties of the individual atomic actors that are held in a thermal milieu. The only actors that are initially quite widely separated are those that exist in the very rarified gaseous phase, or the vapor phase.

Suitable reference material for the mechanics of these atomistic particles is contained in (1–5). The configuration and dynamics of individual atoms or molecules that are at equilibrium in a cavity with thermal radiation will be under discussion, although not explicitly in these terms. It is assumed implicitly that such a configuration can be described consistently by quantum mechanics, classical mechanics, thermodynamics, and classical e-m.

It is further assumed that, in the main, only three entities make themselves evident—electrons, nucleons (made up of protons and neutrons), and photons.

> If one examines the long-lived stable particles of physics circa 1969, they consist of the photon; 6 leptons[1]—the electron and positron and the electron-coupled and μ-particle-coupled neutrino and antineutrino; and 2 baryons, the proton and neutron. The shortest-term particle is the neutron (with a 1,000-second–free-space lifetime); the others are stable. However, practically only the three entities named show up in ordinary low-energy chemistry and physics.

One kind of electron is found. Its diameter is 5.6×10^{-13} cm. Its rest mass is 9.1×10^{-28} gm. Its electric charge is 4.8×10^{-10} esu. Its atomic weight is 5.5×10^{-4} amu ($0^{16} = 16$). Its spin is one-half. The magnetic moment of one Bohr magneton (the unit magnetic moment) is 0.93×10^{-20} erg/gauss.

There are approximately 800 nuclei with lives greater than an hour (i.e., relatively stable isotopes). For the nonphysicist, we can offer the flavor of current atomism by providing the kind of elementary statement given to the nonatomic specialist in physics. Typically, an elementary statement of organization (6) is:

---

[1] There is no need for the nonphysicist reader to "freeze" at these terms. They may be regarded as a few of a large number of names and faces that are presented to him at a crowded cocktail party in the opening scenes of a complex play. As the story unfolds, he can sort out the players that concern him.

On modern theory, the atom consists of a central core (or nucleus) of diameter about 10^{-12} cm, surrounded by a number of electrons. These electrons move round the nucleus in orbits whose diameter is about 10^{-8} cm, and these determine the size of the atom. The nucleus contains two kinds of particles: protons, which are particles roughly 1,836 times as heavy as electrons, but with a positive electric charge $+e$, and neutrons, of very nearly the same mass as protons, but with no electric charge. The number of electrons surrounding the nucleus is equal to the number of protons, and each electron has a negative charge $-e$, so that the atom as a whole is electrically neutral. The physical and chemical properties of the atoms are determined by the number of electrons they contain, and hence the number of protons in the nucleus is characteristic of a particular element. The number of neutrons is roughly equal to the number of protons in light elements but is over 1.5 times as great in the heaviest elements. The mass of the nucleus is determined by the total number of protons and neutrons, and a given element may have several stable forms of different nuclear mass, corresponding to nuclei with different numbers of neutrons, but the same number of protons....

It is now established that the electronic charge is the fundamental unit of charge, and all charges are integral multiples of $+e$ or $-e$. It is, therefore, assumed that the electron is indivisible and is a fundamental particle of matter; so, also, is the proton.... Since charges of opposite sign attract one another, the electrons are bound to the atom by the electrical attraction of the protons in the nucleus. The forces which hold the nuclei together are of different character, and operate only at very short ranges, of the order of the nuclear diameter [i.e., atomistic entities exist—these are their properties].

Thus, nuclei are basically identified by their (long-range) chemical properties which are associated with their atomic number Z, which is identified as the number of protons that have gone to make up the nucleus. In increasing order of atomic number, the nuclei may be arranged in a matrix of chemical groups within periodic rows. There is a tendency for analogous chemical properties to emerge within each columnar grouping.

The electrons are locked up in quantized states with their parent nuclei from a minimum diameter of the order of 10^{-8} cm to somewhat larger in physical atoms; or they can be involved in interpenetrating orbits with other atomic species in chemical molecules; or nuclei binding nonneutralized numbers of electrons can dart around as ions. The object of theory is to account for all of these solar-systemlike arrangements and for their interactions and exchanges.

Assuming that the systems are sparse within the conceptual container and somehow at radiation equilibrium with the walls of that container and that their average cg motion is zero over periods of time comparable to or longer than those required to set up radiation equilibrium, the concern then is only with the interparticle forces. In particular, quantum mechanics was designed, principally, to account for the configuration of the hydrogen atom (the quantization rules of gas spectroscopy had been phenomenologically determined earlier).

The solution to such problems, in effect, offers the character and shape of the potential well that binds the electrons and nuclei together and their absorption and emission characteristics. We will subsequently discuss to what extent this descriptive solution may be approximated in some sort of general scheme.

Thus, to some extent, the assumption of atomistic particles on one particular level, the level of nuclear particles, can be followed by a derivation of their stable orbital configurations at the next level. The nontechnical flavor of modern quantum chemistry and molecular physics may be savored in an article by Mulliken (7).

The detailed technical state of theory is really a different matter. Hirschfelder (8) is an up-to-date reference source. Speaking of intermolecular forces as responsible for most of the chemical properties of matter, he points out that though the exact nature of the forces is complicated, the binding into molecules is not beset by conceptual difficulties. The fundamental theory was developed during the first decade of modern quantum mechanics, with few advances in the past 30 years. The subject has remained semiquantitative, although more current experimental work has revived theoretical interest.

Nevertheless, the gap between the theory of intermolecular forces (8) and the nature of the chemical bond (2, 9) or the kinetics (10) and physical chemistry (11) for the molecular configurations of chemistry is still quite abysmal. The complex many-body problem of the fundamental atomic entities is not a satisfactorily solved system as far as fundamental means is concerned. However, a very large body of empirical literature exists on what configurations may be expected and how they may be synthesized. All of these problems are of concern to the chemical specialist.

SUMMARY

1. The "atomistic" nuclear particles—electrons, nuclei (essentially protons and neutrons), and photons—coalesce to form configurations of atoms, ions, and molecules.

2. Although both levels, nuclear and "atomic," are quantized, they are separated by a continuum field, the electromagnetic, mechanical, and perhaps quantum-mechanical continuum, in which the nuclear particles are embedded. From this continuum interaction, the higher hierarchical level—of atoms, ions, and molecules—emerges.

QUESTIONS

1. Offer a description in graphic and geometric terms of some significant chemical combination that takes place with some difficulty in the gas phase (producing yields in the range of perhaps 1 to 10 percent). Sketch out the quantum mechanics of the compound.

2. Discuss the formation of some significant chemical bond in a sparse collection of molecules (e.g., mean free path as large as the enclosure) in enclosures of different temperatures.

BIBLIOGRAPHY

1. Pauling, L., and E. Wilson: "Introduction to Quantum Mechanics," McGraw-Hill Book Company, New York, 1935.
2. Syrkin, Y., and M. Dyatkina: "Structure of Molecules and the Chemical Bond," 1950 trans., Dover Publications, Inc., New York, 1964.
3. Rossini, F.: "Thermodynamics and Physics of Matter," Princeton University Press, Princeton, N.J., 1955.
4. Hirschfelder, J., C. Curtiss, and R. Bird: "Molecular Theory of Gases and Liquids," John Wiley & Sons, Inc., New York, 1964.
5. Slater, J.: "Quantum Theory of Matter," McGraw-Hill Book Company, New York, 1968.
   ———: "Quantum Theory of Molecules and Solids," vols. 1–3, McGraw-Hill Book Company, New York, 1963, 1965, 1967.
6. Bleaney, B., and B. Bleaney: "Electricity and Magnetism," Clarendon Press, Oxford, 1965.
7. Mulliken, R.: Spectroscopy, Quantum Chemistry and Molecular Physics, *Phys. Today*, vol. 21, p. 52, 1968.
8. Hirschfelder, J.: "Intermolecular Forces," Interscience Publishers, a division of John Wiley & Sons, Inc., New York, 1967.
9. Pauling, L.: "The Nature of the Chemical Bond," Cornell University Press, Ithaca, N.Y., 1960.
10. Kondrat'ev, V.: "Chemical Kinetics of Gas Reactions," Addison-Wesley Publishing Company, Inc., Reading, Mass., 1964.
11. Moelwyn-Hughes, E.: "Physical Chemistry," Pergamon Press, New York, 1961.

6

The Physical–Chemical Continuum of Molecules

6-1 TRANSPORT COEFFICIENTS AND PHENOMENOLOGICAL EQUATIONS OF CHANGE

Now that sparse collections of atoms, molecules, electrons and ions in radiative equilibrium with photons and the constant temperature walls of enclosures have been conjured into existence out of the substances of everyday experience, the problem is to describe the equilibria and equations of change for the denser collections of such systems. At low densities (for conditions in which only a limited number of such atomistic systems have been introduced into the container), the ensemble will consist of an almost perfect gas $pV = nRT$. It may be surmised that elementary statistical mechanics will apply and the Maxwellian distribution (of velocities, momenta, etc.), will describe the collection. According to Gibbs' ergodic theory (or its modifications), the ensemble of all such systems will have comparable statistical measures. [A considerable literature exists for the physics and chemistry of near-ideal gases. (1), (2), (3), and (4) are excellent sources with which to begin.]

At more realistic densities—in the domain of what is generally known as imperfect-gas theory [see for example (5) or (6) and (3)]—the interatomistic force laws must be taken into account. For example, the value of contributions to the terms in the theorem of the virial can be derived. This represents the equation of state of the system.

In the next step, the equations describing the motion of the collection can be derived in a domain which is not at equilibrium while disturbed by forces outside of the collection. [Essential arguments are covered in (6) or in abbreviated form in (7).] The Boltzmann equations (from which the equations of change may be obtained) are treated in various approximations by the development of a formal theory for the transport coefficients.[1] The notable work of Enskog and Onsager led to the development of the thermodynamics of irreversible processes. In the application to coupled mechanisms within a very general statistical mechanics based on the concept of microscopic reversibility of the local processes, Onsager (8) probed at their reciprocal relations (with regard to the coupling). This, combined with his linear law which assumed that the velocity of reactions tending toward equilibrium is proportional to the displacements from equilibrium, gives the contribution to the change of entropy due to irreversible processes in terms of transport coefficients describing the possible couplings.

Thus, without providing a great deal of detail to the specific form of the interatomistic forces, the form of the phenomenological equations of change of not-too-dense atomic-molecular collections can be specified; the number of transport coefficients that the various flux couplings will permit (momentum exchange leading to viscosity, energy exchange leading to thermal conductivity, molecular species exchange leading to diffusion, and chemical affinity leading to chemical reaction rates) can be determined; and a theory of these transport coefficients can be begun.

To obtain precise results for the transport coefficients and equation-of-state parameters from first principles, on the other hand, is still not possible [note (9) pp. 260 and 370, or (10)].

What can we determine from these equations of change? First of all, it may be shown that there exists a range for the variables over which the equations describe essentially continuum processes (if the statistical mean free path is small compared to the size of the field, and the statistical relaxation time is small compared to the process time). At this scale, the atomicity is only detectable as "noise" in the field. (The field is thus bounded from below in size and time.) Second, the characteristics of these continuum systems can be discovered by specifying various boundary conditions from above.

At this point in our description of this system, we have conceived of a near-continuum gas (although we can consider it more generally a fluid).

---

[1] The transport coefficients are physical parameters that characterize degradative processes in a material system. Well known are the coefficients of viscosity, thermal conductivity, and diffusivity.

Although its atomic constituents may have been chemically reactive initially, we may assume that the components of the fluid have come to equilibrium, and we are now involved with only one molecular species in a single phase. We face a variety of transport coefficients—if isotropic, two coefficients of viscosity, one of thermal conductivity, etc.

An interesting argument is outlined in (7) concerning the second coefficient of viscosity, the so-called bulk or dilatation viscosity. Whereas the shear viscosity deals with the statistical mechanical transport of momentum from molecular cluster to molecular cluster at the relaxation time scale of the translational degrees of freedom, the bulk viscosity deals with the transport of momentum at the relaxation time scale of all other internal degrees of freedom, e.g., for the intermolecular degrees of freedom such as vibrational or rotational. The formalism for relating the bulk viscosity to these degrees of freedom was developed by Tisza [see (11)]. In (7), a criterion is suggested for the threshold at which the continuous medium changes from fluidlike to something more plastic or solid. There appears to be a point at which an interference with rapid relaxation and equipartitioning of the energy takes place among these internal degrees of freedom. The possibility of frozen-out degrees of freedom thereby exists. We have thus proposed that this criterion represents a more certain distinction between gases, vapors, and simple liquids on one hand and thixotropic liquids, plastics, and solids on the other than is contained in common elementary physical descriptions.

It seems evident that the equations of change, for states removed from local equilibrium, will differ for these two classes. The equations of change for the "fluids" (gases, vapors, and simple liquids) will be the hydrodynamic equations. They can describe the continuum relaxation toward equilibrium or the maintained motion of irreversible processes. The equilibrium states are states like the Maxwellian, in which energy is distributed among the easily accessible degrees of freedom in the molecules and there is "no" (or very little) exchange with frozen-out degrees of freedom.

It may be useful to provide some flavor of the derivation of equations of change.[1] We can quickly illustrate the phenomenological derivation of equations of change for a single molecular species "transparent" fluid (radiation and absorption are neglected) in its continuumlike regime. The derivation can also be accomplished by a more complex statistical mechanics (3). The singular virtue of this example is that it illustrates quite elegantly that the same result can be solicited both by macroscopic mechanics and by statistical mechanics. It seemed quite clear in the mind of one of the greatest of mathematical physicists, Kirchhoff, that the predictions of these equations represented the most stringent test of the validity of a statistical mechanics.

[1] This book contains a very minimum of mathematical symbolism and derivation. Most non-physical scientists will, of course, likely complain of too much. However, the little which was provided and which can be scanned casually, was meant to illustrate a few mathematical physical processes that are "essential" should any nonphysical scientist want to come to grips, ultimately, with the "physics" of his own field.

In Newton's macroscopic laws of motion on a portion of continuous medium (cg motion of a system of particles)

$$E_{rs,s} + \rho F_r = \rho f_r$$

E_{rs} is a symmetric second-order tensor, the stress tensor (in Cartesian tensor notation the comma notation denotes covariant differentiation), that characterizes the internal forces of one part of the medium on the contiguous part. It is symmetric because the force in the r direction on the s plane must be the same as the transposed set, so that in the limit the fluid may not be able to generate self-rotations.

F_r is external body-force vector per unit mass.
f_r is the acceleration vector.

Disregarding body forces (gravity is ordinarily the most important body force of concern; we will simply consider pressure forces),

$$E_{rs,s} = \rho f_r = \rho \frac{DV_r}{Dt}$$

where V_r is the vector velocity field.

Phenomenologically, we can consider that the stresses in a medium might be functions of position, of relative displacements, and of any order of derivatives of such deformations. We can imagine that substances (fluids) exist in which the stresses are only functions of deformation rates (i.e., that the stresses quickly relax if the deformation rates disappear).

To first (linear) order, we might assume for the most general form of a symmetric second-order tensor

$$E_{rs} = -p g_{rs} + \gamma_{rs}^{mn} (V_{m,n} + V_{n,m})$$

where g_{rs} — spatial metrics (e.g., $g_{rs} = 1$ for $r = s$; $g_{rs} = 0$ for $r \neq s$)
γ_{rs}^{mn} = a mixed tensor representing the possible coefficients of viscosity
p = the mean scalar pressure

The "justification" for this linear assumption is taken to be Onsager's work, which proposed that there likely will be a linear range for conditions of change not too far removed from equilibrium. Further, if the medium is isotropic, there are only two coefficients of viscosity (denoted here as a and b).

$$E_{rs} = -p g_{rs} + a g_{rs} V_{m,m} + b (V_{r,s} + V_{s,r})$$

there being only three types of symmetric terms that can satisfy isotropy.

This expression states that the stress tensor (force per unit area) of one portion of an isotropic fluid on the adjacent can only depend on the mean pressure, dilatation at the point ($V_{m,m}$, the divergence of velocity, measure dilatation), and a symmetric combination of shear forces (the trace $V_{m,n} + V_{n,m}$).

In terms of the shear and bulk viscosity $-\mu, \lambda$

$$E_{rs} = -p g_{rs} - (\tfrac{2}{3}\mu - \lambda) g_{rs} V_{m,m} + \mu (V_{r,s} + V_{s,r})$$

Thus

$$\rho \frac{DV_r}{Dt} = -p_{,r} + [\mu(V_{r,s} + V_{s,r})]_{,s} - [(\tfrac{2}{3}\mu - \lambda) g_{rs} V_{m,m}]_{,s}$$

the Navier–Stokes equation of motion for a generalized inhomogeneous isotropic fluid.

From the conservation of mass, the equation of continuity is derived. Let m be the mass of fluid in a fixed volume V.

$$m = \iiint_V \rho \, d\tau$$

$$M = \frac{dm}{dt} = \iiint_V \frac{\partial \rho}{\partial t} \, d\tau = -\iint_s \rho V_r(v_r \, d\sigma)$$

where M = mass rate of flow
v_r = the unit normal

By Green's theorem, the rightmost expression may be transformed to a volume integral.

$$M = -\iiint_V (\rho V_r)_{,r} \, d\tau$$

Thus

$$\iiint_V \left[\frac{\partial \rho}{\partial t} + (\rho V_r)_{,r} \right] d\tau = 0$$

or since the volume was arbitrary

$$\frac{\partial \rho}{\partial t} + (\rho V_r)_{,r} = 0$$

This equation, or

$$\frac{D\rho}{Dt} + \rho V_{r,r} = 0$$

represents the equation of continuity.

The energy equation is represented by the statement that the change per unit volume of internal energy in an infinitesimal volume portion of a field is given by the net heat flux that flows into the volume plus the net rate at which the surface stress does work on the volume.

$$\rho \frac{Du}{Dt} - q_{r,r} - E_{rs} V_{s,r} = 0$$

where u = internal energy per unit mass
q_r = vector heat flux

Phenomenologically, one might expect

$$q_r = -K T_{,r}$$

where K = thermal conductivity.

The combination of these results with the equation of continuity

$$\rho \frac{Du}{Dt} - \frac{p}{\rho} \frac{D\rho}{Dt} = (K T_{,r})_{,r} + \mu(V_{r,s} + V_{s,r}) V_{s,r} - (\tfrac{2}{3}\mu - \lambda)(V_{r,r})^2$$

represents one form of the energy equation. From the first and second law of thermodynamics,

$$T\,dS = du - \frac{p}{\rho^2}\,d\rho$$

where S = entropy per unit mass.

Substituting this relation

$$\rho T \frac{DS}{Dt} = (K T_{,r})_{,r} + \mu(V_{r,s} + V_{s,r})V_{s,r} - (\tfrac{2}{3}\mu - \lambda)(V_{r,r})^2$$

results in a more compact form. This result exhibits the irreversible production of entropy.

The set is completed by adding two thermodynamic relations for a pure substance.

$$dS = \frac{c_p}{T}\,dT - \frac{\alpha}{\rho}\,dp$$

$$d\rho = \frac{\gamma}{c^2}\,dp - \alpha\rho\,dT$$

where c_p = specific heat at constant pressure
α = volume coefficient of expansion
γ = ratio of specific heats

The thermodynamic variables are functions of only two state variables for a single phase and species. Here p and T are chosen. All of the coefficients are not independent. The following relation of compatibility can be derived.

$$\alpha T = \frac{(\gamma - 1)c_p}{\alpha c^2}$$

Thus the set

$$\rho \frac{DV_j}{Dt} = -p_{,j} + [\mu(V_{i,j} + V_{j,i})]_{,i} - [(\tfrac{2}{3}\mu - \lambda)V_{i,i}]_{,j}$$

$$\rho T \frac{DS}{DT} = (K T_{,i})_{,i} + \mu(V_{i,j} + V_{j,i})V_{j,i} - (\tfrac{2}{3}\mu - \lambda)(V_{i,i})^2$$

$$\frac{D\rho}{Dt} = -\rho V_{i,i}$$

$$dS = \frac{c_p}{T}\,dT - \frac{\alpha}{\rho}\,dp$$

$$d\rho = \frac{\gamma}{c^2}\,dp - \alpha\rho\,dT$$

$$\alpha T = \frac{(\gamma - 1)c_p}{\alpha c^2}$$

is a complete continuum hydrodynamics set for a single fluid component driven by pressure. Such processes as gravitational convection, radiation exchange, diffusion, or chemical change have simply been omitted from the sample derivation.

In (7), the following criteria were obtained under which the hydrodynamic field would act fluidlike and continuumlike:

$$\Gamma\left(1 + \frac{\lambda}{\mu}\right) < 0.1$$

$$\beta\left(1 + \frac{\lambda}{\mu}\right) < 0.001$$

where $\Gamma = v\omega/c^2$ (a continuum parameter—the ratio of relaxation time to the period of any wave frequency of interest)

$\beta = v/cD$ (a continuum parameter—the ratio of mean free path to field dimension)

$\lambda =$ bulk viscosity

$\mu =$ shear viscosity

$c =$ velocity of sound

$D =$ characteristic field dimension

$v =$ kinematic viscosity

$\omega =$ angular frequency $(= 2\pi f)$

$f =$ natural frequency

Another criterion that should be added is that the ratio of bulk viscosity to shear viscosity should not be too high. It is possible that the value of the ratio of the order of 5 may be considered an upper limit to assure fluid mobility. If it is higher, the material may begin to show frozen-out degrees of freedom, and tend toward a more plastic state.

At high temperature, there remain problems in these equations of change that have not been mentioned. As chemical disassociation of the molecules into atomic constituents looms near, then additional frozen-out degrees of freedom may appear during the hydrodynamic process. The same thing is true in a pure liquid if cavitation takes place by energetic release of the vapor phase. What these examples intend to stress is that frozen-out degrees of freedom and easily accessible degrees of freedom refer to the questions of the energy barriers or "walls" that limit the ranges of the degrees of freedom. They refer to issues of penetrability and nonholonomic constrainability that we have not brought into focus.

6-2 THE STATES OF MATTER

To bring the relaxation issue into a slightly better focus, we may examine for a moment the thermodynamic variable of the bulk modulus, say $V(\partial p/\partial V)_T$. This parameter can be defined for any state of matter. Its isothermally defined nature means that it can be obtained experimentally by a very slow process within a small domain around the operating point.

Now the ratio of bulk viscosity λ (g/cm sec) to bulk modulus B (dynes/ cm^2) defines a time. This time constant, essentially, is the effective relaxation time of internal modes. (The ratio of shear viscosity to bulk modulus defines the translational relaxation time.) If the relaxation time is small compared to the observation time, then those degrees of freedom involved are freely available to relax. This is the simple molecular gas showing near-ideal-gas properties. However, if near a "critical" temperature, the isotherm has a horizontal derivative in the $p–V$ plane, and then the relaxation time becomes "infinite." There no longer is a unique equilibrium. The material is "thixotropic" for very long times.

In fact, a barrier begins to make itself evident—the "surface tension." Explanations may be sought within the scope of a theory of the liquid state. A modern theory of liquid structure, such as a cell theory (3), is based on the idea that with increased density the motion of any given molecule is restricted to a cell consisting roughly of the near neighbors.

With the formation of the liquid state, solely through sufficient density below a temperature which is "critical" for each molecular species, there arise changes in various parameters—the appearance of surface tension and thus of a free surface, change in bulk modulus, change in the magnitude of transport coefficients, and change locally to "... a nonperiodic state of matter whose density is too large to allow a rapid convergence of the virial expansion, i.e., too large to allow an imperfect gas description" (3). However, what also arise are metastable and nonequilibrium states. At rest, a liquid "ultimately" may reach equilibrium with its vapor, from which it is separated by a free surface. It is not assured that it will. (Example, the cubic instability that is commonly known for a van der Waals gas below the critical temperature can possibly persist indefinitely in a container in its metastable state—although we do not know this for an experimental fact.)

6-3 FLUID BEHAVIOR AS BRIDGE BETWEEN CHANGE AND STATE

What is represented here is another common property of complex coupled nonlinear systems. This nominally continuous hydrodynamic field can operate with various disequilibria (within limits) for indefinitely long periods. (To illustrate with a chemical example, common atomic species exist that do not combine except when exposed to catalysts.) Yet, weak forces exist which may restore the equilibrium or may maintain the disequilibrium indefinitely. Subject to some cautions, however, the thermodynamics of irreversible processes is an apt description for the equations of change within each homogeneous phase. Problems remain, generally at thin interfaces between such phases. The interface is often marked by jump phenomena (examples; free surface, a reacting surface as in a flame, shock waves, the boundary between city and country).

Thus, each phase in the fluid state, whether gas or liquid, may be characterized by continuous equations of motion and near-equilibrium distributions of atomistic particles. Additional phenomena—surface properties, etc.—are required to match physical conditions at the boundary. Certain specialized phenomena (including such matters as solubility) may be meta-stable within the body of each homogeneous phase. The essential parameters that characterize the state are temperature (i.e., equilibrium with the wall; note that one still is dealing with the mixed m/e-m field that emanates from the wall) and molar density (which determines the closeness and magnitude of atomistic interaction). In the simple fluid state, the atomistic entities are mobile and permit free equipartition of energy.

6-4 ELASTICITY-PLASTICITY

Further increase of molar density or the degree of interaction of complex molecules or of complex aggregations tends to increase the bulk viscosity so that the material becomes more fixed. Density alone will essentially fix the plastic-elastic state. [Discussed in an unpublished report (12).]

An adequate theory of the plastic-elastic state is still in process of being written. References (13), (3), (19–20) are useful source material. [In addition, we will make reference to ideas contained in our own unpublished work, (12) in 1954, (21) in 1955, and (7).]

Some major characteristics that emerge in the molecular collection which is being held in a container, as we reduce temperature, are the following:

1. A change in the character of the viscosity-temperature relation. For the near-ideal gas, the viscosity varies nearly directly with temperature. As the temperature is reduced, the relation turns, and now the viscosity increases with decreasing temperature. As the temperature approaches the critical temperature, the viscosity increases very rapidly (7). It is the extra activation energy, the energy necessary for the creation of new holes in the liquid struc-ture, one of the concepts that Eyring and coworkers developed in his theory of rate processes (15, 22), that accounts for this viscous increase.

2. An increase, of even greater magnitude, in the bulk viscosity.

3. A more moderate change in the magnitude of the bulk modulus. [One must note that the bulk modulus has to change from its gas value—which is the applied pressure itself (this being the consequence of an active atomistic system that is constantly playing a stream of changing linear momentum on any wall) to its value as a liquid. The change is moderate at the point that the gas or gas-liquid mixture is ready to be liquefied.]

4. The appearance of a shear modulus, which then changes even more rapidly than the changing magnitude of bulk modulus.

5. The possible appearance of a "latent heat" (here of fusion), indicating a phase change to greater ordering.

Although not so commonly thought of in this way, it is the small magnitude of the ratio of shear viscosity to bulk modulus that dominates the fluid state (liquids and gases) and lets it "flow" near (but not at) thermo-dynamic equilibrium, whereas it is likely the large magnitude of bulk viscosity to shear modulus that dominates the slow volume relaxation of the near-fluid and nonfluid state (fluids showing phase changes, supercooled fluids, metastable states, amorphous solids) and crystalline solids. More certainly, it is the large magnitude of bulk viscosity to bulk modulus that dominates the slow relaxation of the solid state.

The complexity of the functional dynamic characteristics of the solid state (beyond the elementary assumption of the existence of Hooke's law) should be noted at least in Nadai (17) and Alfrey (15). It is not sufficient to think of the state as marked by such linear properties as are implied by the Voigt or Maxwell model of stress and strain relaxation. (These are linear relations between stress and strain and their rates of change that are commonly used to illustrate dynamic material properties.) Instead, phenomenologically, one must note the need for description of both small and large strain effects found associated with:

Elastic deformation
Creep
Hysteresis
Age hardening
Work hardening
Failure
Endurance limit
Nonlinear elastic deformation (it is not even certain that a linear Hooke's law always holds)

in various materials, particularly plastic.

It is no wonder that the problem has become the subject of engineering mechanical study rather than theoretical mechanical study, simply because of the complexity of the dynamic states that seem to arise.[1] Yet, by this time, certain characteristic mechanical properties seem to be known:

1. As the density of molecular particles is increased, the motion of particles in a freely available universe—in which all states are ergodically traversed—then begins to resemble a continuously filled universe (averages over any space-time interval quickly converge to the mean over a broad frequency spectrum with characteristics lying between a flat "white noise" spectrum and a single-moded Gaussian spectrum), and particles begin to be crowded and impeded by local neighbors.

2. With a slower relaxation time, this neighboring cell of atoms begins to form a grosser atomistic structure which, for liquid densities, still retains considerable motility.

3. For still-higher densities, this grosser atomistic structure begins to lose mobility and is finally locked into place. (The necessary condition for

[1] A most vehement commentator in the field of the theoretical mechanics of the solid state is Truesdell.

locking the structure is not just the density but also the temperature. At sufficiently increased "temperature," which means roughly that there is an appreciable energy that can still be carried by the translational modes of the molecules as compared to the total energy available to the molecule, the mobility of a near-ideal gas can always be obtained.)

However, the degree of fixity of this local cell or "lattice" depends on the regularity of this structure. As a picturesque description, in talking about metals Hume-Rothery (16) points out, "You may ... regard the assembly of atoms in a metallic crystal as forming a gigantic molecule. This molecule has a series of electronic states or energy levels, and at the absolute zero of temperature, the electrons occupy the lowest energy levels, subject to the Pauli Restriction Principle...." The theory of possible symmetry groups begins to arise among these regular lattice arrays, and the energy per unit lattice begins to have significance with regard to the phase transition from liquid to solid. In the amorphous solid, this is much less well defined.

4. However, the regularity of this local lattice structure cannot persist. It extends out to "dislocations." Dislocation characteristics are illustrated in Zwikker (18), Hume-Rothery (16), and *Scientific American* (23). It is such dislocation domains, and the grained structure they form, which tend to make up the atomistic elements of the solid state. Such structure can only be treated phenomenologically at present. [However, see (24).]

> To provide some flavor as to how the solid state, say in metals, is regarded, we may extract some comments from Cottrell in (19). By etching a solid, one finds a "honeycomb of boundaries that partition the metal into small polyhedral cells called grains, typically about .01 inch across....
>
> "Clearly each grain, however regular in shape, is a single crystal, and a piece of metal consists of a mass of differently oriented crystals joined together along common boundaries."
>
> Regarding the grain boundaries, "gradually people came to the view that when a metal was cooled below its melting point, adjoining grains would crystallize as many as possible of the atoms that lay between them and so reduce the boundary to a mere interface, only about one or two atoms thick, across which the crystallographic orientation changed abruptly from that of one grain to that of the other."
>
> To test the state of knowledge about the structure, one could compute the strength of the material for its observed crystalline bonds. A discrepancy up to a factor of 1,000 in strength is found. The discrepancy in strength arises from the dislocations in the crystals. Irregularities allow atomic planes to slip much more easily than they would in a perfect crystal. The dislocations tend to pile up at grain boundaries. The appearance of dislocations and grains thus spoils an elementary view of how matter may be regularly arrayed.

5. Only a limited number of its properties can be described by a near-equilibrium "thermodynamics" of the solid state. It is the limitations in theoretically modeling structure that still confront us here.

Yet, one must remember that it is such a solid-state material which makes up the isothermal envelope in which we conducted our thought experiments regarding the e-m field. Such materials exist. They can be highly impervious. They can persist for a long time. They can be brought into thermal and radiative equilibrium with an outer heat source enclosure. Except for some possible properties at very low temperature [see, for example, Mendelssohn on superfluidity and superconductivity in (25) or in *Physics Today*, April, 1969], the material is made up of irregular domains that contribute electrical resistivity and thermal resistivity.

Thus, our general systems theory modeling cannot be described as simple, smoothly alternating -*A*-*C*-*A*- levels. We find, more commonly, that the appearance of a nonlinear instability in the motion of a system of atomistic particles exhibiting coupled phenomena, which is sufficiently extended in space and large enough in number to act as a continuum, instead likely must be ordered as a sequence of nested atomistic clusters, each of increasing stability, until a stable field continuum is finally found. (Electrons and nuclei form atoms; atoms form molecules; molecules form local cells; local cells form dislocation-governed domains; dislocation-governed domains form grains; grains form the continuum field that ultimately represents the extended solid state. While this may appear confused, at first sight, what we are saying is that stabilization of the solid state at the atomic level does not fully occur until the atomic dislocation is reached. Later, we will consider that biological cells form organs and organs form complex organisms; or, that humans pair, pairs form families, and families form tribes.)

It is not that the original scheme is defective but that the association that stabilized the atomistic motion was not stable enough to permit indefinite growth of such entities. Professor Gamow provided excellent lecture illustrations in how the building of structures took place. A sequence of orderly blocks may be piled one on another, but they disorder at some number and tumble down. If many such structures are built a somewhat more helter-skelter arrangement may provide increased stability to permit continued building of the structure. At some point, a structure of sufficient stability—or with some strong-enough cement added—will permit the gain of very many courses of blocks, so that the tower of blocks appears indefinitely continuous. Ultimately its own crushing strength, the "structural strength of mountains," may govern how far that structure may be continued.

It must be noted, however, that each atomic level may be quite stable under some conditions or for some limited length of time. Thus, it is possible to maintain a field of electrons or nuclei, of atoms, of molecules, of cells, of dislocation domains, of grains. In addition however, there may always be conditions under which metastable (or marginally stable) states can be maintained. Thus, a completely free ergodic hypothesis is not possible; only the quasi-ergodic hypothesis is. It is still approximately true that the time average, over the ensemble members of all systems that have been formed with the same constraints (i.e., temperature and potentials of the thermal shell, energy, and other summational invariants), will be like the spatial average of a sample configuration—for states that can be reached. (That is, there also may be frozen-out patches that cannot be reached with ease.)

6-5 CHEMICAL UNIT PROCESSES

Chemical change and combination are even more complex. They may be viewed as combinations that take place in gas phases in the presence of other materials, that take place in liquid phases, or that take place in solid phases. In reasonable truth, we might say that anything that changes the atomic-molecular configuration of an aggregate, that is sufficiently repetitive in extent to form a "finite" entity lies within the content of chemistry. Pauling (26) states that "there is a chemical bond between ... atoms ... in case that the forces acting between them are such as to lead to the formation of an aggregate with sufficient stability to make it convenient for the chemist to consider it an independent molecular species."

Note what we are driving at, and also are driven by the chemist to accept, is that the repetition of an "atomistic" structure is sufficient to create a spatial-temporal field that can be regarded as nearly continuous. Thus the physics and chemistry of not extremely dilute fields of ions, atoms, molecules, and cells tend to merge. It is only by emphasizing particular instabilities that specialized problems emerge. To illustrate—a collection that is chemically stable but acted upon by unstabilizing momentum (pressure) forces demonstrates hydrodynamic motion toward equilibrium. A collection that is not chemically stable demonstrates chemical motion toward equilibrium. A collection that is chemically stable but acted upon by unstabilizing energy (temperature) sources demonstrates heat-transfer motion toward equilibrium, and so on.

The specialization that represents chemical interests is well borne out in such a textbook as Moelwyn-Hughes (27). Other specialized sources may be mentioned for a theory of chemical change. The kinetics of gas chemistry may be examined in Kondrat'ev (28). A much more specialized example is Porter (29). For a classic source, Pauling (26) presents structural chemistry, although some readers may find another older book source (30) more directed. Streitwieser (31) illustrates an introduction, not complete, to modern organic quantum chemical theory.

The chemical changes that are most often of concern in our imaginary thermal container are relaxation processes. This arises because the prototype of reaction kinetics have been first-, second-, etc., order reactions [see, for example, (27) Chap. 22]. The literature on chemistry, particularly with the significant contributions of Eyring, has a considerable body of material—

from cookbook recipe to highly abstract theory—for dealing with such questions for complex chemistry. Beyond the nature of bonds[1] it is interesting to note the chemical engineer's view of unit processes and reactions. Very loosely, this may be categorized as:

Breaking bonds	*Making bonds*
solution	synthesis
raising temperature (decomposition, distillation, reflux)	lowering temperature (condensation)
electrolysis (decomposition)	surfaction
surfaction	crystallization
Exchanges—breaking and making bonds	*Physical separations*
catalysis	thermal diffusion
exchange reactions (redox, simple exchange, etc.)	osmosis (physical, electrical)
chemadsorption	electrophoresis
absorption	Knudsen flow

These questions can be explored within the body of chemical and chemical-engineering literature.

6-6 THE BIOCHEMICAL MILIEU

Although this gross description has hinted at a path toward much of current continuum (batch) chemistry (although no descriptive path for catalysis has been indicated), this material is not satisfactory for biochemistry, the chemistry of biological systems. Loosely speaking, the hydrogen bond is the major organizing element in biochemistry. To characterize the bond, Pauling states, "It is now recognized that one hydrogen atom, with only one stable orbital .. can form only one covalent bond, that the hydrogen bond is largely ionic in character, and that it is formed only between the most electronegative atoms."

The basic problem posed by biochemistry of living systems is how, in a continuous (batch) field of molecules, certain replicating "atomistic" processes—at the molecular level, both in space and time—can come into being. This involves the characteristics of two fundamental classes of biological polymers, nucleic acids and proteins. The quasi-statics of the spatial problem, for nucleic acids, was opened by Watson and Crick, in 1953, by the novel structure they proposed for DNA as a double helix of intertwined polynucleotide chains; for proteins, it was first illuminated by Gamow (1954) as the numerological (coding) problem of reproducing amino acid order in

---

[1] Pauling offers, loosely, electrostatic, covalent, and metallic bonds. Organic chemistry, for example, is largely an exercise in covalent bonding. More commonly, others distinguish molecular bonds and hydrogen bonds as distinct.

genetic material. The field of molecular biology has since burgeoned. Illustrative references are (32), (33), and (47).

> To provide the nonchemist reader with any brief introduction to biochemistry is, of course, out of the question. A very compact book, perhaps useful, is by Ingraham (34).

However, a missing ingredient has been some description of the dynamics of limit-cycle performance, of "oscillatory" reactions, that could not only be replicating and reproducing but also reliably hereditary, within the near-continuum biochemical field. This hope has settled upon the domain of enzymatic (catalytic) reactions.

The problem has received a great deal of current impulse starting in 1961 from the work of Monod [e.g., see (35)] and from Goodwin's book (36). More recent exposition may be found in Gander and Goodwin's work in (37), in Higgins (38), or in (39). The basic ingredient that has been proposed is a shift of velocity (reaction rate)–concentration curves by allosteric enzyme accelerators or inhibitors. This provides "negative-resistancelike" characteristics that make oscillations possible in the biochemical chains.

Nevertheless, as Pattee voices doubts in his article on the origin of life (40), there is little guarantee that such discoveries as the significance of polypeptides and polynucleotides as precursors to proteins in the static Watson-Crick model of DNA replication will lead to "spontaneous generation" of a reliable hereditary cyclic process in the batched continuum of a primordial ocean or a present-day laboratory solution. No sequence of steps can be specified in a dynamic program by which a hereditary process can arise or evolve by physical interactions.

> The physical dogma which appears challenged by a mechanistic foundation for the living process is the second law of thermodynamics. If stated in the form that as a result of natural processes disorder in the universe increases, the living system with its emergent order seems to violate this principle. However, this is not necessarily the case, if it is understood that "ordering" has to do with the height of the potential levels available for conversion. In the "early" stages of the universe, the major potential available was temperature. No chemical reactions were possible. At later stages, with parts of the universe cooled off, chemical processes and even chemical oscillators could come into existence. The same situation exists with regard to nuclear processes and even nuclear oscillators (stars). The detailed processes of biological systems—we would believe at present—can only exist within limited physical environmental conditions. Thus, complex oscillator structure is potentially only a phase of the running down of even larger-scale systems.

Slurring over the details, Pattee really poses the quantization problem (of dynamic biochemical cycles that "oscillate"; that is, that tie each epoch to the next by an invariable pulse of behavior) all over again. He decides:

We may therefore conclude that the normal classical laws of motion as derived from energy principles, with completely specified initial conditions and only holonomic constraints, do not lead to dynamical behavior of matter consistent with the concept of hereditary evolution. Nor is the assumption of template memory replication alone sufficient to account for natural selection. On the other hand, if we postulate that the energy-determined structures or conformations of matter introduce nonholonomic, finite *time-delay* constraints *inside* the system, there arises behavior which is consistently described as memory with hereditary propagation. If we couple such a system with an *outside* environment to cause mutation and selective interactions, we may expect some form of hereditary evolution.

The rest of his article on automata theory, although quite significant to our ultimate purpose, can be skipped now. Instead, we prefer to point out that the apparent connection between Pattee's concepts and ours is, as his statement suggests, that, in order to obtain "hereditary propagation" in the quantization of radiation or atoms in our imaginary enclosure, we had to have a nonholonomic constraint; namely, conditions at the wall involving some kind of temporary nonpenetrating binding and a finite time-delay. We wonder whether we and Pattee are not saying the same thing about the need for some specific nonlinear coupling to the boundaries of or substrate in the milieu.

The basic point, we believe, that Pattee is leading toward is that it is not simply the template—"kinematic"—action that can create a hereditary propagation in living systems, but the need for a dynamic coding element furnished by enzyme action. This dynamic nonlinear chain, when (or if) shown capable of happening in solution, will furnish a clear demonstration of biochemical activity in the watery-solution milieu for living systems and will represent a possible source or origin of prebiological systems, as the subject matter has recently been referred to and discussed (41).

An adequate description of transport phenomena of continuum biochemistry is thus nearly ready to blossom.[1]

That understanding about chemical and biochemical oscillators is coming into focus was well borne out in a session on the origin of life at the 1969 Biophysical Congress. The papers of Katchalsky, Prigogine, Morowitz, and Pattee clearly enunciated conditions for start-up of life systems essentially identical to those discussed here. These conditions included dissipative

---

[1] The flavor of the biochemical problem in the living system, comparing its current molecular status to the needs for future dynamic modeling, can be tasted in the descriptions of an enzyme molecule and its action (42) and an active protein carrier, hemoglobin, and its action (43). However, as the editor's afterthought in (44) points up, there may still be a gap in understanding the catalytic action of some enzymes.

structures, rapid flow processes (as in hydrodynamics), nonequilibrium thermodynamics, nonlinear stability, macroscopic inhomogeneity both in space and in momentum space, and heterogeneous catalysis. Chemical oscillators have been demonstrated by Zhabotinsky and Katchalsky. It appears increasingly clear that although the formation of oscillator chains by direct chemical-chemical coupling may be remote, it is not so difficult to achieve by chemical-mechanical, chemical-electrical-chemical, and other intermediate couplings. For example, a chemical-thermal oscillator (a reaction in a high-pressure catalytic cell) was discovered a number of years ago by Imperial Chemical. [See (45); on the chemical oscillator, see (46).]

We can summarize the growth of ideas about systems in this section, even if we did not describe the systems in detail. As our attention expands in size, there is a nesting of atomistic complexes of electrons, nuclei, atoms, molecules, molecular cells, and macromolecular complexes—possibly ending with dislocation domains—with varying degrees of stability (as the number of nesting levels increase, the certainty of stability increases). These complexes, which are enclosure-temperature and number-density dependent, form active elements whose relaxation times give rise to a wide variety of transport coefficients (viscosities, conductivities, diffusivities, activities, etc.). The transport coefficients are noted in grosser space and time, in which the system of such repetitive complexes acts like a continuum, according to near-equilibrium dynamics, but with these phenomenological coefficients representing conversion losses (i.e., according to the thermodynamics of irreversible processes). Thus, we can imagine (and in many cases, write) the equilibrium distributions and the new equilibrium continuum equations of change for such ensemble classes as the equations of states of matter, the equations of hydrodynamics, of heat and mass transfer, of chemical change, and—in projection—of chemical and biochemical oscillating processes.

"Catalysis," in the broadest possible sense, as the dynamic process at the atomistic level by which the metastable or frozen-out states are released and possibly stabilized into a much more rapid nonlinear cyclic process, is the main mystery at this hierarchical level.

SUMMARY

1. A collection of atomic-ionic-molecular entities, at increased number density, begins to act like a continuum.

2. Because of the interparticulate forces, as we previewed in the section on statistical mechanics, an equilibrium statistics emerges from and among these active elements.

3. More specifically, four states of matter emerge in the continuum—the gaseous, the liquid, the solid, and the reactive chemical state.

4. If these dense collections of active elements in dynamic equilibrium are disturbed so as to be driven into motion, either the methods of statistical mechanics or phenomenological physics can lead to equations of change.

5. For temporal rates or field gradients that are not too large, one can develop the statistical mechanics or statistical thermodynamics of near-equilibrium but nonequilibrium irreversible processes. Its major property is that although every local relaxation—e.g., collisional interaction of particle to particle—may not be at equilibrium, averages over a small number of such relaxation epochs quickly approach equilibrium.

6. Phenomenologically, a number of transport processes and parameters develop from these nonequilibrium relaxations. Well known are the coefficient of viscosity, which measures the relaxation of momentum; the coefficient of thermal conductivity, which measures the relaxation of energy; and the coefficient of diffusion, which measures the relaxation of segregated mass species.

7. At low-number density or high temperature, the particles have near-ideal gas mobility. At increased density, the interparticular forces impede the motion to neighboring cells. Liquid-state mobility emerges. In addition to the shear viscosity which is a measure of translational momentum, a second relaxation parameter, the bulk viscosity, emerges which measures the relaxation of momentum of all (internal) degrees of freedom other than the translational. The changing magnitude of bulk viscosity is the gateway to the emergence of form out of function. It increases from zero for the noble gases at low density, to a few tenths of a percent of the shear viscosity for near-ideal diatomic gases, to a number of times larger than the shear viscosity for liquids, toward hundreds and thousands times the shear viscosity for high-polymeric near-plastic "thixotropic" liquids, to essentially infinity for solids.

8. Form (or structure) as contrasted with function may be noted by comparing the ratio of bulk viscosity to bulk modulus (change in pressure per fractional change in volume), which defines a relaxation time of internal modes, with a time of observational interest. If the ratio of such times is large, the degrees of freedom are frozen out and may be regarded as form. If small, a functional relaxation may be viewed. In solids the time ratio is generally large (unless one is interested in observations covering inordinately long periods of time). Thus, the solid state is most often considered to be the carriage of form or structure.

9. However, even the solid state is not fully stabilized in its form by its interparticulate forces (it is generally high-number density itself which is sufficient to bring the solid state into existence). There are a series of structural levels—from a regular lattice array in the nuclear constituents that make up atoms to dislocations in the array—which "key" the structure together. Thus, even before the continuum extends very far, issues of stability, leading to further quantization, quickly arise.

10. The theoretical details of the chemically reactive state are hardly in focus today. Yet it must be imagined as part of the prebiological state. We may surmise that the search for active biochemical processes (as opposed to passive relaxational steps) will expose conditions for its existence.

11. The common chemical continuum, well-known to physical chemistry, is characterized by degradative reaction kinetics (although explosive exothermic reactions are, of course, also known). In principle this is like the viscous degradation of a fluid in which its motional state decays to rest. However, not so well known is the possibility of near-"resonant" or sustained states. In the fluid state, it is the fluid inertia that contributes to near resonance. How to get chemical systems to do this is not completely clear. Coupling with other phenomena, typically mechanical or thermal, is one known path.

QUESTIONS

1. Assuming only three broad phenomenological concepts: *a.* that the medium possesses an elastic modulus $[V (\partial p/\partial V)_T \neq 0]$; *b.* that the interatomistic elements exhibit some kind of closeup attraction and repulsion, perhaps like a potential well, perhaps not; and *c*, that the atomistic elements are active, discuss the kinds of states that you can see emerging as the number density is increased by squeezing the elements into a diminishing enclosure.

2. Attempt to verify the criteria proposed for a continuum field $[\Gamma(1 + \lambda/\mu) < 0.1; \beta(1 + \lambda/\mu) < 0.001, \lambda/\mu < 5]$. Attempt to show that this covers both gases and liquids.

3. Show by a number of examples how physical boundaries are matched by jump equations in such cases as *a.* shock waves; *b.* surface phenomena at gas-liquid interface; *c.* work functions at solid-gas interfaces; *d.* biochemical membranes; and *e.* gel-liquid interface. That is, indicate the governing equation set that holds within the volume on each side and that governs the jump.

4. Discuss the linear modes permitted by the hydrodynamic equations in the small-amplitude case (quadratic terms in field parameters are vanishingly small).

5. Consider a filament of material (e.g., metal, ceramic, or plastic) which is mechanically stressed in a given thermal environment. Discuss very carefully all of the classes of observed force-displacement characteristics that you can obtain.

6. What is a theory of the engineering strength of polycrystalline materials?

7. Examine the possible foundations for a theory of biochemical oscillators.

BIBLIOGRAPHY

1. Tolman, R.: "Statistical Mechanics with Applications to Physics and Chemistry," Chemical Catalogue Company, Inc., New York, 1927.
2. Kennard, E.: "Kinetic Theory of Gases," McGraw-Hill Book Company, New York, 1938.
3. Rossini, F.: "Thermodynamics and Physics of Matter," Princeton University Press, Princeton, N.J., 1955.
4. Eyring, H., D. Henderson, B. Stover, and E. Eyring: "Statistical Mechanics and Dynamics," John Wiley & Sons, Inc., New York, 1964.
5. Chapman, S., and T. Cowling: "The Mathematical Theory of Non-uniform Gases," Cambridge University Press, New York, 1952.
6. Hirschfelder, J., C. Curtiss, and B. Bird: "Molecular Theory of Gases and Liquids," John Wiley & Sons, Inc., New York, 1964.
7. Iberall, A.: Contributions Toward Solutions of the Equations of Hydrodynamics, Part A, The Continuum Limitations of Fluid Mechanics, *Contractors Rep. Off. Nav. Res., Wash., D.C.*, Contract No. Nonr 34-5(00), December, 1963.
8. Onsager, L.: *Phys. Rev.*, vol. 37, p. 405, 1931; vol. 38, p. 2265, 1931; *Ann. N.Y. Acad. Sci.*, vol. 46, p. 241, 1945.
9. Hirschfelder, J.: "Intermolecular Forces," Interscience Publishers, a division of John Wiley & Sons, Inc., New York, 1967.
10. Ernst, M., L. Haines, and J. Dorfman: Theory of Transport Coefficients for Moderately Dense Gases, *Rev. Mod. Phys.*, vol. 41, no. 2, p. 296, 1969.
11. Herzfeld, K., and T. Litovitz: "Absorption and Dispersion of Ultrasonic Waves," Academic Press, Inc., New York, 1959.
12. Iberall, A.: "Phenomenological Basis for a Generalized Elastic Theory of the Solid State for Application to Structural Design," unpublished study, 1954.
13. Fowler, R., and E. Guggenheim: "Statistical Thermodynamics," Cambridge University Press, New York, 1952.
14. Mott, N., and H. Jones: "The Theory of the Properties of Metals and Alloys," 1936 ed., Dover Publications, Inc., New York, 1958.
15. Alfrey, T.: "Mechanical Behavior of High Polymers," Interscience Publishers, a division of John Wiley & Sons, Inc., New York, 1948.
16. Hume-Rothery, W.: "Electrons, Atoms, Metals and Alloys," 1948 ed., Dover Publications, Inc., 1963.

17. Nadai, A.: "Theory of Flow and Fracture of Solids," vols. 1, 2, McGraw-Hill Book Company, New York, 1950, 1963.
18. Zwikker, C.: "Physical Properties of Solid Materials," Interscience Publishers, a division of John Wiley & Sons, Inc., New York, 1954.
19. Piel, G., et al. (eds.): "Materials," A Scientific American Book, W. H. Freeman and Company, San Francisco, 1967.
20. Wannier, G.: "Elements of Solid State Theory," Cambridge University Press, New York, 1960.
21. Iberall, A.: Bumper Stop Design, unpublished study, 1955.
22. Glasstone, S., K. Laidler, and H. Eyring: "The Theory of Rate Processes," McGraw-Hill Book Company, New York, 1941.
23. Dash, W., and A. Tweet: Observing Dislocations in Crystals, Sci. Am., vol. 205, p. 107, 1961.
24. Mura, T. (ed.): "Mathematical Theory of Dislocations," American Society of Mechanical Engineers, New York, 1969.
25. Beiser, A. (ed.): "The World of Physics," McGraw-Hill Book Company, New York, 1960.
26. Pauling, L.: "The Nature of the Chemical Bond," Cornell University Press, Ithaca, N.Y., 1960.
27. Moelwyn-Hughes, E.: "Physical Chemistry," Pergamon Press, New York, 1961.
28. Kondrat'ev, V.: "Chemical Kinetics of Gas Reactions," Addison-Wesley Publishing Company, Inc., Reading, Mass., 1964.
29. Porter, G.: "Progress in Reaction Kinetics," vols. 1–4, Pergamon Press, New York, 1961–1967.
30. Syrkin, Y., and M. Dyatkina: "Structure of Molecules and the Chemical Bond," 1950 trans., Dover Publications, Inc., New York, 1964.
31. Streitwieser, A.: "Molecular Orbital Theory for Organic Chemists," John Wiley & Sons, Inc., New York, 1961.
32. Vogel, H., V. Bryson, and J. Lampen: "Informational Macromolecules," Academic Press, New York, 1963.
33. Rich, A., and N. Davidson: "Structural Chemistry and Molecular Biology," W. H. Freeman and Company, San Francisco, 1968.
34. Ingraham, L.: "Biochemical Mechanisms," John Wiley & Sons, Inc., New York, 1962.
35. Monod, J., J. Wyman, and J. Changeux: On the Nature of Allosteric Transitions: A Plausible Model, J. Mol. Biol., vol. 12, p. 88, 1965.
36. Goodwin, B.: "Temporal Organization in Cells," Academic Press, Inc., New York, 1963.
37. Fischer, R. (ed.): Interdisciplinary Perspectives of Time, Ann. N.Y. Acad. Sci., vol. 138, p. 367, 1967.
38. Higgins, J.: The Theory of Oscillating Reactions, Ind. Eng. Chem., vol. 59, p. 18, 1967.
39. Mesarovic, M. (ed.): "Systems Theory and Biology," Springer-Verlag New York Inc., New York, 1968.
40. Pattee, H., E. Edelsack, L. Fein, and A. Callahan: "Natural Automata and Useful Simulations," Spartan Books, New York, 1966.
41. Fox, S. (ed.): "Origins of Prebiological Systems," Academic Press, Inc., New York, 1965.
42. Phillips, D.: The Three-dimensional Structure of an Enzyme Molecule, Sci. Am., vol. 215, no. 5, p. 78, 1966.
43. Perutz, M.: The Hemoglobin Molecules, Sci. Am., vol. 211, no. 5, p. 64, 1964.
44. Bernhard, R.: Enzymes—The Moment of Truth, Sci. Res., vol. 4, no. 4, p. 21, July 7, 1967.
45. Bush, S.: The Measurement and Prediction of Sustained Temperature Oscillations in a Chemical Reactor, Proc. Roy. Soc. A, vol. 309, p. 1, 1969.
46. Iberall, A.: Comments on the Problem of the Biochemical Oscillator, NASA Contractors Report CR 1806, 1971.
47. Morowitz, H.: "Energy Flow in Biology," Academic Press, New York, 1966.

7

Continuum Instability; Physical and Biological Examples

We have arrived at a continuum level in which the underlying atoms are in complete or near equilibrium and have formed whatever stable clusters the interatomic binding will permit. One vaguely expects the possibility that the field of these tightly bound clusters can now grow indefinitely in size. Nevertheless, with the slight disturbances that the extended field is exposed to in its surrounding bounded environment, it becomes unstable in the large.

The following continuum systems illustrate such coarse instability, which is somehow related to the transport coefficients associated with the field material. These coefficients result from the underlying molecularity of the system; yet the instability is not at the molecular level but at the continuum level of the aggregation. Examples are:

Turbulence in the body of the hydrodynamic field
Surface phenomena in the hydrodynamic field
Point and surface phenomena in the solid field

Point and surface phenomena in the chemical field
Cellular processes in the biochemical field

The effect of such macroscopic instability is to form "superatoms" within the continuum field. (The superatoms are each rich enough in atoms so as to make up a continuum segment.) Our problem now is to discuss the stability conditions that create these emergent macroscopic atomistic "forms." In principle, we have (or can imagine having) equation sets that describe change in the continuum and—independently—an equation set that describes the statistical mechanical equilibrium state in terms of atomistic fluctuations. We will discuss these illustrations, separately, in outline form.

7-1   TURBULENCE IN THE HYDRODYNAMIC FIELD

A set of continuum equations for a nonequilibrium physical hydrodynamic field—under the action of surface and body forces but with no chemical changes taking place (i.e., chemical equilibrium has been assumed to have occurred first)—were sketchily derived in the preceding chapter. The set consisted of an equation of motion, an energy equation (e.g., a statement of the phenomena, both reversible and irreversible, changing the entropy), an equation of continuity (conservation of mass), and thermodynamic relations. Illustratively, in decreasing order of restriction, various problems that can be undertaken are: no body forces present, isothermal walls and sources, motion under pressure gradients; gravitational forces added, isothermal conditions; accelerational motions, such as rotation, added; motion under large temperature gradients, such as a difference in source and wall temperature; added coupling of electrical or magnetic forces, etc. Well-known sources, Lamb (1) and Rayleigh (2), illustrate the classical methodology for the mathematical physics of hydrodynamic fields. Goldstein (3) provides experimental content.

It may be first noted upon examining the equation set that, by permitting the driving forces to approach zero, a rest state for the system emerges. This represents one singular state of motion of the set. There may be remnant stationary potential fields that persist (e.g., the Laplacian temperature field of pure thermal conduction). Second, by changing the driving force, a motional field emerges. One such motional field (small velocities) will be essentially linear in velocity. It may be described by solutions of the equation set for various boundary conditions. These solutions illustrate a laminar or creeping flow field. Finally, above some critical value of velocity (for each boundary-value problem and type of driving force), these apparently stable solutions break down. The field breaks up into eddies and flashes. It becomes turbulent. Since Reynolds' (1883) highlighting study, the search for an adequate explanation for such turbulence has persisted.

Our present concern is not with developing a theory of turbulence but simply with pointing up that this illustration appears to be a clear example of instability, in which a continuum field (the velocity field in a fluid medium) cannot be extended indefinitely without an unstable motion ensuing. (The critical parameter, the Reynolds number, which governs the transition is DV/v; D = a characteristic dimension of the field, V = an average velocity in the field, v = kinematic viscosity. It is thus clear that indefinite extension of a hydrodynamic field, by growth in D, is not possible without instability arising.)

A fluid field with any driving source can exhibit the instability. (We are assuming that the bulk viscosity of the fluid is reasonably bounded, else we get properties that resemble more nearly a supercooled fluid. We have seen movies of the flow of hot polymers under high pressure through orifices. It is very dubious that turbulence could be demonstrated in such materials—according to our ideas, not because of the high shear viscosity of the material and the required high power to drive the field, but because of its high bulk viscosity. Other internal relaxation modes could carry the energy off.) Some illustrations of the instability are: Richardson's vertical-density gradients (3); G. I. Taylor's flow between rotating cylinders; von Kármán vortices; Bénard cells; pipe flow instability, tidal oscillations in the atmosphere, etc.

The clearest example that a nonlinearly stable ("quantized" limit cycle) pattern can emerge was demonstrated by G. I. Taylor for rotating cylinders (3). An extensive discussion of hydrodynamic stability may be found in (4) and (5). It is our contention, although not widely accepted as yet, that our study (6) illustrates, somewhat more generally than Orr–Sommerfeld theory, that a classical mathematical physical path likely can be found for the discovery of nonlinear limit cycles, using describing function techniques. Thus, at present, we believe and are willing to emphasize that techniques, albeit crude, are available for determining whether stable limit cycles may exist in the hydrodynamic field. Characteristically, their structure is exhibited as stationary or moving eddies. Since these are at continuumlike dimensions, we may regard them as atomistic rather than atomic. They form macroscopic superatoms as compared with the underlying molecular structure. Important strictures defining the applicability of the near-equilibrium continuum field and its describing equations (6, 7) are a spatial constraint (that the mean free path is small compared to scale dimensions in the field, including boundary-layer dimensions) and also a temporal constraint (that internal relaxation times are short compared to the shortest observational periods or periodicities of interest).

7-2 SURFACE PHENOMENA IN THE HYDRODYNAMIC FIELD

We have already alluded to surface tension, in the case of the equation of state, as representing an energy barrier that requires a discussion of a "jump" phenomenon between liquid and gas [see Adam (8), or (9), and (10)]. There is an additional property of systems which is suggested both by this example and the atomistic vortical cells that show up in turbulence. It appears quite generally that the instability in a continuum system is grossly relieved at points or surfaces; that is, that material which is involved in the relief mechanism lies within restricted zones. (This is partially obvious in that otherwise the mechanism would be distributed throughout the body and no continuumlike integrals would have existed. The material field would exhibit the character of white noise. Nevertheless, this does not guarantee that such relief surfaces could not be quite skittish or evanescent. Triple-point phenomena come closest to exhibiting such skittishness. We are suggesting, however, that not only do such points and surfaces exist but that they also have a much more permanent character in time.)

In this problem category, we may consider shock fronts in compressible flow (apart from the front, the separated phases may be treated by the continuum equations; the front may be treated by jump relations); cavitation; metastable states of droplets in the vapor phase, or vapor bubbles in the liquid phase; nucleating lattice-formation in the liquid, particularly near freezing, etc. Thus, instability here seems to exist on a global scale which, for a given temperature, is intensified by largeness of dimension, speed of reaction, number density of sub-atoms, and sub-atomic jitter velocity (namely, by all the factors that make up the Reynolds' number. One additional means of viewing the instability is that if the kinetic energy which the system sweeps into the local spatial field exceeds the relaxational energy which the atomistic elements within the local field can absorb, then the field becomes globally unstable.). By breaking up the large field to smaller, more quantized size, it can be stabilized. The largest degree of stabilization can occur with more specialized arrangements of these superatoms. However, once a marginal degree of local stability is set up (i.e., not fully thermodynamic equilibrium), it is hard to state over what sort of epoch this pattern will persist. Theoretically, one may surmise that when the perturbation time interval of convergence (in this case for a continuum field, not for the perturbation time interval of point particles as in the n-body problem) breaks down the field solution cannot be guaranteed except by another extension from the new condition. This process is likely too difficult to describe step by step. Thus, a statistical mechanics is once again in order, but for continuous modalities.[1]

Historically, it has always been convenient to use such concepts as van der Waal's long-range attractive forces between finite-sized molecules as models for phenomena associated with states differing from the ideal gas. However, the theory of phase changes, notably near critical phase points, is in process of review currently. [(11) is illustrative of the discussion in progress.]

---

[1] If the sub-atomic particles also emerged as the result of continuous modalities in their m/e-m "ether" then the character of instability would be the same in these two cases.

7-3 POINT AND SURFACE PHENOMENA IN THE SOLID FIELD

[See (8)] With high number density, the molecules in the solid field are not capable of much translational motion. Relative motion proceeds very slowly. (Yet it proceeds. Slip, slow creep, and other anelastic properties assure this.) By careful design (namely, choice of crystal symmetry, "impurities," alloying material, etc.), "heat" treating, and "work" hardening, possible structures can be obtained that are quite stable. [They are complex poly-crystalline, no longer simple, structures. The single crystals—"whiskers"— do not provide the stable answer unless they, too, are immersed in a complex matrix. It was one of the purposes of our 1954 unpublished report (12) to discuss this problem.] The simple, metastable states of a newly frozen-out solid are thereby eliminated, and instead, a highly stable structure is arrived at.

In this stable arrangement, the molecular array extends regularly out to its dislocation. The dislocation entities tend to "pin" and "stabilize" the extended structure. The dislocation domains extend out to grain boundaries. The structure—as a whole—is thus inhomogeneously stabilized. (Later, we will attempt to identify such dislocations as analogous to human elites.)

How did this structure come about? Obviously, the final very stable structure did not come about from an elementary thermodynamic process. [See Jackson in (13), for example. We are up to the same kind of questions that were raised by Pattee in (14). His concern was how to provide a stably repetitive temporal process that could assure reliable hereditary propagation. Here, in trying to illustrate the process, we need only a theory of automata, since we, the human craftsmen, are willing to invest all of the necessary constraints—of space and time—to bring about stepwise the desired end of a "hereditary" structure; namely, a structure that repeats a coded description, a "cookbook" recipe for a reliable extended material. See, for example, discussion in (15). Yet, we would be hard put to bring into being the replica-tion as a temporally hereditary process by which good material proceeded to replicate good material.] Nevertheless, as in Watson–Crick coding, we can pretend to carry through a quasi-static description. Cool the material, with imperfections, from the melt. Around a uniform distribution of nucleating centers, a very regular local cell of neighbors begins to form and grow. At some point, there is sufficient instability that dislocations form to stabilize the regular arrays by only small mismatch. These dislocations may be accentuated by the distribution of impurities to make their density optimal. Through a large number of research trials, we learn what constitutes good ingredients to put into the melt and what are good heat-treatment steps in the cooling process of formation.

Thus, because of their finite size, it is the dislocation domains themselves that are the first superatom in the solid state. The array of such domains is extensive enough to be a continuum. It is the dislocations that bring the theoretical yield strength of the material down to the order of its low experimental value rather than to the magnitude of the theoretical crystalline-bond strength. [As source material on dislocations, see (16) for an elementary description or Friedel (17) for more technical discussion.]

Beyond this, the subsequent levels which are used to stabilize the solid state so as to prevent motions during appreciable mechanical stress are, at present, still matters for engineering mechanics rather than fundamental science. Loosely speaking, it is a chemical metallurgy of macroscopic "molecules." While this thought may appear to be a weak summary, it really means to suggest that there is considerable geometric concern with how to wedge and prepare a matrix so that only the dislocation-domain instability may govern. Chemical and physical boundary attack is used as the tool (typically, to protect the "whiskers").

7-4 CELLULAR PROCESSES IN THE BIOCHEMICAL FIELD

The potential existence of oscillating reactions in a soup is not the end of the biochemical problem. At the present time, various investigators are looking for oscillating systems examples [(18–20), Katchalsky, Prigogine]. The nonlinear conformational changes in allosteric enzymes suggested by Monod, much like snap-diaphragm characteristics or tunnel-diode characteristics, have held some of the most promise in suggesting a workable mechanism. Nonbiochemical examples have been demonstrated. The experimental fact is that evolutionary steps "quickly" ($1–3 \times 10^9$ years?) may have proved sufficiently unstable to develop the biological cell.[1] Currently a sequence of steps to organize this viable cell-making process may still appear horrifying to physicist, chemist, and biologist. Commoner (21) clearly states one major dogma quite well. "For example, the proposal that DNA is a self-duplicating molecule conflicts with one of the basic principles of biology—the cell theory —which holds that living substance cannot be reduced to any unit smaller than a single cell and yet exhibit the properties (such as self-duplication) characteristic of life." Yet, these are the issues that concern Oparin and Pattee (22).

In the cell, we are most familiar with the current work of Goodwin (15, 23). According to our understanding, Goodwin believes that it may be possible to describe the 200 to 300 oscillating enzyme chains in some simple

---

[1] It seems appropriate to us to identify formally, really only to stress what we had said before, yet another principle of GSS. Whereas it is common to think of evolutionary processes as somehow bucking the tide of entropic degradation, we propose an alternate view. In the larger-scale degradative process, which lingers in passing at various stability eras at which new limit-cycle complexes emerge, the closeup view is that of evolutionary emergence. The actual process is only a step in global degradation. Life—we suggest—is such an example.

bacteria within some finite time. [See (24) for a beginning of such a description.]

> We use the term "chain" in our biological descriptions rather the electrical term "network," the physical term "degree of freedom," or the chemical engineering term "unit process" to suggest that we are speaking of causal chains of linked processes and mechanisms. The processes are not so fixed in space or time nor so restricted to one hierarchical level as to fit the connotations of these other terms.

Ideally, the biochemical cell would be described by a rather precise chemical characterization of its essential constituents, of its fluid matrix, of its internal structural elements, of its external membrane, of its dynamic nonlinear operating modalities, and of its exterior fluid sphere of influence. This cannot be done today with any real success. If no other single incident proved impressionable to us, it was time-lapse photographs (by A. Bajer, University of Oregon) at the fractional micron scale showing the nature of mitosis in a cell (25). It seemed clear that no set of hydrodynamic equations could be brought to bear on the infinitely complex nonlinear process that takes place.

A second, equally impressive demonstration may be furnished by photography under phase microscopy, indicating a large sphere of influence well beyond the membrane with which the cell is coupled.

Thus, there are two methods that one might use in describing the atomistic cell. One is that of static descriptions—the cell as histological entity. The second is to assemble some questions, some primitive block diagrams, and some hypotheses on the dynamic cell and to discuss these for various primitive cellular forms.

The former does not meet our inclination or purpose. Instead, the reader can be referred to (26) for a popular introduction; for more detailed descriptions, to general physiology texts (27–31); or to more specialized sources (32–36). Also see (57) for introductory cellular energetics.

Cell complexity (in the animal kingdom, particularly the human) extends from the red blood cell, muscle cell, capillary-wall cell, nerve cell, to the single-celled animal (amoeba, etc.). The cell is presented in elementary form as possessing a nucleus, a nuclear membrane, cytoplasm, and a cell membrane. A summary of the major characteristics of the cell is found in (26). (For example, the current visualization of the geometric structure of the cell is depicted on page 9 by Brachet and page 52 by Robertson.)

For a background of some details about the cell, one may examine the nature of enzymes given in Dixon and Webb (37) or in elementary form in Bernhard (38), and the nature of organelles described in Morrison (39). The latter two are from paperback series that may serve most introductory purposes. Also Bernhard gives many excellent references to the current status of enzyme kinetics and catalysis. For the flavor of organelle origins, see (55).

For a rudimentary dynamic picture of the cell, one might emphasize the linked elements listed in Fig. 7-1.

The black-box active element, we propose, is the active membrane-enzyme link. In our view, it is likely the membrane in most cells that reacts to the outer aqueous milieu with some electrochemical specificity or to conductive electricity in the sheath in the case of nerve cells.

In addition, we are inclined to believe that water may be bound to the membrane and may be partially or totally organized within the cell. The membrane-enzyme combination itself presents active "engine" characteristics. Its mechanisms are not understood, although, following Danielli, many sources consider that a bimolecular lipid leaflet determines the form of the membrane and that a "sodium pump" illustrates a functionally active unit. However, many controversies raging in this field of membrane theory make it foolish for an "outsider" to take sides.

A tour through classic unit-membrane theory—lipid and protein layering—may be found in such references as Davson's excellent textbook (27); Davson and Danielli's classic work (40); (26); (41), in which Robertson presents a review of his unit-membrane theory; DeLuca's discussion in (42), a thoroughly enjoyable unresolved dialectic on the biochemistry and physiology of mitochondria, the membrane-bounded energy, and calcium power packs within the cell; (43); Lucy's discussion on membranes in (44)—this article, on steroid hormones, goes much further into biochemical questions than most readers may wish to delve; and (45). References (43) and (46) were suggested to us as being representative of the current position regarding the structure of the membrane, which Danielli has been so instrumental in advancing. Stein (45) has written an extensive review of current knowledge on the membrane. Another view, somewhat more dynamic, of the cell periphery is by Weiss (48). Another interesting article is by Racker (47). There are

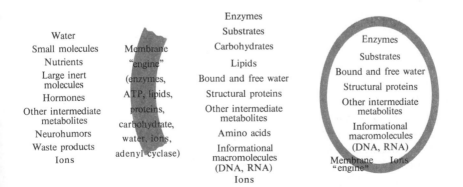

Fig. 7-1   The transport-influencing system (or systems) in a cell, with (or without) a nucleus.

innumerable other symposia which can easily be found that illustrate the considerable interest and difficulty in elucidating membrane function. Hoffman (49) and Bell and Grant (50) are further examples.

On the other hand, there are investigators who are even more critical of the current modeling of the active processes in membranes [see for example, Ling's article in (43) and (51), or an article regarding the views of Green (52). Ling's thesis, based on nmr evidence, is that water within the cell is polarized; Green's that conformational changes in the membrane will mechanically link its properties to transport.]. However, see Solomon (56).

> It is clear that a vigorous dialectic is going on in this field. We disclaim taking any stand, not because of any timidity, but because of lack of familiarity with sufficient data obtained in normal operating biological systems on cell "permeability" and transport. We will cite our own red-cell-flow work to underline reasons for our uncertainty.

We have been quite impressed by the preliminary nonlinear modeling of K. Kornacker for nerve cells (23). Our own growing concern has been with the exchanges taking place between muscle tissue, the capillary membrane, serum, and blood cells. We are in process of trying to determine the characteristics of the metabolic exchange at this level, which seems to involve chemical-electric control of the axial file of red cells itself which we find to be periodic. (This work is done in vivo in muscle tissue in normal, unanesthetized animals.)

Some other current work that we find experimentally impressive is that of Silvio Baez. We believe that he has developed techniques that begin to show the spatial dynamics, in vivo, of the capillary wall, too [e.g., see (53) as a techniques reference].

We mention these pieces, not as being in the established main line of membrane science, but because they each emphasize the active nonlinear or limit-cycle nature of two major membranes—the nerve cell and the capillary membrane within the complex biological organism.

The main ingredient, which we and Kornacker seem to stress—he for nerve cells, we for capillary cells—is the need for active mechanisms within the membrane itself with specific dynamic characteristics that fit actual transport processes, not simply a passive porous membrane or microfilter action, regardless of its specialized character (thickness, layer constituents, holes, active sites, tight binding) or a simple specialized active transport carrier. In our opinion, it is even the in vivo observations of the actual transport-process dynamics in the complex organism that are lacking. (For example, observations are missing starting from just a simple description of dynamic water transport across capillaries.)

> We can suggest reasons for seeing complexity in the problem of membrane transport by introducing our own speculative views of one biological cellular system's dynamic response involving membranes. In three separate experimental

assays (54), we have shown that the axial flow of red blood cells possesses a periodic flux in their number through the lumen of small capillaries. In these small capillaries, in our opinion "nutrient" capillaries, there is a considerable mechanical interaction between the cell membranes and the capillary membranes. We find periodicities of the order of 0.3 sec (as a cell jitter) and of the order of 30, 100, and 300 sec.[1] There was no evidence that this cyclic flow was governed by any sphincteric action. Thus, the only alternative is a complex chemoelectric gating, involving the two interacting membranes of red cell and capillary wall, combustion by-products of the tissue surround—here muscle—possibly circulating hormones within the capillary, and the cellular contents within the membranes. Such a membrane-mediated model may be even more complex than pump models.

Thus, similar to the automata replication problem of the solid state (how can one make a solid system reliably reproduce its own structure) but with the added need for hereditary replication, in time, of the life process, the question arises whether the evolutionary development took the path of oscillating chemical reactions (such as DNA template chain or a precursor with an enzyme escapement for life replication) in watery milieu to form a biochemical continuum. For the present, we require that the reaction in the milieu be chemically active enough to help form an active hydrophobic lipid-protein engine (the closed biochemical chain-enzyme escapement life process).

Conceptually, in a science-fantasy sense, we might take note of the antagonistic sense of water and oily polar bonds. If there were strong hydrophobic bonds surrounding the atomistic oscillating life-replication reaction, these bonds could, in imagination, have suppressed the reaction. (A second developmental step of the aggregation of fats and oils as well as amino acids in prebiology is imagined.) Thus, only hydrophobic surrounds that had some dynamic permeability could be selectively viable, particularly if the surround enhanced motional instability, etc. Then an active "automobile" system (i.e., a living system) would emerge. The chain described in imagination is fantastically long to be able to specify with any confidence.

Even more fantastic is the real coding of electrical and electrochemical events that produce mitotic division. Thus, we may put down a black box of outside chemical-active–membrane-enzyme chains; but such a description is still incredibly naive. [For a suitable primitive carrier in attempting to swallow or reject such speculative pills, we suggest (27).]

Thus, with mystery no less than atomic quantization out of the continuum of active subnuclear electrical stuff, we have the biological cellular quantization out of the continuum of active biochemical stuff.

It is a fact today that a number of groups are seeking to find a path to develop a mobile, materials exchanging, self-replicating "cell-like" structure.

---

[1] Vasomotion in bat wing capillaries, involving a periodic contraction in the range of seconds has been noted in reports by Wiedeman (1968) and D'Agrosa (1969).

It is a fact that they have not yet succeeded. Nevertheless, they have been able to take C, O, H, and N as atomic material, energize it, produce elementary molecular groups like CO, CO_2, NH_3, H_2O; further produce formaldehyde, ammonium compounds, amino acids; and even produce protocells. The latter are cellular structures that may have one or more kinds of internal actions or structures. These investigators, in public statements, are quite enthusiastic in their belief that synthetic biocellular structures will in the near future be achieved. We premise that they are right. Thus, the existence of cells vis-à-vis physical-chemical processes permits consideration of the next level of the collection of cells or colonies.

To summarize the thoughts in this section, although presented with very little detail, we note that the molecular continua do not persist indefinitely. A gross Reynolds' numberlike criterion determines a nonlinear instability by which the field breaks up and becomes stabilized into continuumlike, surface-bounded, volume-compartmentalized domains. The suggestion of a generalized Reynolds' number instability means that it becomes more unstable with size, with relative field convective velocity, with number density of molecular particles, and with the jitter velocity of the molecular particles (i.e., "thermal" agitation, Brownian motion). "Thermodynamic" equilibrium only governs in the same sense that it always does. There is internal equipartition only among those degrees of freedom that are not frozen out. It is not yet clear what freezes out some degrees of freedom. Generally, it appears associated with considerable number density. The local cell seems to be capable of preferentially locking up some nonlinear state instead of others. Among those that are accessible, the quasi-ergodic hypothesis essentially applies. Instability is relieved by a nested sequence of stabilizing sublevels. The complexity of superatoms which emerge from the continuum field is fantastic. The reason, suspected from hydrodynamic sources, is the rich number of modalities that are available. There are nonlossy propagative modalities of mechanical and electrical (and one may suppose ultimately, gravitational) nature. These are momentum diffusive and energy diffusive (the propagative modes may be mass diffusive or electric-charge diffusive). Generally, one might say that the propagation is either nonlossy (wavelike) or diffusive. However, they arise from all the tensor couplings that Onsager theory permits (10).

This point is worth some elaboration. The number of degrees of freedom in an electromagnetic wave guide, in the radiation field within a blackbody enclosure, or in a continuous turbulent spectrum is infinite, but not many modalities are involved. Thus the patterning complexity, the "color" of the response, appears limited. On the other hand, adding chemical coupling begins to increase the coupling modalities considerably. The patterning complexity—as the biochemical life forms indicate—increases considerably. When we come to the complex biological organism with its many internal systems, we will see even greater patterning "coloration" because of its many more modalities.

SUMMARY

1. All of the continuum states whether gaseous, liquid, solid, or chemical become unstable at some size and then break down into superatoms.

2. The instability is essentially relieved at surfaces or points. The regions in between remain the density-determined continuum.

3. Examples are the breakdown of the turbulent gas atmosphere into a cell structure or the turbulent fluid into cell-like vortices. In the solid, the latticelike organized structure of nuclear particles—that extend as atoms or molecules—ends as domain-organized grains at dislocation-point imperfections that define slip planes. In compressible flow, the domains are relieved by shock fronts. In two or more phase substances, the regimes are separated by surfaces.

4. In addition, the chemical state has exhibited a stabilized state in cellular processes of a bio-chemical nature. This is not the only possible stabilization of the chemical field. There are many geochemical complexes known. The complex geophysical earth exhibits many other such oscillating processes. However, the biochemical oscillator is a striking one that has evolved in time.

5. In the case of the biochemical system, which is responsible for life, the "remarkable" surface active system, the membrane, helps maintain these processes. Its action is not really understood. Its action should perhaps be named dynamic catalysis.

6. Dynamic catalysis may perhaps be labeled as the system by which an active surface—by virtue of geometric properties and by virtue of possessing an active "engine" property, which however utilizes little energy—can regulate and regularize other properties involving relatively large amounts of energy transfer. The process—particularly associated with regulating chemical as well as biochemical reactions at the molecular level—is not particularly well understood. It can loosely be called a dynamic gating.

7. Although evolution of active mechanisms is commonly considered as running counter to the entropic tide, it is possible to view the entire cosmological history as a degradative process. However, at various epochs in its degradation, local instability permits the formation of limit cycle oscillators for limited eras. The conceptual model is that of a global pinball machine.

QUESTIONS

1. Examine the nonlinear equation set that includes the G. I. Taylor formulation of rotating cylinders, Bénard convective cells, and the treatment of parallel plate flow of (6). See if you can set up the decomposition into a mean flow state and first-order fluctuating equations in general so that you can show how autonomous oscillations in the nonlinear set can be traced to an eigenvalue problem associated with the first-order fluctuating set. Pursue the secular equation associated with satisfying boundary conditions to determine conditions when the allowed steady-state spectrum is discrete, continuous, nonexistent, or combinations of these.

2. Attempt to examine when and why field instability in the general inhomogeneous equations of hydrodynamics will lead to thin boundaries separating molecular configurations with different binding. Keep in mind shock fronts, free surfaces of liquid and solid, the separation surface of a submerged jet (liquid in liquid, or gas in gas), critical phase points, and a burning wave front.

3. Discuss at considerable length the "pinball machine" theory of history mentioned in summary, in which emergent evolution of new limit cycles is the characteristic property of a globally degradative system (i.e., the big bang theory of cosmology and a subsequent cooling-off in an expanding universe).

4. Examine in some systematic way means for creating chemical oscillators in a batch process and in a continuous flow process. This may involve chemical-catalytic substrate means, chemical-mechanical coupling, chemical-thermal coupling, chemical-electric coupling, chemical-mass diffusion coupling, etc.

5. Describe in considerable detail the systems characteristics of some simple cell, say a simple bacterium. First sketch out its current structural organization (i.e., all of its organelles), then indicate its functional chains, improvising where necessary. Attempt to arrange the description so that the bacterium's two temporal modalities of motional food-ingestion cycles and growth and division cycles can come off.

6. After reviewing various recent sources of material on membranes, attempt to write a systems description of some well-defined membrane structure and formation that will bring its active limit-cycle characteristics into focus.

7. Attempt a (science-fiction-historical) theory of forming an active oil-water oscillator in some cellular structure. Make use of the strong polar bonds.

8. Examine the theoretical or experimental base for a protocell that has characteristics similar to a simple living cell. (Find time-lapse motion pictures on a cell colony whose motion and division characteristics are clearly periodic and attempt a good analogue for this system.)

BIBLIOGRAPHY

1. Lamb, H.: "Hydrodynamics," Dover Publications, Inc., New York, 1960.
2. Rayleigh, L.: "The Theory of Sound," Dover Publications, Inc., New York, 1945.
3. Goldstein, S.: "Modern Developments in Fluid Mechanics," Dover Publications, Inc., New York, 1965.
4. Lin, C.: "The Theory of Hydrodynamic Stability," Cambridge University Press, New York, 1955.
5. Betchov, R., and W. Criminale: "Stability of Parallel Flows," Academic Press, Inc., New York, 1967.
6. Iberall, A.: Contributions Toward Solutions of the Equations of Hydrodynamics, Part B, Primitive Solutions for the Fluctuating Components of Turbulent Flow Between Parallel Plates, *Contractors Rep. Off. Nav. Res., Wash., D.C.*, Contract No. Nonr 4559(00), October, 1965.
   ———: A Contribution to the Theory of Turbulent Flow Between Parallel Plates, *Seventh Symposium on Naval Hydrodynamics*, Rome, Italy, 1968 (in press).
7. Iberall, A.: Contributions Toward Solutions of the Equations of Hydrodynamics, Part A, The Continuum Limitations of Fluid Mechanics, *Contractors Rep. Off. Nav. Res., Wash., D.C.*, Contract No. Nonr 34-5(00), December, 1963.
8. Adam, N.: "The Physics and Chemistry of Surfaces," Dover Publications, Inc., New York, 1968.
9. Hirschfelder, J., C. Curtiss, and B. Bird: "Molecular Theory of Gases and Liquids," John Wiley & Sons, Inc., New York, 1964.
10. Rossini, F.: "Thermodynamics and Physics of Matter," Princeton University Press, Princeton, N.J., 1955.
11. Domb, C.: Thermodynamics of Critical Points, *Phys. Today*, vol. 21, p. 23, 1968.
12. Iberall, A.: "Phenomenological Basis for a Generalized Elastic Theory of the Solid State for Application to Structural Design," unpublished study, 1954.

13. Jackson, K., in H. Reiss (ed.): "Progress in Solid-state Chemistry," vol. 4, Pergamon Press, New York, 1967.

14. Pattee, H., E. Edelsack, L. Fein, and A. Callahan: "Natural Automata and Useful Simulations," Spartan Books, New York, 1966.

15. Waddington, C. (ed.): "Towards a Theoretical Biology, 1, Prolegomena," Aldine Publishing Company, Chicago, 1968.

16. Piel, G. et al. (eds.): "Materials," A Scientific American Book, W. H. Freeman and Company, San Francisco, 1967.

17. Friedel, J.: "Dislocations," Addison-Wesley Publishing Company, Inc., Reading, Mass., 1964.

18. Monod, J., J. Wyman, and J. Changeux: On the Nature of Allosteric Transitions: A Plausible Model, *J. Molec. Biol.*, vol. 12, p. 88, 1965.

19. Fischer, R. (ed.): Interdisciplinary Perspectives of Time, *Ann. N.Y. Acad. Sci.*, vol. 138, p. 367, 1967.

20. Higgins, J.: The Theory of Oscillating Reactions, *Ind. Eng. Chem.*, vol. 59, p. 18, 1967.

21. Kasha, M., and B. Pullman: "Horizons in Biochemistry," Academic Press, Inc., New York, 1962.

22. Nord, F.: "Advances in Enzymology," vol. 27, Interscience Publishers, a division of John Wiley & Sons, Inc., New York, 1965.

23. Waddington, C.: "Towards a Theoretical Biology, 2, Sketches," Aldine Publishing Company, Chicago, 1969.

24. Datta, P.: Regulation of Branched Biosynthetic Pathways in Bacteria, *Science*, vol. 165, no. 3893, p. 556, 1969.

25. *Symposium on the Dynamics of Fluids and Plasmas*, University of Maryland, October, 1965.

26. "The Living Cell, Readings from *Scientific American*," W. H. Freeman and Company, San Francisco, 1965.

27. Davson, H.: "A Textbook of General Physiology," Little, Brown and Company, Boston, Mass., 1964.

28. Best, C., and N. Taylor: "Physiological Basis of Medical Practice," The Williams & Wilkins Company, Baltimore, Md., 1966.

29. Ruch, T., and H. Patton: "Medical Physiology and Biophysics," W. B. Saunders Company, Philadelphia, Pa., 1965.

30. Guyton, A.: "Textbook of Medical Physiology," W. B. Saunders Company, Philadelphia, Pa., 1966.

31. Mountcastle, V.: "Medical Physiology," vol. 1, The C. V. Mosby Company, St. Louis, Mo., 1968.

32. Giese, A.: "Cell Physiology," W. B. Saunders Company, Philadelphia, Pa., 1957.

33. Brachet, J.: "The Cell," 5 vols., Academic Press, Inc., New York, 1960.

34. McElroy, W.: "Cellular Physiology and Biochemistry," Prentice-Hall, Inc., Englewood Cliffs, N.J., 1964.

35. Fawcett, D.: "The Cell," W. B. Saunders Company, Philadelphia, Pa., 1967.

36. Bloom, W., and D. Fawcett: "A Textbook of Histology," W. B. Saunders Company, Philadelphia, Pa., 1968.

37. Dixon, M., and E. Webb: "Enzymes," Academic Press, Inc., New York, 1964.

38. Bernhard, S.: "The Structure and Function of Enzymes," W. A. Benjamin, Inc., New York, 1968.

39. Morrison, J.: "Functional Organelles," Reinhold Publishing Corporation, New York, 1966.

40. Davson, H., and J. Danielli: "The Permeability of Natural Membranes," Cambridge University Press, New York, 1952.

41. Purpura, D. (ed.): Current Problems in Electrobiology, *Ann. N.Y. Acad. Sci.*, vol. 94, p. 339, 1961.

42. DeLuca, H., in A. Budy (ed.): "Biology of Hard Tissue," New York Academy of Sciences, New York, 1967.

43. Loewenstein, W. (ed.): Biological Membranes: Recent Progress, *Ann. N.Y. Acad. Sci.*, vol. 137, p. 403, 1966.

44. Lucy, J., in L. Peachey (ed.): "Conferences on Cellular Dynamics," New York Academy of Sciences, New York, 1967.

45. Stein, W.: "The Movement of Molecules Across Cell Membranes," Academic Press, Inc., New York, 1967.

46. Korn, E.: Structure of Biological Membranes, *Science*, vol. 153, p. 1491, 1966.

47. Racker, E.: The Membrane of the Mitochondrian, *Sci. Am.*, vol. 218, no. 2, p. 32, 1968.

48. Weiss, L.: "The Cell Periphery, Metastasis, and Other Contact Phenomena," North-Holland Publishing Company, Amsterdam, 1967.

49. Hoffman, J. (ed.): "The Cellular Functions of Membrane Transport," Prentice-Hall, Inc., Englewood Cliffs, N.J., 1964.

50. Bell, D., and J. Grant (eds.): "The Structure and Function of the Membranes and Surfaces of Cells," Cambridge University Press, New York, 1963.

51. Ling, G.: A New Model for the Living Cell, *Intern. Rev. Cytol.*, vol. 26, 1968.

52. Bernhard, R.: *Sci. Res.*, vol. 3, p. 33, 1968.

53. Baez, S.: Recording of Microvascular Dimensions with an Image-Splitter Television Microscope, *J. Appl. Physiol.*, vol. 21, p. 299, 1966.

54. Ehrenberg, M., and S. Cardon: Dynamics of the Microcirculation, *Proc. Ann. Conf. Eng. in Med. and Biol.*, vol. 8, 1966.

   Cardon, S., C. Oestermeyer, and E. Bloch: Effect of Oxygen on Cyclic Red Blood Cell Flow in Unanesthetized Mammalian Striated Muscle as Determined by Microscopy, *Microvasc. Res.*, vol. 2, p. 67, 1970.

55. Raven, P.: A Multiple Origin for Plastids and Mitochondria, *Science*, vol. 169, no. 3946, p. 641, 1970.

56. Solomon, A.: The State of Water in Red Cells, *Sci. Am.*, vol. 224, p. 88, 1971.

57. Lehninger, A.: "Bioenergetics," W. A. Benjamin, Inc., New York, 1965.

8

The Cellular Colony as Continuum

We have noted the quantization in the continuum field of an extended molecular array into "stable" atomistic superatoms—illustratively the solid-state–physical-chemical grain, the fluid state vortex, and the biological cell. Now in the larger spatial domain, subject to particular boundary-value problems, these atomistic elements may begin to cluster. They begin to form "colonies" or ensembles.[1]

The phenomena which emerge at this level are likely not so different from those at the previous molecular continuum level, but they require some careful thought. At dilute concentrations—small-number densities of super-atoms—it is not so clear what provides interparticle binding force. In fact, if

---

[1] We suggest the terms "colony" and "colonial" organization as being literally most appropriate. We view the ensemble as a group of similar entities living or growing together but remote from the state having nominal control of it. The specialized meaning that we will give to the latter aspect is that the governing algorithm, control function, or "purpose" behind the colony will be somewhat remote or not immediately apparent. Thus it is "colonial."

no binding force exists, colonies do not form. Ideal gas, or Henry's law of behavior, emerges from the statistical ensemble. However, what are commonly found are weak forces which organize the colony. We cannot discuss these in general. The difficulties are at least as great as accounting for the forces at the intermolecular level. In colonies, the diversity of binding forces is likely greater. There may be a great variety of reasons why any particular colony is bound together. However, one illuminating insight can be provided as hypothesis for the colony type of organization.

There is a quantum mechanical force, the exchange force, which is commonly presented as having no classical analogue. We wish to suggest a nonlinear analogue. If in stable dynamic configurations involving quite similar particles which involve many types of coupling forces (e.g., mechanical, e-m, others), it is possible to interchange particles, momenta, energy, internal state configurations, etc., without involving a great deal of energy difference, such exchanges may, if phased correctly, provide a binding "force." Note the exchanges do not have to be of a direct type involving exchange in the translational states of the cg of the system. They may involve exchanges among the internal degrees of freedom. The exchanges, further, may not have to involve equipollent energy transfer. They may be communications type or catalytic exchanges in which a small amount of propagative energy can trigger internal active modes that have amplifying capabilities (e.g., an e-m signal triggering an internal relay. The classical comic-book representation is the "force" of a "thought" turning on an internal "light" in the brain with an internal dynamo implied therein).

Shorter-range forces likely make themselves more evident with increased density of particles.

At the present time, we cannot fruitfully illuminate this section with any more general comments. It is based on the structure of engineering concepts that are found in alloys [for example, see (1, 2, 3)]; in the hydrodynamic field [for example, see (4)]; in wave systems in extended atmospheres [for example, see (5)]; or, in the cell colonies, in the current work that B. Goodwin or B. Chance's group is doing [of which (6) or Hess' article in (7) are examples]. It is clear in Goodwin's work that the processes of growth and reproduction in bacteria and higher forms, such as hydra, are synchronizable and suggest a complex communications process. It is clear from the synchronization of events in the motional hydrodynamics of the cell or from mitosis within the cell that the communications system is made up of mixed signals involving electrical, chemical, and mechanical processes. Chemical communicationslike coupling within the colony seems quite reasonable.

The flavor of the problem may be found in an article by Moscona (8). Loewenstein's paper in (9) and, more recent, his paper at the February 1968 Biophysical Society meeting or at the August 1969 Biophysics Congress illustrate work in progress on intercellular communication that is or will soon be available. See, for example, (10). A fascinating illustration from the more specialized literature on biology, including discussion on its aggregation, is a monograph on the slime mold (11). The full dynamic complexity possible in a simple "colony" organization, such as the slime mold, was elegantly presented in time-lapse photography by H. Rusch, "Some Biochemical Events in the Life Growth Cycle of Physarum Polycephalum," at the Biochemistry Society Symposium, April 14, 1969 in Atlantic City. The complex periodic subprocesses found in a simple living structure, as complex as anything that might be found in a 20-acre chemical manufacturing plant, yet devoted to only two or three modalities—namely ingest, grow, divide—could easily serve as an object lesson in systems' science for all physical scientists. Also see (14).

The net effect of coupling is that the colony motions do not consist of random, independent actions of each individual but possess a considerable degree of coherence. It is such processes of limited coherence which produce a coupling of pulselike phases rather than the simply, unitarily caused, everlasting sinusoidal oscillation, which confronts us here and at every other level.

It was Huygens, in about 1665, who sought explanation for the entrainment of two clocks on a wall. "Explanation" has been in terms of the small-amplitude acoustic signals which eventually couple the nonlinear escapement phases.

We have attempted a preliminary discussion of the mathematics of sequences of pulses unfolding in time [in Sec. 6 of (12)] as distinguished from sustained sinusoidal oscillations.

The basic idea is that each pulse of action, in a nonlinear system embedded in a real universe, emerges as a new creation out of its past. It is the sustained linear instability in the local environment that ensures the repetitive quality of the action. On the other hand, in the idealized lossless linear isochronous system, with its characteristic sustained sinusoidal oscillation, causality for the action would be yoked irrevocably to the endless past and to an unending future.

In the elementary case of the theory of the clock system, a simple explanation for designing its sustained motion can be proposed. Choose a nearly linear device—pendulum, spring, crystal, atomic system—whose performance looks "clean" and linear for limited periods of time and provide an energy escapement per "pulse" or small number of "pulses." This assures the sustained coherent, timekeeping motion. It is likely that this is the philosophy inspiring the work of the biological space-timekeepers such as Pattee, Goodwin, and Chance or circadian scientists such as Halberg, Pittendrigh, Bünning, and Sollberger. Although we believe that it is "obviously" nonlinear properties that produce the sustained dynamics within the colony in all cases, we cannot isolate them always. Because of the difficulties of identification, the search for "binding" forces, the "Morse" potentials for colonies, will be some time in progress.

Plausible areas to seek information about colonial organization are developmental biology, geochemistry, and the plastic-elastic characteristics of solid-state materials. Many other colonial organizations, including those in higher hierarchical levels such as in human societies, are, of course, quite well known. However, we suggest that explaining these small organization problems is at least as difficult as explaining larger organization problems. In biology, this has been recognized by the group of investigators working on what they have facetiously referred to as "virtual" animal development (that is, of simpler animal forms) in defense against their colleagues who have identified themselves as studying big-brother "real" animals (generally complex mammals).

To illustrate these difficulties of accounting for organizational character, we can point to the following: It is Goodwin's thesis that biological development takes place by a logic of sequencing temporal events rather than by a logic of thresholds.[1] We have no competence to master all the background, issues, or facts in this field, in which the embryonic modalities arise. Besides, the problem becomes seriously involved with the start-up phase of a system rather than its life phase; that is, how the epochs of order temporarily emerge in the cosmological pinball machine. The story of genesis, not of a single individual, but of a coupled colony will be fascinating to watch as it unfolds during the next decade. As one further illustration from its current state, one must examine (13).

Similar remarks can be made about the genesis of salient geochemical complexes found on and in the earth. This is particularly aggravated in the discussion of the older structural forms. The geological problem is also too specialized for our consideration.

The third illustration is that of the plastic-elastic characteristics of the solid state. An engineering-mechanics problem of interest is to discover the bulk characteristics of a material from a sample, i.e., from a colony of grains. As (2) discusses in the theory of the tensile test, it turns out that determining the properties of the bulk material is a flow process, not a static process. Thus the results are sensitive to the form of the test specimen and how the test is made.

The author has been concerned, involving considerably more details in fully characterizing the relationship between bulk properties of the material and of the structure, with the case of threaded nuts and bolts. The reader will often find the nut and bolt dismissed in physics texts as an elementary application of the inclined plane. In engineering mechanics texts, a few additional geometric design factors and mechanical performance factors are touched upon. The great number of aspects that are actually involved in the mechanics of a nut and bolt highlights once again that there is a large amount of specialized detail in the colonial organization that depends upon its form and functional application and its previous history. These characteristics are not solely the characteristics of the bulk continuum, and generally, their specialized nature is beyond the scope of our interests.

SUMMARY

1. The "superatoms" that formed from instability in the ionic-atomic-molecular continuum now begin to coalesce into colony-type organization.

2. The available organizing forces are weak, having been largely employed in binding the sub-atomistic levels. It is likely that the remnant forces are best viewed as "communications"-type bonds.

[1] He characterizes four components for the state vector of a cell in its differentiated state.

3. It does not follow that every collection of atomistic elements will bind. However, if there is a large-enough field and active wall boundaries, ultimately some kind of colonial organization will come into being. If the superatoms develop into active entities instead of being dissipative and feeding only upon the communicated fluxes, then it is likely that a colony develops which becomes organized through these fluxes by modes which are best described by a more-specialized sense of the word "communicated." Thus, if the entities are "near-resonant" or "near-threshold" so that only minor energy is required for changes in their state, then the transforming fluxes act with a coding character in how they change state, i.e., the entities "communicate."

4. This does not happen in the hydrodynamic field alone. There vortices are dissipative. However, add the couplings that create an earth's atmosphere, and weather emerges from the "colonies" of cellular air masses. (Appropriately, hurricanes are named in human fashion.) When noted in the earth, as the geochemical complexes come out of the melt, the solid-state grains are organized into complex colonies. More appropriately when other dynamic processes are coupled with chemical changes—moving up the line from such coupling as ionic interactions, to ion exchange, to polar bonding (reactions move from simple solubility, to ion exchange of material in the ground-water system, to liquid coacervates, toward sol-gel organization, to lyophobic-lyophilic bonding)—the emergence of "communications coding" becomes more definite. Finally, in the biochemical element, the surface-active membrane, this process comes into its own. The jump from there to language, which is technically horrifying, is conceptually simple.

QUESTIONS

1. Explore the literature and pose theory or conjecture on the nature of the organizing force structure in such organizations as a colony of protozoa; or simple metazoa; or geochemical complexes; or air mass movements by cellular colonies.

2. Run through the dynamic structure of how some biochemical form within the complex living organism provides its organized colonial step; examples, protein replication, enzyme synthesis, virus replication, organelle function.

3. Run through a theory on how one might test all the small- and large-strain plastic-elastic characteristics of a material sample which can be related to the bulk properties of the material (including fatigue, failure, combined stress characteristics, creep, strain rate effects, etc.).

BIBLIOGRAPHY

1. Hume-Rothery, W.: "Electrons, Atoms, Metals and Alloys," 1948 ed., Dover Publications, Inc., New York, 1963.
2. Nadai, A.: "Theory of Flow and Fracture of Solids," vols. 1 and 2, McGraw-Hill Book Company, New York, 1950, 1963.
3. Zwikker, C.: "Physical Properties of Solid Materials," Interscience Publishers, Inc., New York, 1954.
4. Goldstein, S.: "Modern Developments in Fluid Mechanics," Dover Publications, Inc., New York, 1965.
5. Fultz, D.: in T. Malone, "Compendium of Meteorology," American Meteorological Society, Boston, Mass., 1951.
6. Higgins, J.: The Theory of Oscillating Reactions, *Ind. Eng. Chem.*, vol. 59, p. 18, 1967.
7. Mesarovic, M. (ed.): "Systems Theory and Biology," Springer-Verlag New York Inc., New York, 1968.
8. Moscona, A.: in "The Living Cell, Readings from Scientific American," W. H. Freeman and Company, San Francisco, 1965.

9. Loewenstein, W. (ed.): Biological Membranes: Recent Progress, *Ann. N.Y. Acad. Sci.*, vol. 137, p. 403, 1966.
10. Loewenstein, W.: in L. Bolis and B. Pethica (eds.), "Membrane Models and the Formation of Biological Membranes," North-Holland Publishing Company, Amsterdam, 1968.
———: in M. Locke (ed.), "The Emergence of Order in Developing Systems," Academic Press, Inc., New York, 1968.
Ito, S., and W. Loewenstein: Ionic Communication Between Early Embryonic Cells, *Develop. Biol.*, vol. 19, p. 228, 1969.
11. Bonner, J.: "The Cellular Slime Molds," Princeton University Press, Princeton, N.J., 1959.
12. Halpern, M., E. Young, and M. Ehrenberg: "Oscillatory Behavior of Heart Rate, Ventilation and Skin Temperature in Resting Humans: Toward a Spectral Analyzer," Aerospace Medical Res. Labs., AMRL-TR-67-2281, January, 1968, Clearinghouse Federal Scientific Technical Information, Dept. of Commerce, Springfield, Va.
13. Mitchison, J.: Enzyme Synthesis in Synchronous Cultures, *Science*, vol. 165, no. 3894, p. 657, 1969.
14. Keller, E., and L. Segal: Initiation of Slime Mold Aggregation Viewed as an Instability, *J. Theor. Biol.*, vol. 26, p. 399, 1970.

part three

Detour—The Line of "Man" Systems

9

Functional Organization of Simple Organisms, Organs, Parts

Biological systems are "minor" atomistic systems within the framework of the universe. However, we are one of them. Thus our myopic concentration is on their epigenetic unfolding.[1] Beyond this restricted interest, the living system is quite impressive in its internal complexity, and we have much to learn from its organization. Thus, rather than presenting a complex nested sequence of stabilizing biosystem structures as part of the main line of the development of the hierarchy of systems, we will take a detour through the

---

[1] We may distinguish, concurring with Goodwin (1), three biospectroscopic levels. The first, metabolic or motor, is the high-frequency domain at which the operating chemical chains occur. The second, "epigenetic," is the medium-frequency domain in which the fixed genetic coding unfolds with links that can only be formed from the content and experience derived from the external ecological milieu. The third, "genetic," is a frequency domain which is rate-governed by a long time relaxation phase (the life of the parent system) and a short time escapement phase (its reproductive phase). At this level, chemical coding exists for the dynamic gating of catalysis, by which structural-functional hereditary chains of reliable reproduction emerge.

systems that embed our species. If we were a virus or the total "consciousness" of a planetary bioecology, we might have a different view. But we are not. We are man.

The indefinite though cooperative colonial ensemble that extends continuumlike finally ends at boundaries, often through specialized boundary conditions. (We always imagine our systems immersed in some large iso-thermal and constant potential container.) These create an organization. As an example of the bounding of a simpler system, a fluid made up of vortices can end at its boundary as a liquid droplet or a planetary atmosphere. At this point, we would like to consider, as part of the biological line of systems, the colonies bounded in the form of simpler organisms, organ systems, or man-made functional parts.

Beyond the one-cell biosystem and the coordinated colonial ensemble, the next level of organization is of the simpler organisms which show some specialization within their bounding surfaces. It is as if sufficient repetitive binding existed to form a colonial unit (like a polycrystalline aggregate of atoms) and then a number of slightly different colonies were put together so that these functions were cooperatively linked to make a somewhat-specialized structure of such colonies (e.g., the North American colonies that joined to form the United States). Such organization likely holds true for the organ systems. Introductory material to the organ systems includes such references as (2–4). To compare widely different levels of simple organisms, see such interesting sources as (5, 6). To compare widely different levels of organization, see such sources as (7, 8).

Although excellent descriptive material, these or any other contemporary sources do not adequately describe the dynamic action of any of the organ systems or their organization into organisms. Yet, there are many current, exciting beginnings such as work on the dynamics of the eye, the kidney, the heart, the adrenals, etc. What is still lacking is the full dynamic causal chain by which the processes take place. [For an introduction, see (27).]

However, to capture the imagination of the reader, we can suggest a sampler of a few organ-system analyses. The list is not meant to be complete. Apologies are presented in advance to the many excellent investigators whose names should be mentioned at this point.

For vision, Lettvin et al. (9), Lipetz (10); for the olfactory sense, Lettvin (11); for the kidney, Smith (12); for the heart, van der Pol (13), Rushmer (14), particularly Chaps. 2 and 10, and Berne and Levy (15); for endocrine control systems, Harris (16). Also, in (17) the papers, in particular, by Yates and Brennan [adrenocortical function; also see (18)], Cahill (glucose homeostasis), Brown-Grant (the pituitary-thyroid system), and Schwartz (gonadal function). Also, (19) and (20) to illustrate a brain control system and (21) for gastro-intestinal control, in particular the paper by Lepkovsky (on the regulation of food and water intake).

We cannot explain the detailed and complex processes by which biological organisms or their operative parts organize. Nevertheless, we hasten to bring one final item into the discussion—the nature of functional parts made by man. Our purpose is to attempt to clarify the characteristics of organization and self-organization of simpler organisms, i.e., what perhaps makes colonial assemblies into organ systems.

If one examines a mechanistic part made by man, it is quite obvious that it could be made by the application of automata theory. There are many transformation machines today that make parts automatically. There are also programmed machines; one can visualize the Turing machine. Yet a functional part cannot really have come into existence through such automata processes. It takes—as yet—the nonlinear process known as a man to create a Gestalt of a functioning entity, to visualize a collection of independent functional parts that can realize the functioning entity. Then transformation machines, tools, and material can fashion these parts by coded routines.

A functioning part is more than a geometric structure that spans a spatial dimension or fits constraints. It must operate, with sufficient life, within a space-time domain. Thus, there is not really a great deal of difference in principle in the creation within boundaries of the specialized entity that is a simple organism; an organ; functioning parts created by internal reproductive processes by the combination of man, tool, and material; or functioning parts created by external productive processes by the combination of man, tool, and material.

The salient element which is required is informational feedback.[1] A clue to what this means is the following: The act of "creation" of the next pulse-in-time is something that is required even to propagate time. This should not be taken to mean that there is much practical chance for losing "time" from pulse to pulse. The ordering of the various hierarchies (cosmos, galaxy, star, planet, etc.) ensures the unfolding of time, pulse by pulse.

> We do not wish to be accused of crackpot mysticism. We are indebted to D. Bohm (22) and the inspiration derived from reference (23) for the original kernel of the idea. There is a hard search, in many minds, for the source of quantization of both spatial and temporal structures. Here Bohm has offered a source for the very act of "creating" quantized time. Time, as was discussed at a New York Academy symposium (23), is not a universal unity for all levels of organization. Yet levels are nested within one another and, within limits, are referable to each other.

It is very likely that a small, most often undetected, communications-feedback signal provides the coupling. We have no theory for this. It is only a

---

[1] A simple machine or network may rigidly play out its dynamics for an epoch. However, in a complex system, it is "information," not generated within its internal links but communicated from outside, that cues the system over and over again. To illustrate: when you have finished eating, it is not yet determined with certainty when you will eat next.

metaphysical speculation. Yet it is the same class of phenomena as the exchange forces.

Some amplification of the physics of such a problem as self-organization of a system may be helpful. We are discussing at this point a cooperative ensemble of parts, each of which itself may be an ac active atomistic system which may have quite a few colonial subassemblies. The question is what is the foundation for its dynamic operation. We propose the following general systems structure.

The individual atomistic system may be regarded as built up from nonlinear oscillators. The stricture is that these nonlinear oscillators may be built as desired; namely, by any causal chain that can create a limit cycle in the phase plane [see for example (24)]. The theory of the clock furnishes one scheme. According to its theory, choose a linear oscillator. Add an escapement which will feed in small impulses of potential energy, preferably at the point of highest velocity, so as to maintain the oscillation. (While grasping the recipe for building an oscillator, for guidance we may keep in mind's eye the common design problem of building a complex vehicle or manufacturing factory, etc. What we have just done is laid down a prescription for such items as the power plants.) The electronics literature names two other common schemes for building oscillators—feedback and negative resistance. The design art for the complex system, at this point, is then to be able to mediate the oscillator by one or more inputs—characteristically parametric adjustment. (That is, by making some parameter change, the oscillator characteristics may be changed.) We now potentially have power packs, switches, and amplifiers.

Now, although a simple system may be put together by direct control of these parameters (as complex an example of direct control that is commonly in use might be step switches used in telephone systems), ultimately the crowding and competition suggest the use of propagative phenomena which are coded; namely, the use of information or communications feedback. What is used for communication feedback is not obvious. It may be any kind of sign. A useful introduction to the science of signs—"semiotics"—may be found in Cherry (25). An overall view of the information sciences may be found in (26).

To handle the complex contingencies of the milieu, there must be a code book—an algorithm—and a memory, each rich enough to produce a sustained motional pattern of the system. Let us examine what this might mean at the most primitive systems' level by illustrating one feasible scheme.

A series of particles introduced into a box most commonly will come to a static equilibrium (e.g., sand particles in a box). Even with a more active nature to the particles, it may be quite difficult to obtain any action other than a "saturation" or an unstable oscillation between two polar extremes. Thus, an algorithm, to achieve a more uniformly occupied phase space, has considerable sophistication. One routine with promise is to attempt to achieve a ring oscillator, which is not necessarily rigidly cycled but will statistically manage to ring all of the modal states of the system. Such a ringing system must likely make up a fairly complex information feedback system. We believe that this exists in biosystems and suspect that it is formed by a fairly complex group of biochemical chains. It was such an end in mind that influenced our choice of some of the organ-system studies referenced.

SUMMARY

1. We have begun a detour among the sciences of man. Whereas colonial systems develop and may turn out to be persistent, when finally superorganizations evolve (not out of all colonies, but out of some), the colony may become a functional unit within the larger system. The atomistic living systems—that make up the biosphere—are an important case in point (to us).

2. The major function of a functional unit—its "purpose"—is to propagate a transformation in time. In so doing, it tends to "create" time, epoch by epoch. It is only when its causal chains are sustained in their action in their external links that the sense in which time is mechanistically "created" can be visualized.

3. The means by which a functional atomistic entity made up of colonies achieves its complex action is information feedback.

4. The evolutionary steps by which complex parts and function emerge is part of the start-up of systems and is outside of our present concern.

QUESTIONS

1. Visualize the problem of an artificial organ system to be used as replacement (e.g., sensory system unit, kidney, liver, heart, spleen, G.I. tract, brain, endocrine system unit, motor system) or to provide some new function. Trace out its logical requirements and its physiological requirements, and specify the design of a physically achievable embodiment.

2. Visualize a complex organization involving man-machine links (e.g., a complex factory including its operation within the economic environment). Design a mechanical system that performs a significant function and thus represents a significant functional unit.

3. Carefully trace whether or how information feedback regularizes the functional performance of functional units. Provide a richness of examples.

4. Review the status of intercellular communication to see if a model of a coordinated colonial assembly can be derived. Discuss the model, indicating its unknown areas and filling in plausible steps. (You may keep any level of complexity in mind, from the simple metazoa to the kidney or even to the brain.)

BIBLIOGRAPHY

1. Goodwin, B.: "Temporal Organization in Cells," Academic Press, Inc., New York, 1963.
2. Best, C., and N. Taylor: "Physiological Basis of Medical Practice," The Williams & Wilkins Company, Baltimore, Md., 1966.
3. Ruch, T., and H. Patton: "Medical Physiology and Biophysics," W. B. Saunders Company, Philadelphia, Pa., 1965.

4. Guyton, A.: "Textbook of Medical Physiology," W. B. Saunders Company, Philadelphia, Pa., 1966.
5. Bonner, J.: "The Cellular Slime Molds," Princeton University Press, Princeton, N.J., 1959.
6. Lentz, T.: "The Cell Biology of Hydra," John Wiley & Sons, Inc., New York, 1966.
7. Ramsay, J., and V. Wigglesworth (eds.): "The Cell and the Organism," Cambridge University Press, London, 1961.
8. Fraenkel, G., and D. Gunn: "The Orientation of Animals," Dover Publications, Inc., New York, 1961.
9. Lettvin, J., H. Maturana, W. McCulloch, and W. Pitts: What the Frog's Eye Tells the Frog's Brain, *Proc. I.R.E.*, vol. 47, p. 1940, 1959.
10. Lipetz, L.: "Information Processing in the Frog's Retina," Aerospace Med. Res. Labs., AMRL-TR-65-24, February, 1965, Clearinghouse Federal Scientific Technical Information, Dept. of Commerce, AD #614-249, Springfield, Va.
11. Lettvin, J., and R. Gesteland: A Code in the Nose, *Bionics Symp.* 1966, Aero Med. Div., USAF, Wright-Patterson AFB, Ohio.
12. Smith, H.: "From Fish to Philosopher," Ciba Pharmaceutical Co., Summit, N.J., 1959.
13. van der Pol, B., and J. van der Mark: The Heartbeat Considered as a Relaxation Oscillator . . . , *Arch. Neerl. Physiol. l'Homme Anim.*, vol. 14, p. 418, 1929.
14. Rushmer, R.: "Cardiovascular Dynamics," W. B. Saunders Company, Philadelphia, Pa., 1961.
15. Berne, R., and M. Levy: "Cardiovascular Physiology," The C. V. Mosby Company, St. Louis, Mo., 1967.
16. Harris, G.: "Neural Control of the Pituitary Gland," Edward Arnold (Publishers) Ltd., London, 1955.
———: The Central Nervous System and the Endocrine Glands, *Triangle*, vol. 6, p. 242, 1964.
17. Stear, E., and A. Kadish (eds.): "Hormonal Control Systems," American Elsevier Publishing Company, Inc., New York, 1969.
18. Mesarovic, M. (ed.): "Systems Theory and Biology," Springer-Verlag New York Inc., New York, 1968.
19. Jasper, H. et al. (eds.): "Reticular Formation of the Brain," Little, Brown and Company, Boston, Mass., 1958.
20. Kilmer, W. L. et al.: The Reticular Formation, Parts 1 and 2, Contract No. AFOSR-1023-67B, *Interim Report No. 3*, prepared for Directorate Information Sciences, AFOSR, Arlington, Va., February, 1969.
21. Morgane, P. (ed.): Neural Regulation of Food and Water Intake, *Ann. N.Y. Acad. Sci.*, vol. 157, p. 531, 1967.
22. Waddington, C.: "Towards a Theoretical Biology, 2, Sketches," Aldine Publishing Company, Chicago, 1969.
23. Fischer, R. (ed.): Interdisciplinary Perspectives of Time, *Ann. N.Y. Acad. Sci.*, vol. 138, p. 367, 1967.
24. Andronow, A., and C. Chaikin: "Theory of Oscillations," Princeton University Press, Princeton, N.J., 1949.
25. Cherry, C.: "On Human Communication," John Wiley & Sons, Inc., New York, 1961.
26. Iberall, A.: "Information Science, Outline, Assessment, Interdisciplinary Discussion," prepared for Army Res. Off., AD #635-809, June, 1966.
27. Bloch, E., et al.: Introduction to a Biological Systems Science, *NASA Contractors Report CR 1720*, 1971.

10

The Internal Organs and Brain as a Biochemical Continuum

Just as on a lower level the atomistic components of nuclear particles coalesced to form the extensive collection of molecular species, so the cellular constituents of biology coalesce during a long process of evolution to form complexly differentiated organ systems and functional parts. We are concerned now with the extensive internal organization that may emerge from this collection.

Consider then these specialized organ systems, that themselves are built out of colonies of cells. At a lower hierarchical level, the atomistic biological cell motored itself about in a watery milieu. The necessity for this exposure is still stamped within their complex emergent state, for the colonies. the organs, and the simple organisms still have to retain their connection to the water milieu whence they evolved. It is to the credit of a sequence of men like Bernard, Sechenov, and Cannon that the concept of the functional "purpose" of life at all levels emerged as the regulation of the conditions of the internal watery environment of the biological system, independent of the conditions of the external environment. This is referred to as homeostasis. [For a succinct introduction, we can offer references (1, 2, 3) to supplement (4–8)].

As the developmental line of biological organisms proceeds upward hierarchically, the system of cooperative organs becomes more complex. The biochemical-mechanical-electrical collection of atomistic elements grows in number and again a near-continuum organization emerges. The ensemble resembles an extensive community that proceeds to develop "without end" (that is, the community contains transformation rules for extending the ends and repairing structure). The cooperative organ systems involve only a limited class of phenomena—hydrodynamics, plasticity-elasticity, electricity, thermodynamics, biochemistry, mechanics, and their couplings. It is clear that major control systems—the nervous system, the endocrine system, biochemical regulation and control systems—play essential roles. (The ubiquitous and still mysterious membrane plays the decisive role at their foundation. However, internally the complex biological system is continuumlike, as internal system piles up on internal system.)

We have referred to the dynamic regulation by which homeostasis is achieved as homeokinesis (2). It is an operation by which a complex of oscillating biochemical chains and mechanisms are shifted in operating point, mainly by mediating or releasing factors that inhibit the chains. However, a satisfactory state of operation cannot be achieved within the internal environment for an indefinitely extending biochemical continuum. It would become unstable, tending toward a stasis or toward a destructive hunting. It can only be stabilized from the outside. Therefore, there must be a limit in size to the organism, and we must turn to the outside for the regulation.

The physics of this problem deserves some discussion. C. Draper of the Massachusetts Institute of Technology gave a Wright Brothers lecture for the Institute of Aeronautical Sciences in the mid-1950s in which he put forth a very provocative thesis. The Wright brothers did not solve the problem of flight. Both the aerodynamic theory as well as most of the essential experimental results had been laid down by earlier investigators. However, the Wright brothers added one crucial step. Their airplane had no aerodynamic stability, which all earlier investigators had attempted to achieve. Instead, their craft was stabilized by the pilot—the "steersman." It is this function of providing marginal stability in response to the external "winds of the milieu" that is an essential step in achieving an adaptive complex system.[1]

In the first place, as this section attempts to point up, a viable system must be linearly unstable so that it is capable of responsive motion. However,

[1] To illustrate: if the newborn infant were launched into its new environment only with its internal chains and with no mothering assistance in developing external chains, it would soon become unstable and perish. In simpler organisms, of course, the external environment itself contains many cues.

it must then be inhibited ("clutched") if it is not to be "spinning its wheels" or on the verge of "running away" at every instant. The information for latently stabilizing the system must, most plausibly, come from outside. It is not necessarily an ac spectrum that is presented to the organism. It may be the slowly changing dc conditions. (Actually, the ambient environment for both living and nonliving systems seems to be best described as an impulse spectrum. Poetically expressed, as the disturbances in the environment pass in file, they make up the "vicissitudes of the milieu.")

Thus, internally the system is likely not stable. Conversely, when one finds a biological system in process of turning "in" on itself (that is, it attempts to shut itself off more and more from the external signals, which are regarded as too stressful), then one can suspect significant instability. We would place schizophrenia in this category (at least its depressed phase).

It is desirable to keep in mind, even on the most rudimentary level, the dilemma we must face. For the system to be capable of a lively internal response, it must be fairly undamped, i.e., it must possess inductive-capacitive characteristics. However, to prevent it from being wildly resonant, it must have damping characteristics. In order that it be "auto-motive," it must utilize informational fluxes that are hierarchically distinguishable from power fluxes which run the components. Yet if such a system were to run solely from its own internally circulating fluxes, we conjecture that only two possible states would arise—the stasis of death or shutdown (if because of phasing, the degradative fluxes took hold), or the drift into a "meaningless" banging aroung through some or all of its internal modes of action. What else could the system do but "drift" or "bang around" aimlessly if only internally guided (9)?

The theory of the "steersman," that marvelous organ known as the brain, which for this thermodynamically "open" system seems to possess so many wonderful "volitional" characteristics (yet still behaves quasi-ergodically when a number of brains are gathered in ensemble) is still in process of being written. Although it is easily possible to find an extensive biological literature on the brain,[1] some engineering cybernetic sources should not be overlooked. Brodey (10) is a useful source for a number of such references, and the work of Kilmer et al. (11) is worth inspection. An interesting Russian source is Luria (12).

SUMMARY

The continuumlike colonial organization of cells ended abruptly with the atomistic structure of coordinated colonies, simple metazoa, and organs.

---

[1] A provocative characterization of the human brain, namely three brains of different developmental statuses, may be attributed to Paul McLean (13). Although the model may not withstand indefinite scrutiny—an issue for experts—it helps as an elegant orientation toward brain complexity.

1. Continuing our studies with these functional entities, we see that the organ subsystems in the interior of the biological system, including the brain, seem capable of growing out indefinitely without end in number. They thus seem to create a near-continuum internal field.

2. The primary algorithm that seems to govern this internal field is regulation of the internal states of what is essentially a watery environment. This is known as homeostasis.

3. In accordance with our general thesis, if internal subsystems are active, they will likely exhibit instability as a collection. Accentuated by the circulating informational flux, these colonial systems would go into oscillations. "Hunting" patterns would emerge. If an adaptive regulation is to emerge, stabilizing signals likely must arise from the outside. Thus, once more the continuum ends in a boundary. The super-superatom, which is the complex living system, emerges. Now, if it is to be a viable active system, the chains of causality are completed from outside.

4. The scheme by which dynamic stability emerges is referred to as homeokinesis. It assumes that the actions of these subsystems will emerge as limit-cycle chains—oscillators—which are marginally unstable and whose operating points are shifted by the flux of signals from the exterior. It is the pattern of evershifting limit-cycle oscillators, providing a dynamic regulation by which homeostasis emerges, which forms the nature of the biological system.

5. Once again the issue of degradative flux exchange versus near-resonance and the emergence of communications flux arises. In the human being, the internal systems are near resonance. Exquisitely balanced chemical reactions (using the "exquisite" membrane engine) are prepared to trigger the man. As infant, he emerges nearly unstable into the surrounding complex ecological milieu. The infant cries, and if he were to be unattended, he would cry, and cry, and cry, and cry—unto death.

QUESTIONS

1. Choose as complex a biosystem as you can handle. Discuss the scheduling of all of the internal chains that must be regulated in order that homeostasis (or its dynamic equivalent, homeokinesis) may occur.

2. Attempt to set up the general theory of the trajectories of a system made up of many organized subsystems, each capable of functional performance, plus an "intelligent" communications system (i.e., one capable of viably dealing with information fluxes) which has been operating as an open system in a viable homeokinetic mode and which is suddenly closed. (Illustrations to keep in mind are a factory that has been struck and is being effectively picketed, and the neonatal infant, newly born and unattended, who has been provided with prenatal care by its mother.) Although "death" or "destruction" may ultimately arise, what are all the emergent paths?

3. Sketch out a theory of brain function first that is occupied, not with anatomy and physiology, but with mechanistic function (i.e., more with a physical scientific-engineering view of the system). Then review the entire theory and description after a more intensive exploration through detailed literature on form and function within the brain. Are the views correlative?

BIBLIOGRAPHY

1. Cannon, W.: "The Wisdom of the Body," W. W. Norton & Company, Inc., New York, 1932.
2. Iberall, A., and S. Cardon: Control in Biological Systems—A Physical Review, *Ann. N.Y. Acad. Sci.*, vol. 117, pp. 445–515, September, 1964.
    ———— and ————: Regulation and Control in Biological Systems, *Proc. of IFAC Tokyo Symposium on Engin. for Control System Design*, 1965.
3. Waddington, C.: "Towards a Theoretical Biology, 2, Sketches," Aldine Publishing Company, Chicago, 1969.

4. Davson, H.: "A Textbook of General Physiology," Little, Brown and Company, Boston, Mass., 1964.
5. Best, C., and N. Taylor: "Physiological Basis of Medical Practice," The Williams & Wilkins Company, Baltimore, Md., 1966.
6. Ruch, T., and H. Patton: "Medical Physiology and Biophysics," W. B. Saunders Company, Philadelphia, Pa., 1965.
7. Guyton, A.: "Textbook of Medical Physiology," W. B. Saunders Company, Philadelphia, Pa., 1966.
8. Mountcastle, V.: "Medical Physiology," vol. 1, The C. V. Mosby Company, St. Louis, Mo., 1968.
9. Freud, A.: "Normality & Pathology in Childhood," International Universities Press, Inc., New York, 1965.
10. Brodey, W., and N. Lindgren: Human Enhancement: Beyond the Machine Age, *IEEE Spectrum*, vol. 5, p. 79, 1968.
11. Kilmer, W., W. McCulloch, and J. Blum: An Embodiment of Some Vertebrate Command and Control Principles, *J. Basic Eng.*, vol. 91, p. 295, 1969.
———— et al.: The Reticular Formation, Parts 1 and 2, Contract No. AFOSR-1023-67B, *Interim Report No. 3*, prepared for Directorate Information Sciences, AFOSR, Arlington, Va., February, 1969.
12. Luria, A.: "Human Brain and Psychological Processes," Harper & Row, Publishers, Incorporated, New York, 1966.
13. MacLean, P.: in L. Ng (ed.), "Alternatives to Violence," Time-Life Books, New York, 1968.

11

The Complex Biological System in the External Milieu

11-1 THE INDIVIDUAL AS BEHAVIORAL ATOM

The homeokinetic biological system, crammed into the overall envelope that makes up the body, consists of a multitude of biochemical chains, each operative with characteristic rate-governing steps that help establish an essentially periodic character to their occurrence. This large bank of oscillators, in its operation and shifting and changing of operating characteristics, results in the starts and stops of the motor systems, the search for food, the inter-mittent eating, the voiding, the breathing, the blood-coursing, the perspiring, and the larger organized complexes (perhaps best referred to as syndromes) of anger, euphoria, fear, attention, speech, etc. One is hard put not to conclude that in order to attain global stability the large number of only loosely coupled chains must represent an extensive collection of atomistic elements which have to be canonically constrained in accordance with an ergodic hypothesis.

The master "player" of that "musical organ" that makes up the bank of oscillators is the brain. It is "he" who "whistles Dixie and all other tunes." It is the existence of this most complex controlling mechanism that provides, by its combined genetically unfolding, maturational, and adaptively learned algorithms, the overall excitation patterns among these oscillators. This patterned response begins to make up the atomism of the individual's behavior.

How the individual chains are developed and how they act have been studied as part of the physiology of living systems. As such, with their weak coupling to other chains (weak, or at least acting in separated time domains), they represent quantized-state elements—atoms—of the biological system. A heart in an isolated constant nutrient milieu (or even just a more-limited piece of myocardial tissue) beating away for year after year, as Carrel demonstrated, illustrates that these complex systems are "atomistic." As they become more extensive in number, they make up the biological continuum that is the organized internal milieu of the human organism. In cooperative endeavor, they provide an extensive spectrum of internal effects.

However, the continuum made up of an indefinite extension of organ parts, even including the brain, cannot be stabilized from the inside. It must become quantized as an individual. For the successful operation of a total organism (meaning its persistent sustained operation within an extended dc or slowly varying ac ecological milieu over a considerable period comparable to a lifetime), there must be a searching motion for and drawing in of input from the outside. Thus for stability, here the stability to persist from generation to generation, there must be a physical bounding of the system, a "skin." Again, the continuum field fractures, quantizes, and forms a bounding surface. Now stabilizing inputs come in from the surface. Motor systems extend to the surface. Spectral responses must arise. There are classes of spectral response patterns that are satisfactory as well as those that are unsatisfactory. What appears to be the case is that the controlling system, the brain, and the physiological systems, form a global—nearly unstable—system. The organism will not stay put. If at rest, it will ultimately (within a reasonably constant time period) begin to move. If moving, it will ultimately come to rest. We must refer to this as marginal instability and indeterminate gain at zero frequency. Out of these internally self-selected movements come the successful patterns—the external limit cycles of behavior—that sweep the oscillator banks.

Let us look at this behavioral story. The internal fluxes within the organ systems of the human make up a behavioral continuum. Here an external drama unfolds in exciting dynamic boundary-value form. "Once upon a time (i.e., let $t = 0$, in such and such a year), there was a prince ..." the fairy tales state in near-standard form. Stripped of its more poetic elements, the story from close up unfolds as a problem of the patterned interaction from slowly changing outside boundary inputs to the internal systems under the control of the brain.

However, viewed from outside and somewhat more removed in space, a new field comes into focus, the individual as one of a sparse number of interacting entities within a slowly changing ecological field. In this field, the individual appears as an atom—hard (that is, covered and not penetrable), indivisible, compact, with some characteristic interacting qualities for inputs from the outside and internally activated outputs from inside, and a characteristic spectral response.

Although there may be a preliminary tendency to view human behavioral exchanges as belonging to a continuum (for each individual), another round of thought suggests much more that the individual's behavior tends to fall into a very discrete pattern of orbits. It becomes characteristic of social intercourse (namely, the organized behavior of a system of such atoms as an n-body problem) for it to develop rather specialized excitation modes and interchange modes, suggesting that stability-instability considerations govern socially bound systems. The difference with physical systems is that the dimensionality of internal degrees of freedom is so much greater than in simple physically bound molecules (rather than just vibrational and rotational m/e-m modes there may be thermal, mechanical, acoustic, optic, chemical, and electrical modes providing many more degrees of freedom in the interactional spectrum). A complex spectrum, more akin to that for classes of organic compounds, emerges. It is likely that its greater dimensionality shows up in the evolutionary nature of the orbital paths. It is very likely in this complex that man's "volitional" behavior lies.

We would like to provide the nonbiologist reader with some useful biological source material for human behavior. A more complete descriptive picture, recently prepared, of behavior may be found in (1–3). A partial list of some of the major persons who are representatively responsible for developing significant foundations for human behavior includes Darwin, Mendel, Sechenov, Freud, Sherrington, Pavlov, Jackson, Wertheimer, Köhler, Papez, Hebb, McCulloch, G, Harris, Gesell, Piaget, and the modern behaviorists.

It would not be seemly to suggest that it is possible to lay down or to attempt to lay down the entire physical chain by which the behavioral system is run. However, it is possible to offer the reader a personal sampler of writing that has appealed to us. To glimpse the dynamic foundations for behavior, we could propose reading Arieti (4), in particular Part 12; Crosby et al. (5); Magoun (6); Penfield et al. (7); Jasper (8); Hebb (9); McCulloch (10) and Kilmer (11); Luria (12); Harris (13); (14); papers by Lepkovsky, Soulairac, Morgane, and deRuiter (15); dippings into Piaget, for example (16); and a recent New York Academy Conference on emotional behavior (17). There is, of course, a specialist literature on the many structural and functional compartments of the brain—the reticular formation, basal ganglia, limbic system, hypothalamus, thalamus, cortex, etc. A most interesting speculation on memory is offered by Pribram (18). An adequate theory for memory is still one of science's great mysteries. (More recently, the work of E. R. John, reported at the 1969 Biophysical Congress, is in process of creating considerable excitement.)

We are personally encouraged that the descriptions of behavior, as it is found in such a recent meeting as (17), are tending to move toward a homeokinetic model of behavioral patterns, in which the genetic coding of the animal unfolds epigenetically within the ecological field. Since this is a basic concept from ethology, it is possible to feel a very sympathetic urge toward the utilization of physics and ethology as the scientific paths that psychology will have to unfold along within GSS. Physical ethology is an apt portion of GSS to be designed to deal with the behavior of individual living systems embedded in physical fields made up of both living and nonliving matter. As far as the living system under observation is concerned, the field distribution of matter represents boundary conditions.

For example, in (17) the following presentations were quite impressive: Lat's dynamic data on "nonspecific excitability"; Melzack's demonstration of stereotype repetitive behavioral patterns, in a number of species, for animals reared with sensory deprivation, which obviously suggests dynamic instability; Hinde's presentation of dynamic patterns in the mother-child rearing complex; Kaufman's demonstration of markedly different mother-child and group behavior in closely related species (of macaque monkeys); Eisenberg's "easy" ability to trace behavioral patterns through different mammalian species; Delgado's beginning catalog of the fragments of behavioral patterns, found by brain stimulation; Scott's identification of nine behavioral systems [distress, eliminative, agonistic, investigatory, comfort-seeking, sexual, care-giving, ingestion, fear-pain. The listing proposed in (1), although far from identical, had the same intent. We are quite willing to admit that "expert" psychophysiological study will soon enough define the categories of behavior, its fragments, its orbital segments or chains, its modes —whatever one wishes to call them—with some greater determinateness than we are capable of]; Breul's attack on one speaker's talk, pointing out that the behavior of Breul's wild animals (rats) did not at all resemble the behavior of the laboratory animals that were being discussed in an extensive factorial experiment and thereby really attacking most multivariable-factorial-designed psychological experiments; Henderson's fundamental statement that since no useful results were to be obtained from open-field tests these should be dropped; Coudland's demonstration of such diverse species' response as the "emotionality syndrome" of chickens "freezing" in an open field, contrasting with rodents response by "flight"; Randrup's demonstration of strong motor stereotypes in a number of species for periods of the order of hours (clearly a gross system instability) induced by a biochemical precursor (DOPA) to catecholamines; Norton's beautiful demonstration of Markov chains in animal behavior (cats); Flynn's efforts to work out the neural sequence in such a behavioral complex as attack behavior (of rats by cats); Hunsperger's effort to relate the stress of affective or emotional reactions to motor "tone" and heart activity (i.e., to the dynamics shown by these muscle systems—the skeletal muscles and heart); Miller's magnificent demonstration that animals can be trained, by rewards and punishments, to change their autonomic (automatic) responses (e.g., increase or decrease of heart rate, vasoconstriction or dilation in the rat tail, increase or decrease in urine formation. Ergo, visceral learning is possible); Brady's demonstration that the physiological autonomic concomitants of an "anxiety" emotional type conditioning (changes in heart rate and blood pressure) could be changed greatly by its association with a cue (i.e., an impressive demonstration was given of the determinacy of characteristic response patterns of interrelated physiological systems under "emotional" conditioning); the impressive scope, hinted at by Brown, for ethology or, more specifically, for neuro-ethology.

Although this meeting certainly is not proposed as representing all behavior or all the interests and views of behaviorists, it is indicative of a more objective and more physically based search for behavioral descriptions and theory. It would be our contention that the one item missing for a tremendous outpouring in this field is a grasp of the key concept of biospectroscopy —homeokinesis—that it is by dynamic spatial-temporal chains that the system unfolds its behavior.

To this central thesis, we can now suggest an additional quite speculative thesis. Long-term behavior, it seems to us, unfolds from the endocrine system. Thus an electrochemical view of the complex animal's behavior is the following: the winds of the milieu, both internal and external, provide signals. Developmentally, through the learning process in infancy, they form response paths in the brain by which a patterned neuroendocrine response complex emerges. These then key and cue the endocrine systems. It is these patterned brain responses—of the neuroendocrines—that ultimately make up the substance of Freud's superego.[1] It is thus clear that it contains no values, only patterns of behavior. Further, we believe that the complex of endocrine responses that emerges is the foundation sought for by Selye in his search for and identification of generalized stress "syndromes."

---

[1] If we view behavior at a time scale in which its high-frequency response is filtered out [say averaged over 100-sec epochs—see (3)], it seems clear that the system makes characteristic judgments. The very complex and extensive nature of this algorithm, or guide to behavior, must be written down in the system in structural form. We identify the algorithm, we believe appropriately, as Freud's superego. In our opinion, Freud's structure for behavior is more comprehensible as a communications theory of brain function than a ponderous power engineering. However, we suggest that it is based more on chemical signaling than on electrical signaling.

Thus, whereas the patterned inputs to the autonomic system present a more fixed, longer-term regulating pattern for internal functions and organs, the sympathetic system response, epigenetically unfolding chained patterns of function, presents the running, working, responding animal operating and adapting to its changing ecology. Through the action of these internal chemical systems, the animal patterns its way in the external milieu.

In the long run, the animal is driven by his chemical signaling. [See, for example, (3) or Richter in (19).] However, one added thought may provide useful clarification. Since many individuals are hard put to lay aside their belief in their own control of their "volitional" activities even though they are willing to accept a largely deterministic view of their habits, we are compelled to comment on the point.

The internal organ systems, including the many structures of the brain, form biochemical chains which are not all closed in the internal milieu (as an example, there are many inputs that can affect heart rate even though the heart has at the base for its function a primary autonomous oscillator). There are a large number of chains that are closed externally. The detailed problem for physiology and related disciplines is to identify the "network" elements in the external loop that close some of these chains and to identify their network parameters. Because of the many modalities available, the coding is often subtle and not fully understood. Nevertheless, to a very significant extent and with considerable occurrence of Markov processes, the episodes of behavior pour forth from the individual in largely nonvolitional fashion. We recognize it as his style or personality.

Such physiological or biochemical "explanations" of behavior are "deterministic." Our culture has a "free will" bias. In order to better understand this, it is necessary that a person understand his epigenetically acquired cultural biases. A few sources that give some insight into the biochemical and neuroendocrinological rationale for behavior and some that distinguish cultural components are Richter (19), Jenner (20), Sheldon (21), Parnell (22), Sargent (23), and Eysenck (24).

Of course, the thesis that behavior emerges from the coursing of neuroendocrines and endocrines is quite speculative. Its basis, most naively, is expressed by the following syllogism. Behavior either emerges from internal electrical communications or chemical communications. Electrical communications seems to be relevant for events up to 0.1 to 1 sec (or a few at most). Behavior, however, persists for seconds, minutes, hours, days, weeks, months, years. Thus it seems more probable that its source is chemical. The most likely sources are the messenger chemicals, the hormones.

Less naively, "harder" evidence for a chemical foundation for behavior is beginning to be furnished by investigators working in what they refer to as "virtual" animals, i.e., their joke to represent animals that have nervous systems with only few ganglia. For example, in a session at the 1969 Biophysics Congress cochaired by D. Kennedy on "Behavior of Neural Nets as Exemplified by Nervous Systems Having Few Elements," Kennedy's introductory remarks intimated the coming revolution. The following talk by F. Strumwasser was most relevant. Strumwasser pointed out that the nervous system exists for behavior. The most common findings, as unit processes, are action and postsynaptic potentials in the milliseconds range. However, from these rapid processes, it is difficult to understand organism behavior that emerges in the period of seconds, minutes, days, and weeks. He thus proposed to demonstrate long-term processes in aplysia that he could relate back to its simple ganglia. He showed that the parietal-visceral ganglia "produced chemical material for export" and that this synthesized chemical material (identified as a polypeptide) had effects on egg-laying in the range of hours and with variation over the year. He showed autonomous circadian rhythms in the eye (synchronous in isolated eyes). He demonstrated autonomous nervous rhythms at the 4-minute level for ganglia in calcium-free seawater. Although the chain of explanation is long, nevertheless it is clear that gradually a common frame of reference in chemical processes is emerging, as illustrated by work at the single cell level (Chance); in the simple organism (Strumwasser); in the total organism (our investigations); with temporal foundations for the development process itself (Goodwin).

11-2 THE QUANTIZED EXTENSIONS OF MAN

The most marvelous probe of the stability thesis we are describing is the quantizing ability of man himself. Man, the complex behavioral atom, is not fully stabilized in his activities by his own mechanisms. He manipulates the external milieu and produces quantized extensions of himself. The emergent evolution of these extensions is indicative of the richness of his genetic coding, epigenetically unfolding. (In fact, it is common for anthropologists to suggest that it is the use of words and tools that makes man unique.)

Man covers his protective skin with a protective skin. He covers his thermally regulating outer zone with a thermally regulating outer zone. He extends his toollike hands with tools; his motor systems with motor systems; his protective means with protective means; his food-preparatory mechanisms with food-preparatory mechanisms; his sensory metrical instruments with metric instruments; his feedback communications elements with feedback communications elements; his genetic coding with epigenetic coding.

The latter two are possibly the most fascinating and most subtle.

Consider communications and the operation of the biological system. Most commonly this operation is described in terms of feedback theory. However, we have been critical of such application of this theory. Our point of view with regard to feedback was expressed in (25). Namely, feedback is coupling from an output stage, through power amplification with a shaped functional transformation, to an input stage to correct or produce some desired change in the output.

We regard this as different from regulation or compensation which we view as the selection of mechanisms that can provide one or more outputs with a bounded measure over a prescribed range of inputs (i.e., all feedback controllers are regulators, but not conversely. Regulation often can be achieved by compensatory mechanisms). That this view is not unique to us may be noted by examining such a source as (26). Although the origins of the controversy are fairly old, there was and still is good reason for these distinctions.

We have not been ready to accept uncritically the action of every component of the biological systems as a feedback controller. Instead, upon examination, most often we have found dynamic regulation consisting of mediation of the action of nonlinear limit oscillators. Under the stimulus of preparing the last section, we have gotten another clue.

We have stated that it is perhaps quite likely that there is a small (energetic) signal that transforms each creation of a pulse of action in a system into the next act of creating a pulse and that this sequence is orderly compulsive. It strikes us now, more fully, that this feedback signal, which we described as informational feedback, is the prime character of communications. We have provided an extensive introductory discussion about information theory in an earlier report (27). With Cherry (28), this is worthy of review.

In information theory, the concept of entropy has been introduced to provide a measure of information.[1] In (27), we suspended judgment about the virtue of associating entropy with messages, except as a formal device. To many including ourselves, the content of the concept of message negentropy was possibly empty. However, a meaning is beginning to emerge, related to our earlier discussion about the functional part that is created by man, that informational feedback may be associated with the negentropy of the escapement in a nonlinear cycle of causation. If the pulsing cycle of being is routinely forced (by an unchanging single causal factor), then there is no creation in the subsequent pulse. However, at any point that the informational feedback changes or modifies the cycle, then this informational or communications content can possibly be associated with message negentropy. However, message negentropy must be associated with the measure of energy that it controls through the nonlinear amplification process. (These are not complete ideas but a declaration of intended direction.) Thus, the exchanges that pass through central nervous system (brain) and coordinated and localized endocrine amplifiers are what we will be concerned with. This is the part of the extensions of man—his external sounds and signs and internal pulsings —that represents the communications language both outside and inside.

---

[1] Actually the negative entropy, or negentropy, is Brillouin's usage.

With a little more amplification, what we are provisionally stating is that message negentropy is to be measured by the product of the discrete quantity of communications energy contained in the message times the power amplification of the receiver system times the (orbitally cyclic) time of action of the receiver. Since this is hardly measurable, it must be measured by the energy released by the receiver in response to the cueing message.

[Physical entropy change is measured by the change in energy flux per unit temperature, in which temperature is a measure of the translational kinetic energy of the internal atomistic elements. It thus represents the change of energy flux per unit of internal kinetic energy. In probabilistic terms, the entropy change is the number of quantum state changes per (large) number of quantum states of the system. Since we are not in a position to define the change of entropy for an entire universe of receivers, we had better restrict our estimates to the message negentropy for each particular receiver.]

"Go to lunch," or "Wash the dishes," or "Stop the war!" may have little message negentropy if the receiver already has made up his mind to perform the act or not to perform the act. The message triggers nothing. However, "Death to the enemy!" at a particular time may hurl a crusade involving hundreds of thousands, even millions, of people into extensive active states of change.

However, the subtle argument still exists that possibly the apparent newly appearing energetic states would have been cued anyway—perhaps not by the particular message but by some other message. One could argue that the individual receiver will be confronted by such a number of grief-provoking incidents in life (e.g., his parents will die; a probability of death exists with regards to relatives, friends, etc.; the accident rate with regard to other cherished persons, objects, beliefs, is known) that one may expect a certain number of major tragedies, rebellions, or aggressions in life, etc. Thus, the specific instances and times may be shocking for the individual but not for the population. (The doctor and the undertaker are socially useful professionals who are not less human or morbid because they deal with and expect tragedy to be part of life.)

Thus, if message negentropy is to have any meaning (and we are not yet fully convinced it has), then its meaning can only be to the specialized viewer who can identify new ordered states where none existed before. We will illustrate.

A person who puts up a sign "Eat here!" has not succeeded in providing any message negentropy if the number of people who eat in the vicinity has not changed. However, the restaurant keeper who can attract more than his share of customers by a happy piece of "advertising" (by an extension of sensory signaling, say) can speak of the message negentropy of his advertisement. (For example, he could say that just by location he would get 340 customers per 1,000 potential customers who would eat in the neighborhood.

By advertising, he might get 500 customers per 1,000 for a year. If subsequently the business dropped back to 340 per 1,000 because his competitors advertised or his advertising lost its appeal, then the remaining message negentropy is zero.)

If we look at man in the long time scale rather than just a lifetime, then it is fair to say that the apparent behavioral changes are not great. His list of behavioral "foci" has not particularly changed over the ages. Thus, one might question what have the permanent effects been of the message negentropy.

The effects are likely only apparent to the specialist. He can say that they appear in man's extensions.

The doctoring specialist, as extension of the doctoring capabilities in the body, can state that the advertising "messages" over the ages convince man to diet more carefully and to treat debilitation and disease with more tools.

The lawgiving specialist, as extension of the superego capabilities in the body, can state that the "messages" over the ages have convinced man of more states in his relation to man. (The "thou shalt nots" have certainly grown.)

The engineering specialist, as extension of the motor capabilities of the body, can point to the "permanent" gains in these capabilities (of technology).

Whether the concept of message negentropy is convenient or not, nevertheless the dynamic analysis of complex systems must have two levels—the "informational" small-power signaling level, which then can act to regulate or control the large-power level of the system (i.e., it certainly involves power amplification, and may involve feedback), and the large-power motor level.

What does not have much message negentropy is the oral output—the verbal chains that pour forth from the mouth. Since this is such a drastic thesis, as part of human behavior, it must be expanded upon.

First at the high-frequency end. If I say, "Watch out!" I can get an immediate reaction. Thus, it is true that the very high-frequency response of the system to oral output exists. Message entropy does undoubtedly exist in that spectral domain.

If I say, "Watch out, watch out, watch out . . ." (i.e., if I nag), the response becomes much more moot. Thus, it is not certain what information exists as one approaches the dc level. (In fact, it is the central problem in art or in advertising to create works whose repetitive presentation will provide unique focal orbit-forming attention to the viewer so that they will indeed contain some message negentropy.)

If I say, "Give up smoking. It will causally influence your life expectancy through possible lung cancer and cardiovascular pathology"; or if I say, "Also watch your diet, restrict your caloric intake, and increase your activity," this likely will have very little effect, although both statements are demonstrably true and the hearer "knows" them to be true intellectually (that is, in his central nervous system). No amount of added emotional overtone will affect the outcome very much.[1] One can hardly ascribe any (or at least much) long-term message entropy to these communications.

A more revealing illustration was contained in a TV program, *Where is Prejudice?*[2] The program covered a study conducted over a number of days among middle class college students of different faiths and races in which they probed at their own biases and attitudes with some degree of steering by the investigators. (It was not possible to determine from the viewing to what degree the conversations had been steered. On the surface, it appeared limited.)

We cannot review the entire program, but we can point out the conclusions we drew, which overall were "shocking" to us and which have prompted their insertion here.

One conclusion which was articulately drawn by the youthful participants was that they expected their parents to be prejudiced and responsible for having "fouled up" the world, but they expected their own kind—the youth—to be pure and clear and to be responsible enough to ultimately change the world for the better in their turn. However, as this searching experience taught them, and frightened them, they were no different from the adults, their parents. (As adults, we do not find these attitudes surprising.)

After having exhausted the facts, ploys, and all other conversational circuits they could find, they gradually turned their conversations to their childhoods.

Gradually it became apparent that the talking had almost no meaning, in the long run, and its hopelessness also became apparent (shocking both to the youths and ourselves). It may be true that talking spreads out or deintensifies aggressive behavior, but it does not prevent it. Thus, one cannot say that civilized, educated, articulate people will not "cut each other up into bits" one whit less or more than uncivilized, uneducated, inarticulate people.

The children were formed in their parents' image (not shocking), practically independent of the entire noisy intervention of the mass information media, the school, and the religious institutions (shocking).

Although the argument might appear stretched to some, it appears reasonable to us that messages that govern the logic of behavior are not formed out of the logical content of high-frequency messages processed by the brain, regardless of how often repeated.

More and more we find ourself forced to the contrary conclusion that the biological system—including man—is not run by the central nervous system but by the endocrine system. The nervous system provides him high-

---

[1] The author regrets that he could not convince his four daughters not to take up smoking. The cultural influence of their peers was much greater than any rational warning he could give them.

[2] National Education Television, Channel 12, Wilmington, Delaware, December 11 and 12, 1967.

frequency response, a memory of past analogues for future performance, a smoother correlative performance, and a rudimentary algorithm to correlate and integrate deferred motor performance. (We consider this a basic idea —one which is still very new to us—that the correlative performance of the reticular system–thalamic-cortical chain is mainly for immediate motor control and that a somewhat different chain controls the longer-range deferred action performance. The latter, with its memory elements, makes up the content of the primitive superego.) "Morality" serves the longer-range "visceral" (really endocrine) functions of the system. Although apparently cynical, this emerges as the statement closest to the truth that we can make out. The high-frequency nervous "music" provides a busy signaling complex. The slower-moving endocrine follower systems dispose of the longer-range action.

The most directed hypothesis that now seems plausible is that the primitive "superego," the rudimentary guide algorithm that develops within the young child, is "formed" by the endocrine system out of the infant-mother experience (that is, by the action of the endocrines and neuroendocrines back on the nervous system). Thus, its internal logic is an endocrine-system logic, not a nervous-system logic. What this likely means is that the rapid and busy business of living arises from the unstable high-frequency nervous system. However, the slower response to the slower-changing average conditions of the milieu (weather variables, interpersonal variables, etc.) is formed by the waves of endocrine output passed into the blood stream. They act in the "geological"time of the body to weather and form the internal physiognomy of the system and its characteristic response patterns.

We can depict the problem, partially by analogue, as follows: The genetic coding in the animal permits a viable epigenetic unfolding of behavior in the ecological milieu. This perhaps may be descriptively followed by current concepts. Mitotic division transmits the genetic coding, with reliability, to the next generation. (Observing the electrohydrodynamics of the process leads quickly to the conclusion that the process cannot be described in detail with any real meaning. Nevertheless, reliable genetic transfer does emerge.)

In analogous fashion, an epigenetic patterning emerges in the parent as a result of transfer from his parents and some adaptive learning from the space and time and cultural milieu. The genetic child is born. In an analogous "epimitotic"[1] fashion which cannot be described in detail at present (stated as both father, teacher, and scientist), a process takes place in which the child emerges as a "reliable" image of his parents. This is what we have found shocking. His epigenetic patterning may then appear different—this is what makes for social progression for man—but the primitive motor and deferred

---

[1] The word is here coined. It describes the outside "mitosis" by which the child splits off with the experience he has been bombarded with in his family unit.

motor performance relates to his parents. Since he has two parents, there is divergence, hybridization, etc., which tend to be adaptive. However, he is a "chip off the old block." This is the real message that emerged from that program. There is considerable rigidity and topological similarity in transference to and in the biological species.

Specifically, the coloration of the talk of the young and their responses derive from their own culture, but their modes and patterns derive from their parents. Finally, it is even somewhat likely that their "superegos" derive from grandparents (i.e., it is their influence which marked their children in that critical period of childbearing).

Thus the extremely difficult task facing the psychiatrist as "systems" analyst (which is what he is) is really to determine the primitive logic of a system whose patterns must have been formed not during the seven years that Freud originally felt was necessary for the formation of the superego, but more nearly the two years suggested by Melanie Klein. The patterns are elements of organization that the ethologists are just beginning to explore.[1]

If one sought a catchphrase summary, although quite speculative, one might say that the temporal patterns of conduct—long-range behavior— emerge from the manner in which the nonlinear coupling and crosstalk in the endocrine systems form spatial patterns in the central nervous system. It is this character that makes the behavior system so dumb, mute, and abstract.

Thus recording—epigenetic coding—finally represents ways that the system extends its communications, by signs and symbols and "numerology,"[2] in both spatial and temporal pulsings.

In what seem to be such distinguishable ways (thus far), characteristic external extensions of man have emerged. It is interesting that they have become part of man's engineering capabilities. In fact, it is fair to say that essentially all of what should become general systems' engineering likely begins with this topic. This entire document is thus only an introduction to this aspect.

An example of what may be the content of such books of the future is Bekker (30), which gives some introductory idea of the common engineering foundation of vehicles. A measure of the current status of systems' engineering may be inspected in Machol (31), in the IFAC Symposium referenced in (25), in (32), and (33), and the IEEE conference referenced in (3). There is also a whole body of general systems' literature in the *Yearbook of the Society for the Advancement of General Systems Theory*, which was discussed in (34) and which may be compared with this document.

---

[1] Reference (29) is perhaps an interesting addition to such a line of inquiry.

[2] The term "numerology" was used by Gamow to denote the almost-mystical character that one might find in the arithmetization of many processes, most commonly integers that expressed quantization rules.

11-3 EXTENSIONS OF THE INTERNAL DRAMAS

The individualistic human atom is not a stable system. His motional characteristics show this. His associative characteristics also show this. He locks into
orbits. The binding force is the exchange force in which he exchanges his
"body" image (an image of his internal systems and the status of the external
world and an accessible memory) with his foci.

This projection of body image creates dramas involved in internal
"brain" computer time, not real time. Illustrative evidence for this is found in
the dream state. In this section, we will sketch out some of the major "creative"
foci that arise out of this internal drama.

The first organizing one for behavior was the formation of a superego.
This complex internal algorithm represents an ego-ideal which the system
uses to guide the formation of its patterned behavior. The communicated-
recorded representation of this abstraction represents the foundation for
morality. The abstraction of "morality" does not arise only in the complex
that is the human species. It is the abstraction that guides lion, and worm, and
amoeba.

> For example in (17), the treatment of the child by the community in the
> absence of the mother was shown to be quite different in the bonnet as compared
> to the pig-tailed monkey, both very close species of macaques. One may validly
> state that these two species, or the specific ethological groupings under test,
> showed different "moral" conduct. The world image of the human child in our
> culture is well presented by Gesell and Ilg.

The second thought that arises is a broadening of the concept of
"creativity." In (1), the question of creativity in men was discussed with the
background idea that it represented newly emergent patterns. We have since
been led to a catchphrase: Creativity in a species is the capability to develop
novel colorful patterns in those modalities that are within the capability of
the species. When exposed, in verbal exchange,[1] to Bohm's thought that each
instant in time was an act of creation, we were confused but then quickly
convinced ourselves that the two concepts were compatible. It is the internal
ordering of the states in systems—including the more primary systems such as
stellar and solar systems—that unfolds time as a generalized displacement
cycle per period and thereby presents time. However, it really presents time
by this newly emergent aspect of a displacement cycle. Since these patterns
are colorfully novel within the capacity of the species (some of the processes
are colorfully novel by the very fact of being unique), the "creation" of
patterns of time pulses or the creation of patterns of modality cycles have a
common foundation for their kinematics.

---

[1] Bohm's concepts are covered in (2). These discussions took place during the 1967 conference
at Lake Como.

Let us expand on this idea to some extent. "Creativity" in a species may or may not be a desirable thing. It depends on the "purpose" that is served. (This question is deferred. The "purpose" of biological systems will be discussed elsewhere.) However, we are concerned, in our defining catchphrase, with the concept of "modalities." A simple system—the elastic earth spinning —has only limited modalities. Its creative pattern is only the ordered spinning-out of its orbits. As its modalities become more numerous, for example a plastic-elastic earth with volcanic action, its modality richness has increased. The earth may wobble and continents drift. Some day the earth may split. All of these emergent patterns may be "creative."[1]

"Creativity" has to do with the unexpected appearance of branching lines, wherever applicable, and the richness of patterning that emerges as a result of these branchings. The implication is that the richness of branchings may be adaptive to changes in the environment, which in turn may redirect the system motion.[2]

The second great capability of the human biological system consists of being able to create internal extensions, those which make the human's esthetics or his fancies.

As extensions of the capability of communication and recording—that is, of the existence of an internal language and an integrative center that permits describing all internal modalities by an external language known as speech by means of a vocal system, and motor systems that can correlate external symbolic language with external symbolic records—newly created modalities have come into existence. These are patterns of externally or internally recorded and communicated material. They have no immediate "purpose" other than using up excess internal computer capacity. They ultimately develop the function of attracting attention (as a key computer function). This act of attracting attention is used as a leisure-time focal imperative to give the observer's or user's life patterns more color. This is the foundation for literature, music, art, and sports. The motor extensions and the recorded abstractions become quite specialized and should be the subject of a science of esthetics or of physiological energetics.

However, its extensive exploration by us at this point would not be seemly. An elegant example of material capable of stimulating beginnings of a general systems' "engineering" for esthetics is presented by Eric Bentley (36).

---

[1] We are indebted to Dr. Paul Siple's data on the fluctuations of the earth's wobble for helping break us away from the too common preoccupation with rigid cycles, and to Dr. H. Wald for the alternate concept of flexible cycles. It seems clear that econometrics faced these issues before physics.

[2] A most provocative discussion, with which the author does not wholly agree, but which nevertheless is at least as important, if not more so, than the discussion in the 1930s on the aims and ends of education has been begun by Paul Goodman (35). What creativity may or may not mean to our society and the young is brought into the problem.

SUMMARY

1. The human individual is born. He is an active atomistic system. Why? Because his internal organistic chains are only marginally stable. His brain and endocrine systems interact and produce patterns of behavior. The threading of all his internal system "hungers" by these unstable cyclic chains is called physiological and behavioral homeokinesis.

2. Many of these chains are closed only through the external milieu. The animal's external behavior tends to mimic or augment many of its internal chains. Much of the internal signals are of an informational feedback nature. In the case of the human, a large augmentation in both communications and external extensions takes place. He speaks with many words, and he makes many tools.

3. From his many internal modalities and the many extensions he can create there appears a rich internal and external drama. These dramas, emerging episode by episode, illustrate the essential creative character of each "turn of the wheel" in limit-cycle behavior.

QUESTIONS

1. Examine (1), (11), and (17). Attempt to identify those modes that the human operates in, each as independent of the others as possible. Sketch out the physiological concomitants.

2. Attempt to discuss behavior as derived from brain structure and the CNS in the time domain longer than 0.3 second, i.e., in the domain of seconds to weeks.

3. Note the character of the Markov chains derived by Norton (17). Attempt to sketch out significant Markov chains in human behavior.

4. Discuss long-term behavior as it might be viewed from the endocrine system.

5. Sketch out the format of a human's behavior (model it on someone's real behavior). Attempt to sketch out a parallel physiological functional picture of how the human was so driven.

6. Derive a formal theory for message entropy in human communications. Apply it to the real record of a day's worth of a human's communications. Indicate to what extent the system dealt in significant information and how much was garbage.

7. Structure the internal logic of a human (say yourself) as formed during infant-mother experience and peer-peer experience in childhood. Abstract the response patterns as primitively as possible, preferably not in cultural terms.

8. Examine the *Yearbook of the Society for the Advancement of General Systems Theory.* Attempt to abstract as many principles for a general systems theory as you can find.

9. Sketch out an abstract theory of creativity.

10. Sketch out a systems science for esthetics—say for the theater or modern dance.

11. As a companion piece to the second question in the previous chapter, attempt to show how the open (or reopened) system attempts to return the upset system (if not too badly upset) back to a normal viable life. In particular, attempt to show how the brain is capable of doing this in detail.

12. Read the Kilmer modeling of the reticular formation (11). Attempt your version or an alternate version of how a command-control system in the complex biological organism will take the system through its entire daily performance.

13. On the basis of a survey of the ethological literature, attempt a crude foundation for a physical ethology theory of the individual. (Examine the bibliography in the next chapter for some preliminary source material.)

BIBLIOGRAPHY

1. Iberall, A., and W. McCulloch: 1967 Behavioral Model of Man, His Chains Revealed, *Currents in Modern Biology*, vol. 1, p. 337, 1968.
2. Waddington, D.: "Towards a Theoretical Biology, 2, Sketches," Aldine Publishing Company, Chicago, 1969.
3. Iberall, A., and S. Cardon: Hierarchical Regulation in the Complex Biological Organism, *IEEE SSC Conference*, Philadelphia, Pa., Oct., 1969.
4. Arieti, S. (ed.): "American Handbook of Psychiatry," vol. 2, Basic Books, Inc., Publishers, New York, 1959.
5. Crosby, E., T. Humphrey, and E. Lauer: "Correlative Anatomy of the Nervous System," The Macmillan Company, New York, 1962.
6. Magoun, H.: "The Waking Brain," Charles C Thomas, Publisher, Springfield, Ill., 1963.
7. Penfield, W., and H. Jasper: "Epilepsy and the Functional Anatomy of the Human Brain," Little, Brown and Company, Boston, Mass., 1954.
8. Jasper, H., et al. (eds.): "Reticular Formation of the Brain," Little, Brown and Company, Boston, Mass., 1958.
9. Hebb, D.: "The Organization of Behavior," John Wiley & Sons, Inc., New York, 1949.
———: "A Textbook of Psychology," W. B. Saunders Company, Philadelphia, Pa., 1966.
10. McCulloch, W.: "Embodiments of Mind," The M.I.T. Press, Cambridge, Mass., 1965.
11. Kilmer, W., W. McCulloch, and J. Blum: An Embodiment of Some Vertebrate Command and Control Principles, *J. Basic Eng.*, vol. 91, p. 295, 1969.
——— et al.: The Reticular Formation, Parts 1 and 2, Contract No. AFOSR-1023-67B, *Interim Report No. 3*, prepared for Directorate Information Sciences, AFOSR, Arlington, Va., February, 1969.
12. Luria, A.: "Human Brain and Psychological Processes," Harper & Row, Publishers, Incorporated, New York, 1966.
13. Harris, G.: "Neural Control of the Pituitary Gland," Edward Arnold (Publishers) Ltd., London, 1955.
———: The Central Nervous System and the Endocrine Glands, *Triangle*, vol. 6, p. 242, 1964.
14. Stear, E., and A. Kadish (eds.): "Hormonal Control Systems," American Elsevier Publishing Company, Inc., New York, 1969.
15. Morgane, P. (ed.): Neural Regulation of Food and Water Intake, *N.Y. Acad. Sci.*, vol. 157, p. 531, 1969.
16. Piaget, J.: "The Language and Thought of the Child," George Routledge & Sons, Ltd., London, 1926.
———: "The Origins of Intelligence in Children," International Universities Press, Inc., New York, 1952.
———: "The Construction of Reality in the Child," Basic Books, Inc., Publishers, New York, 1954.
———: "Six Psychological Studies," Random House, Inc., New York, 1967.
17. Tobach, E. (ed.): Conference on Experimental Approaches to the Study of Emotional Behavior, *N.Y. Acad. Sci.*, vol. 159, p. 621, 1969.
18. Pribram, K.: The Neurophysiology of Remembering, *Sci. Am.*, vol. 220, no. 1, p. 73, 1969.
19. Michael, R.: "Endocrinology and Human Behavior," Oxford University Press, New York, 1968.
20. Jenner, F.: in "International Reviews of Neurobiology," vol. 2, Academic Press, Inc., New York, 1968.
21. Sheldon, W.: "The Varieties of Human Physique," Harper & Brothers, New York, 1940.
22. Parnell, R.: "Behavior and Physique," Edward Arnold (Publishers) Ltd., London, 1958.
23. Sargent, W.: "Battle for the Mind," Pan Books, London, 1963.

24. Eysenck, H.: "Fact and Fiction in Psychology," Penguin Books, Inc., Baltimore, Md., 1965.
25. Iberall, A., and S. Cardon: Control in Biological Systems, A Physical Review, *Ann. N.Y. Acad. Sci.,* vol. 117, pp. 445–515, September, 1964.
    —— and ——: Regulation and Control in Biological Systems, *Proc. of IFAC Tokyo Symposium on Engin. for Control System Design,* 1965.
26. Brown, G., and D. Campbell: "Principles of Servomechanisms," John Wiley & Sons, Inc., New York, 1948.
27. Iberall, A.: "Information Science, Outline, Assessment, Interdisciplinary Discussion," prepared for Army Res. Off., AD #635–809, June, 1966.
28. Cherry, C.: "On Human Communication," John Wiley & Sons, Inc., New York, 1961.
29. Pines, M.: Why Some 3-Year-Olds Get A's—And Some Get C's, *N.Y. Times Magazine,* p. 4, July 6, 1969.
30. Bekker, M.: "Theory of Land Locomotion," The University of Michigan Press, Ann Arbor, 1956.
    ——: "Introduction to Terrain-Vehicle Systems," The University of Michigan Press, Ann Arbor, 1969.
31. Machol, R. (ed.): "Systems Engineering Handbook," McGraw-Hill Book Company, New York, 1965.
32. Optimal Systems Planning, *Proc. 1968 IFAC Symposium,* IEEE, 1968.
33. Dewan, E. (ed.): "Cybernetics and the Management of Large Systems," 2nd Annual Symposium of the Amer. Soc. for Cybernetics, Spartan Books, New York, 1969.
34. Iberall, A.: "Advanced Technological Planning for Interdisciplinary Physical Research," for Army Research Office, Final Report AD #467-051-L, June, 1965.
35. Goodman, P.: The Present Moment in Education, *N.Y. Review,* vol. 12, no. 7, April 10, 1969.
36. Bentley, E.: "In Search of Theater," Random House, Inc., New York, 1953.

12

Steps to a Social Continuum; From Person to Culture

The homeokinetic human exists as a behavioral atom. He is bound chemically and electrically within his interior and to his exterior. The communications-like fluxes that he can use create bonds both to himself and to others. The drama of the social continuum is ready to begin. The internal content of each man's chains is not so rigidly fixed that new emergent patterns cannot arise all the time. Social history enters upon the stage.

12-1 SOME PHYSICAL IDEAS BEHIND A THEORY OF SOCIAL HISTORY

The problem that we have with social history, namely, boundary-value problems in the dynamics of society, is the same as the boundary-value dynamic problem of the solid state. Thus when we talk about society, we will use and keep in mind the nested sequence that was required to stabilize the molecular systems in the solid state.

What brings the solid state into being among entities which exhibit interparticulate forces is (molar) density alone. In the past (1), we have

188

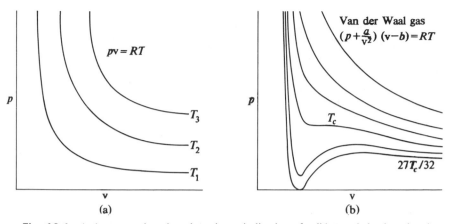

Fig. 12-1 A demonstration that there is an indication of solid-state behavior already suggested in a van der Waals gas. (*a*) Ideal gas; (*b*) first isotherm for which $dp/dv = 0$, at $p = 0$. We might assume this to be the triple-point isotherm, i.e., an isotherm at which the pressure is found to be indifferently near zero. This happens at $T = 27T_c/32$, a crude estimate of the triple-point temperature.

examined the thought that even the most rudimentary deviation from ideal-gas behavior (for example, a van der Waals gas) already shows primitive solid-state characteristics as far as thermodynamic bulk properties are concerned, although statistical mechanical theory can relate the van der Waals constants only to primitive fluid transport coefficients (at that, only to order of magnitude estimates of gas properties, not liquid).

The bordering problem in systems with intermolecular forces appears to be that if there is not too much slowly relaxing energy tied up with internal "atomistic" degrees of freedom, then the bulk viscosity [as modeled by Tizsa (2, 3)] compared to the shear viscosity (i.e., the transport coefficient associated with translational degrees of freedom) is not overly large. The molecular array is then mobile. Further, the array acts fluidlike for all time scales which are long compared to the translational relaxation time or space-scale large compared to the mean free path. With low density, the first criterion regarding the bulk viscosity is satisfied, so that such media are mobile. This is true whether for material systems of molecules or social systems of living organisms.[1] However, two situations occur—one in which the time scales are long compared to translational relaxation, in which case the medium acts continuumlike, and one in which the time scales are short, comparable to the relaxation time, in which case the medium acts "grainy," or atomistic.

---

[1] Mobility shown by a sparse number of trees per acre in a field should not be confused with expecting a whirling-dervish motional performance in the limited time of a casual observation. Trees move slowly in their adaptive responses. The observing human is too impatient. The problem we are concerned with is the freedom to explore the total phase space in process-relaxation time.

In the second case, the application of statistical mechanics to such hydrodynamic systems—the so-called nearly free molecular flow phenomena —exhibits two regimes. In a near-continuum region in which the density is low but not drastically so, there is a fairly clear near-linear superposition of both kinds of effects (Knudsen). For a specific illustration, if at high density a laminar flow is given by

$$Q = a \, \Delta p$$

$$a = a(\mu, d, l)$$

where Q = volume flow
         Δp = pressure drop in some long flow line configuration, which drives the flow
         a = a constant of proportionality which is determined by fluid parameters such as shear viscosity (a transport coefficient) and geometric field parameters such as length l and diameter d

in the near-continuum region, the flow of low-density gas is given by

$$Q = a\left(1 + \frac{b}{p}\right) \Delta p$$

where p = mean absolute pressure
         b = coefficient that represents the near-free molecular flow

In the second very low-density free molecular flow regime, it is likely that

$$Q = \frac{c}{p} \Delta p$$

but

$$c \neq ab$$

At the present, we will exclude from our consideration the free molecular flow region (certainly from our social consideration). It is the superposition range that interests us, in which the statistical near-equilibrium state (represented in molecular systems by such distributions as the Maxwell distribution) is significantly visible and its effects can be superimposed upon mean continuum results. In hydrodynamics if we wanted to be able to treat such a near-continuum, we would have to make use of the Burnett equations of motion, not the Navier–Stokes equations, and modification of boundary conditions from continuity of velocity with the velocity of rigid walls and continuity of temperature with wall temperatures to jumps at the wall (namely, to such phenomena as slip at the wall and two kinds of temperature jumps at the wall). The Burnett equations simply represent an extension of the Navier–Stokes equations to the near-free molecular flow regime.

If we have trouble in the organization of the dynamics of the social system, the problem will not be exclusive to the social system but will exist in all fields for the related problem. Thus, in each case, we will be striving to clarify the same issue.

As a first impression, the issue is the following. At a very fast time scale (e.g., in molecules, 10^{-10} sec), the translational fluctuations are large and considerable and very much random.[1] Many of the collisions seem wild and irreversible. The spectrum of effects thus seems chaotic. Yet, averaged over a number of such time scales, a persistence of type and form of collision emerges (4). It is only very simple "translational" collisions that are involved.

The critical question is now whether there is much interparticle force. Generally, these are forces that attract but fall off (or cut off) at remote distances and repel at short distances. If they did not, likely only one state of matter would emerge, a collapsed saturated state in which particles were pulled together down to a critical density and all motion ended.

If we assumed weak remote forces, then at low density the near-ideal noninteracting system seems to arise. This is like the free molecular flow regime which we have excluded. Thus, the density we must postulate is low to medium (as opposed to very low). Forces of configuration are now likely to be involved.

The forces of configuration tend to be involved in temporal configurations that are much more enduring than the translational fluctuations. Thus the configurational fluctuations are correspondingly slower (e.g., in molecules, 10^{-9}, 10^{-6}, 10^{-3} sec are some typical levels, and in larger molecules—involving many configurational steps—the time may increase to seconds).

Now, it will be found, when we have integrated over the translational fluctuations to obtain some picture of near-equilibrium, that the configurational fluctuations may either be far removed from our observational time scale or near in that time scope. The ones that are far removed in time scale (i.e., long) are formed or formal in character and tend to represent structure. (In man, we will consider these to represent "posture.") For the ones that are close in time scale, on the other hand, their relaxation times can be similarly integrated over to determine their equilibrium. This integration then results in a series of energy-equipartitioned equilibria, in which at each "temperature" (a measure of the internal kinetic energy available) the energy is properly distributed among the configurational and translational modes. This description leads to classical continuum mechanics, and there always is a size and time domain over which such a description holds.

However, the criterion for this status may be violated in one of two ways—either the spatial or the temporal fluctuations may be of the order of magnitude of the observational focus. If the spectrum of temporal configurations is not too densely populated, then the near-continuum results hold, and

[1] At this point, we are still discussing low-density gas behavior.

integration of the Boltzmann equation of change leads in successive approximations to such equations as the Navier–Stokes continuum approximation, the Burnett near-continuum approximations, and, with greater or lesser success, to higher approximations (5).

These approximations are sufficient to deal with moderately dense atomistic populations which may be crowded near their translational fluctuations and which have configurational fluctuations that are far removed in space or time scale. We are in trouble when we look at times or spaces of the order of the configurational fluctuations.

The second point at which we are in trouble is when there is little break, but considerable density of all configurational times as shown in Fig. 12-2.

Some added explanation of this figure may be in order. It is given in the following homely example. A boy, operating within the observational time scale of his day, might be concerned, within the daily time scale of minutes to hours, with his house, his trousers, and a piece of gum. Within that time framework, his house is a formed structure and his chewing gum becomes involved in the more rapid function of being chewed. However, what of his trousers? If they are on the verge of tearing or are torn so that their relaxation properties make themselves evident during that daily time scale, then form and function are confused. As husbands know, torn pants or shirts then become rags around the house.

In another circumstance, the house may become function. For example, viewed in the period of hundreds of years of a civilization's time scale, the 60-year house is just a functional entity in a constant process of change. For example, the house might burn down the next day or furnish fuel or weapons in a riot. Also, the gum may become form at another temperature, when used to plug up a leak.

As a good example to illustrate that temperature and time are interchangeable in their effect on a relaxation spectrum, consider glass. The glassblower plays on the glass with heat, controlling the yielding time scale and bringing it within his manipulating time range. With a few degrees change, the glass has the function of a flowing material which can be deformed as he wills it. [Viscoelastic time and temperature interrelationships are most easily seen in the plastic state. See for example (6).]

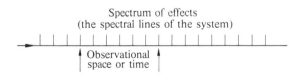

Spectrum of effects
(the spectral lines of the system)

Observational
space or time

Fig. 12-2 Illustrating a somewhat dense spectral region which far exceeds an observational scale toward both the low and the high spectral ends. The scale is the spectral measure, e.g., the logarithm of space or time, such as the logarithm of increasing size scale or of increasing time or frequency scale.

Thus, in this second region, we find quite a few configurational fluctuations that are below and above the observational space-time. How shall we describe such a system?

Let us remember physically what we mean by these systems. In Tizsa's formalism, their bulk viscosity is large compared to their shear viscosity. In Eyring's formalism, such a system consists of a series of many relaxing units. Thus, at every time scale one finds (or may search for) the neighboring "rate governing step." Thus, in our tentative view, these systems show super-atomistic properties. At each observational level, a configurational binding and configurational fluctuation can show up. There arises a spectrum of relaxations to plague us. For well-described systems, one can perhaps estimate these configurations. (Those who have dealt even with such an "elementary" problem as estimating just the elastic modes of some complex physical system know how difficult it is a priori to be certain of all the modes.) For other systems of incomplete specification or understanding, one can only experimentally explore for these modes. This is why loudspeakers, aircraft, and other building structures, molecular complexes, etc., are tested for their relaxational units in their frequency response by empirical spectroscopy.

However, we must remember that the relaxation steps we are discussing here are active ones. Thus we have a system which may become unstable for any change in the dc milieu. It is possible to excite such systems into showing sustained limit-cycle oscillations potentially at all sorts of time or space domains by changes in the ambient dc or ac milieu. (It is obvious in spectroscopy that this can be done by excitation with temperature. We have tried to develop the same general thesis for the biological system and have begun now to realize the all-pervasiveness of the principle. We have begun its discussion in the physical system as part of stability. Now as one further extension, we are proposing to project it for the social system.)

What appear as "superatoms" are frozen-out formal degrees of freedom. We will illustrate with examples. The stirred high-polymeric liquid shows flecks. Even the normal liquid shows very limited near-crystalline formations (as found in x-ray diffraction). The field near the critical temperature shows changing forms. The near-solidifying liquid shows beginning solid organization. Crystallite domains form out of the melt. Silly Putty changes its character with strain rate. A supercooled liquidlike glass shows a changing relaxation spectrum with changing temperature. Water may show anomalous structure. Most commonly, commensurate with energy, the system can settle down into linearly unstable, nonlinearly stable limit-cycle oscillations.

Now, if the spectrum of independent atomistic modalities is rich and the system complex, then there may be many limit-cycle oscillations in progress during any observation time. The possibility of self-development of such a system is clearly demonstrated by the changing geological earth, by the biological system, by the social system, by stellar systems. It feeds from the dc energies available throughout the universe and from the nested instabilities of atom to continuum to atom, etc.

We now know that most systems have an emergent evolution.[1] It is our individual fortune to live now and to be able to describe a little of the process for social history, for in such emergent processes is the foundation for a theory of social history to be built.

12-2 SOME SOCIOECONOMIC IDEAS BEHIND A THEORY OF SOCIAL HISTORY

Although this volume is dedicated to the study of the steady-state life phase of systems rather than to their start-up or degradative phases, we will expand its scope in a few instances when issues that are clarified by start-up are significant. The social system is a good case to think about.

In approaching the social continuum, we are obliged to note what we intend to discuss. Since we are not professionally involved in the sciences of society, our distinctions may be fuzzy. Nevertheless, we shall be concerned with dealing with the following subject content which we shall view in the subsequent context named:

Physiology—the study of functions and activities of living matter organized internally as systems and of the physical and chemical phenomena involved.

Psychology—study of the behavioral interactions between the biological organism—as man—and its physical and social environment.

Ethology—the study of animal behavior. According to Hebb—behavioral science, specialized within zoology, which is concerned with a broader range of species, more comparative study, and more comparative genetic character than psychology. According to Thorpe—the objective study of behavior, without concern for the state of the mind.

Anthropology—the experimental and synthetic study of the natural history of man, as individual and as member of social groups, starting from his anatomical and physiological characteristics and extending through his institutions.

History—the study of the character and significance of events as steps in the sequence of human activities.

Sociology—the systematic study of the development, structure, and function of human groups conceived as processes of interaction or as organized integrated patterns of collective behavior.

[1] Although we believe this for the universe, so far we have little reason to think that this might be true for "fundamental particles" except for their start-up 10 billion years ago. It could be very daring to postulate it, but, in a Bridgman sense, this would be meaningless at this moment because we know of no operational meaning or test to probe at the concept of the evolution of fundamental particles. Others may, and tomorrow we may, but not today. One can certainly imagine it to be conceivably associated with the evolution of the cosmological system. However, it may ultimately have other meanings.

Economics—the study of production, distribution, and consumption of commodities, which are the durable goods of agriculture, mining, or manufacture; usually restricted to limited resources.

It is also possible to distinguish interdisciplinary areas such as psycho-physics, psychophysiology, sociobiology.

It is clear that these sciences each contain an objective portion which fairly overlaps with the domain of physical science. It is this fraction that we shall be concerned with. It is clear, for example, that there exists significant physical content to physiology and to psychology-ethology (that is, to their objective parts). We would certainly consider the following statement from Thorpe [in (7)] to represent considerable authority for our stand: "So I consider that ethology means the scientific study of animal behavior. And, insofar as the vastly greater part of modern experimental psychology is concerned, not with theorizing about emotions, feelings, and states of mind, but with the objective study of behavior, the term can be justifiably used to cover it, too."

That a New York Academy conference (8) was concerned with emo-tional behavior is true. However, in its identification and measure—in which "the number of defecations per unit time" played a significant role—it was quite clear that most investigators were striving for a more objective—physiological—view of "emotionality" than an earlier generation. We are concerned with anthropology—not for its anatomically based anthropo-metrics—as providing us with the objective physical patterns of an "ensemble" or "society" or "culture" of like interacting biological entities that happen to be human. [That is, inclined more to the content contained in books by Mead (9), or Arensberg (10). Ethology already provides some of the data for non-human populations.] As Tax (11) points out, Paul Broca, the great French anthropologist, put forth the expectation that the ultimate ideas of a general anthropology would be extracted on the pivotal base of anatomy and biology. If we invoke "physiology" as we have defined it, we are concerned with the physiology of an ensemble. The objective history we are concerned with is a description of the salient sequences of distinguishable posture in the behavior of an individual or of an ensemble and the "philosophical" explana-tion of the causes and origins of such behavior.

With regard to sociology, we shall be much concerned with the physical processes of interaction of human groups and of their collective patterns and little with systematic study of its historical development and detailed structure.

Finally, with regard to economics, we reject the more narrow specialized technical aspect of current-day economics (not for specialists, but for our pur-poses). Instead, we are concerned with the more classic concept of economics as the study of processes of interaction of human groups and their wants.

It might be considered that the introduction of the concept "wants" creates a teleological cast to our discussion. We do not believe so. What we are concerned with is an ensemble of biological systems that seem to display some characteristic flux patterns—materials and energies moving in and out of the system—and some characteristic behavioral patterns. It is clear that if these fluxes or displacements are prevented (e.g., if the system is prevented from breathing, eating, defecating, drinking, moving, talking, becoming "angry," becoming "anxious," and showing other identifiable flux or state configurations), either physical death, "compensations" in limited degree, or erratic behavioral patterns which debilitate the functioning system emerge. The only teleology we hold to is that continued persistence in time, "survival," or maintenance of function is the overt character of general systems.[1]

Thus our basic concern, hopefully clearly put, is the development of the elements of a systems' science concerned with biological systems—men— organized into ensembles, that interact with each other and their surrounds— the ecology—in a continuing manner. What is fundamental in the biological system, as well as its ensembles, is that its motional state cannot be stopped indefinitely, no more than can an atom or the interconversions of a galactic system, without implying "death" to the system. Thus, validly, these ever-beating systems fall within the province of this report and its fundamental assumption of the existence of active systems.

However, we cannot discuss the complete content of these social sciences. We will offer some comments on their primary characteristics as they concern us, and we will then refer the reader to the more specialized and expert literature.

Socioeconomic considerations begin with the human atom immersed in an ecology. It is important at any point to recognize that the element that must be added is that the human atom at any time has a past history and a memory. Thus, there is an internal storage of some things. For us to discuss the species that is man today, there must have been a history involving a fairly discernible structured social organization, which at most has been slowly changing.

We do not stress this point for trivial reasons. Its implication is that the human being at all times in his previous history had some physiologically viable course of living. This suggests that he had to have an operative inter-coursing family unit at all times. This conclusion is forced by study of all primitive human behavior. It has become a sociological truism that there exists a basic family unit among men.[2] Yet in economics or in economic texts,

---

[1] Further in this system as in all systems, we recognize three phases of its existence—its birth or start-up, its long middle life of relatively unchanged character, and its deterioration phase toward failure.

[2] To cite simply one example, Parsons (158) states, "In social structure there are many variations, but a constant and fundamental point of reference is the nuclear family, the collectivity constituted by a conjugal pair and their biological offspring."

it appears that it is only since Malinowski that firm belief has developed about the fundamental nature of a familial unit and that a social life among rudimentary cultures must always have existed. Judging from primate group structure, man and his hominid ancestors—as the 2–14 million-year-old biological entity that various investigators such as Leakey have been willing to accept—have likely always been social beings. (This is not meant to imply that primates all have a structure of family units like men.) One can only surmise from anthropological sources that man's small number in pre-Ice Age times is what has prevented discovery of a social or cultural prehistory, although one may make a speculative guess that some reconstructions of this prehistory may happen in time.

Our reason for this statement is that the gradual accumulation of a coherent and consistent biological-ecological history from tree-dating, ice cores, etc., of the past 10,000 years as well as the coarser radioactive dating is beginning to provide enough ecological field background that the dynamics of social movement and change can be projected against such data. We are not so sanguine for a history of a half-million or two million years, except in much grosser terms.

It is clear from our modal description of the human being (12) that there is a pattern of behavior which may be described as internal hungers and as "wants" from the outside milieu. The latter are represented as mass and energy fluxes that are to be abstracted from the milieu, and some that are to be rejected to the milieu. It is clear that man must actuate himself into motion to seek some of these fluxes and that dynamic patterns are involved in their space-time acquisition. It is clear that the system is marginally unstable—it does not stay put when introduced into a milieu, and if moving at any observational time, it will stop its motion given enough time. The adage about the commonness of change and the remaining similarity in spite of change amplifies that there is a considerable repetitious nature in man's behavior and, in fact, leads to the possibility of textbook listings of human institutions. Thus, we are perhaps now ready to begin discussion of the "atomistic" individual in continuum society and his exchange fluxes (i.e., the substances and services that make up economic commodities).

The individual as social atom displays a large number of atomistic orbits. However, what is most clear in observing these orbits is that there is a maturational unfolding of phases.[1]

1. The mother-infant constellation furnishes the first phase.
2. The peer-peer play constellation is the second.
3. Attention to the opposite sex marks the third, adolescent, constellation.

[1] The unfolding of these phases does not imply that the basic personality structure that is a man undergoes equally sharp changes in his orbits. It is mainly the behavioral orbits that change.

4. The fixing of an orbit with a person of the opposite sex, and one other major preoccupation, such as career, marks adulthood.

Thus has the structure of interpersonal relations been described by H. S. Sullivan (13). It is not clear, following twenty years of further behavioral study in the field, how fully understood is the nature of adult relations. Nevertheless, the formation of complex social constellations is certainly a human characteristic, and it is certain that it structures around the small, close-knit group organization of the nuclear family unit, often nested within some very simple clan unit involving birth and relation through birth or at most very coherent group characteristics. The number of men who can live alone is small. If a man tries, he disorients. There is a loss in reality or rationality in his orbits. Thus, interpersonal relations, arising from what must represent interpersonal "forces," exist. [For a sampler of behavior, (7) and (14–24) are interesting; also Piaget's many books of which (25) are representative.]

The problem, as with all such n-body problems, is to determine the stable configurations and the underlying spectrum that govern these constellations. However, as in all statistical mechanical problems, when the number of such atomistic particles (those in which the internal degrees of freedom actively permit quantum energy absorption and emissions) becomes large, the detailed n-body problem gets fuzzy, and instead the next hierarchical level of continuum arises.

From the molecule, three states of matter emerged—gas, liquid, solid—all possessing shear and bulk viscosity (albeit the solid has nearly infinitely larger ratio of bulk viscosity to shear viscosity than the others) and elastic moduli. (The fluids have essentially an infinite ratio of bulk elasticity to shear elasticity as compared to the finite ratio for solids.) The interparticle force of gravity is so small that the gas state is not self-contained in anything less than stellar dimensions. The liquid is self-contained by an additional transport property that permits free shear deformation with near-constant volume—the surface tension (which begins at the triple point of condensation). The solid is self-contained, fixedly, into regular or near-regular constellations (the crystallite constellations or, with less regular constellations, the amorphous state).

The aggregation of human "atoms" leads to a sluggish liquid-plastic organization. It is important to realize why. There are rugged individualists in rare numbers who are sufficiently self-oriented that their internal orbits (mainly physiological) are fairly closed (one, however, must remember that they had to come from a family group, or at least from a splinter). But such rare "atoms" have poor survival value. We do not consider it worthwhile at the present time to chase their characteristics down. They may hallucinate, talk to themselves, read, masturbate, rape, or use substitutes for interpersonal "force" relations unknown to us. It is obvious that such persons, to the extent that they exist, represent a rarefied gas characteristic, with little interactional momentum. Most of their energies must be expended in external motional exchanges (long mean free paths) and considerable patterning within their internal degrees of freedom.

However, most people aggregate, and aggregate rather intimately. Population density, where people aggregate, is high. People cluster rather than spread out uniformly like molecules in a gas. This then is the equivalent of condensation forces or strong attractive close-range forces. Now in the solid and liquid state, it was shown (1) that simply the achievement of an atomic density threshold was sufficient to bring liquid and solidlike properties into being. One surmises that a similar condition likely obtains among humans (i.e., coalescence brings social problems into being).

> This concept was arrived at by its pure logical force and was based on a considerable effort to determine the nature of the physiological system [of which (26) is a sample], on a preliminary effort (12) to determine the nature of the behavior system, and on readings in the current work in ethology by such investigators as J. Calhoun [see, for example, (24)] to expose the emergence and development of cultural patterns with changing population densities. However, a recent lecture at the University of Pennsylvania on "Motion (of the Heart) and Emotion," by Dr. A. Katcher confirmed with considerable emphasis that the idea was medical, psychiatric, and ethological "coin of the realm."
>
> He suggests that it is reasonably known that the major coronary disease of angina pectoris was earliest mentioned by Heberden in 1772. [See (27).] Its status was reviewed in Osler's 1910 Lumleian lecture. It is now believed that there is a considerable complex of "emotionally" associated syndromes and "diseases," particularly of the heart, that can be traced to the time of the Industrial Revolution and the growth of extensive urbanization. This seems to tie in also with animal research. Sources that may be cited for those especially interested are (28–38).[1] The immediate thesis is not well proven or even well discussed in these references.[2] However, the thought is actually underlying in these references, and they are provocative. Within a broader frame of reference, Wynne-Edward (33) on "Self-Regulating Systems in Populations of Animals" presents a most exciting document from the point of view of our central general systems theses.

[1] Although commonly China's is cited as a population that seems to violate the thesis, the specialized nature of a strong authoritarian family structure raises countering thoughts that require much more specialized inquiry than we can afford.

[2] This statement is meant literally. However, it is clear in all systems that densification begins to "create" (sometimes helping, sometimes hindering) problems in organization. In animal societies, the studies of Calhoun were quite interesting in exposing this point. We were forced to propose the thesis and to chase it as far as time permitted. One of the most active pursuers of the thesis, Dr. Ratcliffe of the Philadelphia Zoo, upon his retirement (see *Philadelphia Inquirer*, Oct. 7, 1969), commented about his discoveries, "One of the most significant is that coronary heart disease is related to behavior in almost any species that you name, from gorillas to quite a variety of birds. I see no reason why it shouldn't also apply to man, although at present its relationship to man is much less definite. We lack the facilities for applying the experimental method to man. But sooner or later, it will be done."

The basic character of urbanization that emerges is not so much that of "instant disease," but more of a closer interactional range resulting in "stresses" of more intense or new form (remember that stress, as physical analogue, is a force). The problem of uncertainty of position or status and the subordinate relation to many more authority figures are the form of such typical new stresses that may appear in high-density areas.

The lecturer cited three interesting illustrations:

In the Army or in industrial establishments, it is a common belief that more senior officers are kept from too much exposure to the lower ranks so that their authoritative aura should not be tarnished. Is it not more likely, he raises the question, that the exposure is avoided to prevent stress to the lower ranks?

People who migrate from the country to the city show a considerable rise in coronary disease. Is this possibly due to their new, more submissive position?

The highest coronary disease incidence in current society occurs in young black males, not to executives operating under high pressure. Is it not their submissive role?[1] (Unfortunately, poor urban ghetto diet still cannot be ruled out.)

All in all, the idea that social density itself is a major organizing element for animal—and thus human—ensembles is quite attractive. The specific unanswered question relates to what determines high density. To this end, Mayer (24) is very interesting reading.

A major thesis that suggests itself is that in the ecology in which humans interact quite intimately it is not only the humans that interact but all other involved organisms. Thus, at all levels and between all levels of biological organization, there is a homeokinetic interaction, in which each organism has unsymmetric coupling coefficients to all other animals. It is in this milieu that the game of life is played out. Most often, we "see" only man and the "bug" affecting man instead of the entire interaction which includes man affecting the bug, the bug, and both interacting with the plant.

Whereas one may think, in a classical sense, of geometric degrees of freedom as providing the measure of forces in the molecule (that is, in Newtonian macrophysics it is common to think of forces that are functions of such parameters as displacement or velocity), there is also the possibility of measuring forces in terms of near-neighbor indices (for example, in quantum mechanics such forces as exchange forces having no classical counterpart arise). Such forces are configurational rather than geometric. In order to provide them with some sort of theoretical foundation, we may view them as follows: The configurational forces are related to whatever created the quantized states in the atomic configuration (that is, if in the human being the quantum oscillators are physiological and somewhat dichotomous, such as an alternation of eat and not eat, then the interpersonal forces arise from these configurations).

Since this concept of configuration force is so essential to our characterization of how and why social ensembles are bound together, we must provide some background to the concept. It is not possible by any elementary linear or even

[1] Hopefully, the 1960s will be the last decade in which this status occurred.

unitary description to pin down fully the concept of exchange force. In quantum mechanics, where a probability function exists, it is possible to ask what is the probability that a certain exchange of "particles" takes place, and it can be found that an exchange is likely enough for it to exist and become a significant binding entity.

Loosely, the concept of exchange and resonant forces has the following history. In 1926, soon after the invention of wave mechanics, Heisenberg took up the problem of a system with two like particles such as two electrons. If first considered independent, each electron will be described by a (probability) wave function. Their product (since the wave functions represent probabilities) will represent a wave function. Heisenberg treated an interaction between the two particles as a perturbation. Whereas the two unperturbed product functions make up a degenerate system, in their perturbation there is a difference in energy between the two degenerate states. This difference is represented by what is referred to as the exchange integral and thus by analogy is related to exchange forces. For such time-dependent quantum mechanical phenomena, Heisenberg pointed to the analogue of (linearly) coupled mechanical systems of two lossless harmonic oscillator systems of nearly the same frequency with weak coupling. The total energy oscillates back and forth between the two systems and is commonly described as beat phenomena. The "true" motion is a pure harmonic motion in a normal mode (which is linear combinations of the apparent displacement variables).

Heisenberg's work was followed by the efforts of Heitler and London in 1927 to set up a theory of covalent binding for the hydrogen molecule involving the exchange integrals (i.e., what happens when the probability field of one electron overlaps that of a second electron in binding the system, according to quantum mechanical prescriptions).

Finally, Heisenberg's work inspired Pauling, who in 1931 showed a way of combining the degenerate states of Heitler–London theory for binding in the benzene molecule which he interpreted as resonance in analogy to Heisenberg's original use of the term. This was then the source of Pauling's theory of resonance binding. [Since we are not concerned with laying out solutions to quantum mechanics here, only to setting forth ideas, the reader is referred to (39) and (40) for more extensive discussion.]

The added clue to the importance of exchange bonding must be furnished by the fact that among covalent bonds, the hydrogen bond is a most significant case; that its bonding in water and icelike structures is of salient importance in living structure; and that more generally the hydrogen bond is the key structure in living systems (particularly in the polypeptide chains of protein molecules, and by its contribution to nucleic acids).

Essentially, the analogue being sought—here with a validation in quantum mechanics—is that at remote separation the bound fields associated with ac active systems are fairly independent and likely approximate qualities of linear superposition as the systems are brought closer. However, when the copenetration is extensive, then the possibility of an exchange energy may come into existence. A path to known theory is the essentially linear one of degenerate states (the coupling coefficients themselves may arise from nonlinear reasons), which in the theory of linear differential equations has a well-developed exposition. For example, in an equation with repeated roots

$$\frac{d^2x}{dt^2} + 2a\frac{dx}{dt} + a^2x = 0$$

the primitive solutions are the degenerate pair $\exp(-at)$, $\exp(-at)$. The degeneracy is removed in the solution pair $\exp(-at)$, $t[\exp(-at)]$.

$$x = (A + Bt)\exp(-at)$$

can be shown to be a solution. We do not mean to imply that this linear path is the only possible exposition for exchange degeneracy. However, it indicates that removing the degeneracy introduces a new form.

In the human systems, we have proposed exchange forces. What this likely means, in more detail, is that when two human figures see each other—starting from the mother-child configuration—there is an immediate extensive sensory penetration of the overall body image. The penetration is very intimate, sweeping to every recess of the soul. "Monkey sees, monkey does." One need only watch eye movement, or the second-order scanning of the eye watching the other person's eye movements—see Luria (41) for illustrations —to note how extensive this is. Thus, to first approximation, there is an analogous degeneracy of A in B's field and B in A's field. The question may always be raised as to whether this will create an exchange energy degeneracy. The conditions of symmetric copenetration are necessary but not sufficient. However, in the case of the biological system, an exchange energy does arise. "Affection" or "abhorence" comes into existence. The two systems "bind." The binding may not be great, and it may perhaps be assessed in terms of Berne's "strokes" (42). Nevertheless, such descriptions, whether good theoretical descriptions or not, do report on observations of person-to-person interaction. A person-to-fly or person-to-puppy interaction of some significance may or may not take place. (In the limit, person-to-ketchup-bottle interactions may approach the state of a fetish!) Person-to-person ones are hard to prevent. On any encounter between people, the scan takes place with furious speed.

> Although an appreciable exchange energy is evident in human societies, and it may be so in many primate societies, it is not the case that all animals form extensively bound societies. Some animals huddle or cluster, others do not. It is thereby suggested that particular characteristics must be genetically coded within the brain structure for such exchange forces to come into existence.

In human societies, the condensation forces among small groups seem to be physiological, interpersonal, and economic. Whether a chimpanzee group, a gorilla group, a small clan or nomadic tribe, there is a tendency for mammalian species to wander through the ecological milieu in a broad food-searching pattern. The group is fairly supportive in providing its needs.

Whereas the associations of atomistic men into small socially interacting groups suffice for most needs, there is need for a larger group for economic purposes. As social atom upon social atom gather through the ecological pressure of economic forces (i.e., competition for relatively scarce resources, compared with cooperative division of labor; the choice seems to be an optimalization, by coalition, of a friend-foe polarization), the growing continuum finally tends to fracture into a tribe or city-state, an anthropological-sociological unit with some sufficiency unto itself to be viable in the ecological milieu. The group can attack, and it can defend. Thus, its quantization depends upon its interaction within the ecology. The stability is indeterminate in a sparse world. If the world is benign (lemon-drop trees and honey bees), perhaps a loose family life is possible. If hostile (limited food), perhaps tight little marauding family clans are likely. However, if the world becomes crowded, then competition ensues. Then the question of stable organization really arises. This is the problem for the next section.

In this section we are concerned with the history of this loose collection of basic ethological units in an ecological milieu. History, it must be realized, will always mean the dynamic unfolding of phenomena in time and in space, generally retrospective. This is true whether the problem is the "history" of a particular atom emitted from a cathode; the "history" of a falling body from the Tower of Pisa; the "history" of a particular biological cell through its course in a particular culture medium; the "history" of a young man in London—1600; or the history of the Egyptian civilization. Our present concern is with the "history" or dynamics of a society of men, viewed as a colony. We are considering the problem of the same type of organization as the internal continuum of organs in the body.

It is clear that what organizes men are their interpersonal relations and their needs, their physiological needs, and their motor and central nervous system capabilities, those evolutionary and emergent and those transferable by learning.

A tribe–settlement–city–state–nation is a group of people with an age distribution who reproduce at near-stationary rate (so as to persist, slowly decay, or slowly grow) and who control a cylinder of air space and its bounding earth. In the case of nomads, this space wanders in time as the group ranges. It resembles a mobile colony. When agriculture replaced food gathering, the region began to have more fixity.

We may note what the loosely bounded group wants from this space. For it is in the dynamic emergent patterns of satisfying their periodic hungers that the history of the group is written. We pose the problem in a series of questions that a biological species intrinsically asks.

1. What is the quality of oxygen in that space? The biological group seldom has to question this on earth at sea level. In the case of fish, or man stranded at high altitudes, or in time by man's abuse, this problem can become significant. One should note the change in problem areas as one proceeds from dilute to dense ecological concentrations.

2. What is the quality of other gases in that space? Inert gases, e.g., nitrogen, are not commonly disturbing. In fact, nitrogen is a basic ingredient in the life cycle, and life is really adapted to it. Condensible gases, like water, play a highly significant role that seldom comes into focus.

Noxious gases—ozone, oxides of nitrogen, sulfur dioxide—are deleterious. (However, it is not excluded that sometimes species evolve that are particularly adapted to specific material.)

It is of interest to identify the nature of the damaging effects that can result if the quality with which these hungers are provided for, within the social group, is poor. We believe that the following types of independent effects can be recognized:

Debilitation (i.e., diminished performance)
Mortality (i.e., mortality experience of a population)
Genetic effects (transfer ultimately to subsequent generations)

3. What is the quality of water in that space?
4. What is the quality of food (including various minor organic substances) in that space?
5. What is the quality of other necessary minor metabolites and electrolytes in that space?
6. What is the quality of temperature excursions; sunlight availability, wind, storm; the ecological, geographic, geological characteristics of the land? (e.g., a very rocky, mountainous area could make food gathering or home building quite difficult).
7. What is the density of population?
8. What is the state of transmittable memory of the population?[1]
9. What are the avenues of communication available to the species?
10. What are the patterns around which interpersonal relations are organized, including sexual practice, mating, and the care of the young?
11. What are the patterns around which voiding is organized?
12. What are the patterns around which motor activity—roaming, playing—is organized?
13. What are the patterns around which aggression, fear, anger, insecurity, apathy, and their opposites of euphoria, rational placidity and tolerance, security, alertness, play, and sport are organized?
14. What are the patterns of creativity and "rationality"?
15. What are the characteristics of other interacting species in the space?

A dynamic history of action—written in these terms—would represent an ethological-based-on-anthropology history rather than a sociological-based-on-statistics history. It would trace the dynamic cycling and changing tendencies, and then—if ultimately scientifically analytic rather than simply heuristic—it would proceed to trace the physical causality in physiological-psychological terms. An unfolding culture in this view is a study in ethological history.

The nature of such analysis should be viewed with interest, not perhaps for its novelty or lack thereof, but because it seems, by physical reasoning, to

---

[1] As a popular illustration, Nat Hentoff, in the *New York Times*, Sec. 14 on rock recordings, Nov. 24, 1968, points out in a quotation from Al Cohn (Cohn's Law), "It's what you listen to when you're growing up, that you always come back to. For criteria, for pleasure. Like me, no matter what music goes, Basie and Lester Young and Billie are home to me."

stand in similar relation to the analysis for physical and biological systems.

The question we might expect from a social scientist is: "What is the specific analytic character of the dynamics of a society? You physicists can talk about $F = ma$. What do you propose to use for society?" (We will approach this later.)

For the quality of societies, it had been customary to go to the historians, illustratively, Godolphin's *The Greek Historians*, Gibbon's *Decline and Fall of the Roman Empire*, the Domesday Book, and the Victorians in G. Young's *Portrait of an Age*. Since the turn of the century, serious attention has also fallen on the work of the anthropologists. [See for example (43), (44), and (45) for some background on cultural anthropology.]

The rudiments of an archeological-anthropological description may be derived from (46–61). Malinowski, Boas, and Mead are anthropologists who are well known for having brought anthropological material to a more general reading public. An example is (62) or (63). Another interesting illustration of earlier anthropological study is Junod (64). Also see (65).

In recent years, studies that may be considered to be more nearly anthropological-sociological have come to favor. An author like Lewis (66) has become popular in illustrating much nearer, yet still primitive, populations. {Although from an anthropological point of view such poor populations on the fringe of complex modern cultures may not be considered primitive, we really find little difference between their style of life and the scenes depicted by Linton [(57), p. 73] of aborigines acquiring raw materials from a modern culture.} Besides Lewis, it is also useful to inspect two recent books which have received considerable attention and one of their reviews: Dunn (67), Terkel (68), and their review by Friedenberg (69). Dunn provides sketches of slum life in Battersea, England. Terkel furnishes sketches of the entire spectrum of economic life in Chicago. As the reviewer points out, what stands out in the first book is the poverty of intellectual and emotional life, not poverty of things. What stands out in the second book is that the middle class does not live a life much better than the poor. The individuals live lives of noisy desolation, of quiet desperation. The bad guys consistently beat the good guys, but they don't get much out of it themselves. The thought is expressed that the misery is the kind that de Tocqueville predicted, prevalent in competitive, mercantile, egalitarian societies.[1]

---

[1] It would be unseemly to glorify one essay out of proportion to its purpose or possible effect, but there is a more recent review by Friedenberg (165) of a study by Sennett that may be a turning point on ways to gain insight into man's social problems.

The great polarized issues of what runs human societies, those raised by Weber (and his derivative followers), by Durkheim, Mill, Marx, Parsons, enter in a way that is more perceptive of their significance in this essay than, say, the heroic effort of Gouldner in *The Coming Crisis of Western Sociology*. Gouldner raises the ghosts of the past and present, but the reality of the sociological issues are brought into better view by Friedenberg. The plastic-elastic rigidity of the forces that bind humans together is exposed.

It is not a way of social life that quite a few people live poorly; it only becomes so when sanctioned by society's ideology. One must remember that these last two books are not descriptions of poor, rotten villages of the early days of the English industrial revolution, or the class-divided Victorian age, or even the immigrant slums in New York, 1910. They describe life today, in two advanced democracies. For helping to perpetuate such quality of living, we cannot be judged as very advanced.

> For the reader with great ambition, there is Charles Booth's classic 17-volume study of the life of the poor in London, three-quarters of a century ago. For the less ambitious, there is a recent selection from that study (70).
> Dropping back a century earlier, there is another ambitious study, Soboul's classic thesis (71) on the sans-culottes in the eighteenth century at the time of the French Revolution. Here the quality of life of these craftsmen and shopkeepers, in opposition to the capitalist bourgeoise, is brought forth. There are relatively few books like the ones referenced that bother to bring the life of an age into detailed focus. Their study is not academic for any social scientists who want to become concerned with systems' science or engineering.

The history of the social continuum (i.e., the dynamic unfolding in time of a particle or system) is viewed by one major theory, historical materialism, as the unwinding economic thread of its story. Historical materialism may be defined as the thesis that no autonomous history of any human institution can be written without including the economic factor. Whereas we might have accepted this thought 10 years ago as a significant preliminary conjecture[1] and in fact it is a commonly accepted thought in the characterization of man as economic man, the biophysical study of man leads to a modified view. (Economics is viewed in many textbooks as that study which deals with the satisfaction of human wants. Much of the issue will concern what constitutes human wants.)

Physiological-psychological man, operating in a feasible ecology, has about two-dozen "hungers." Man threads these hungers by establishing patterns of activity that fit the time and space scale in a comfortable fashion. If it were not for the evolutionary extensions of man, he would persist in this pattern in ergodic fashion, operating at a slow scale in competitive fashion with other species. Such patterns have been described by Volterra, Lotka, and more recently by Kerner in (72) and (73). They lead to discernible prey-predator cyclic variations, dynamic equilibria states (as in the oceans) among species, the ultimate forest state, etc. These are known in ecological theory. A quick, perhaps somewhat superficial, insight into the nature of the ecological problem can be gathered from (74), Fig. 3-5. It presents a brief summary of the time scales of the growth phase in which related species (birds) expand into an ecological background and occupy it in ergodic fashion. [A more

---

[1] It is the most common response in the industrial community that its economic welfare and profitability is the primary concern of business.

extended article by Odum may be found in *Science* (75). We would be remiss in not calling the reader's added attention to such references as (12), (32), (33), (34) and (76) as representative of a new wave of physical-physiological-ethological thinking about man, other animals, and society.]

However, man developed tools and ultimately techniques for the control of food production rather than persisting in food-gathering and hunting. He exploded in the past 5,000 years, the revolution beginning perhaps 7,000–9,000 years ago. (In the Museum of the University of Pennsylvania, there is a placard describing their archeological findings that in about 3000 B.C., "The high yield of about 30 bushels of cereal to $\frac{4}{5}$ths of a bushel sown revolutionized man's diet.") With this explosion, changes have taken place within men's societies. (They have not taken place in all human societies. There still remain successful primitive societies based on food-gathering and hunting.) First and foremost, economic change has taken place.

> It is not necessarily the case that toil became either easier or harder after the gain in agricultural productivity. Study of primitive societies can attest to this. However, the combination of greater ease, greater certainty, and less stress was the likely factor that determined its success.

What we have sketched out (12) as the nature of psychological man is the fixing of focal imperatives that fit his hungers. When not so continuously concerned with food-gathering, it has been possible for the human in society to expand the number and complexity of his orbital patterns so as to exhibit greater patterning "color." (Up to the present, such expansion has not been found in other animals.) In modern times he has exploded with newly emerging needs. The complex patterning has also included the creation of a significant number of types of elites in human society. (The elites are defined as those persons whose manipulatory actions, both motor and verbal, focus considerable people-energy.) In classical Marxist terms, the elites would have been those who controlled the means of production. This would have made sense if the only factor in society were the economic factor, and economics meant human wants in the specific sense of food and physical necessities. It would then have followed that the productivity control of elites (the amount of motor activity per unit productive accomplishment such as gathering a day's food supply) would be much greater than that of the nonelites.

However, such great productivity control did not occur till much later. The control was likely much more by other manipulatory actions—the priest who could intercede to change the vicissitudes facing man, the artisan who could make things, the healer, the hunter, the warrior, the ruler. One should note that in various ways these were manipulators of the conditions that influence the "hungers."

On the other hand, toolmaking has been long in the developing, at least a half-million years and more likely a few million years. Thus, it is not a single factor such as a developed cortex, or voluntary speech capability, an opposed thumb, or tools—all involving many of the basic extensions of man—but a coalescence of factors within the human pattern of behavior that produces the coherent social group. One must suspect that it required almost a single fixed attractive focus to pile up the people density and that the interacting density itself might do the trick—if certain unstable capabilities exist. The thought of what is latent in the biological organism is here focused on the physical idea of a "latent" instability, one that does not require so much energy or change to bring the linearly unstable, but possibly nonlinearly stable, state into existence.

It is obvious that a system's trajectory near every linearly unstable singularity is not a limit cycle. Why it should be so, or nearly so, in the biological system is partially a mystery yet, but it is subsumed by the homeokinetic thesis advanced earlier. The living system, consisting of so many active chains, often likely has some such sustained chain to fall upon when pushed away from stability in any particular direction. In our view, the complex living system is viable and adaptive in an ac vicissitudinal environment, an environment whose "vicissitudes" make up an impulse spectrum. It is this property that made the biological system viable and thus capable of adaptive emergent evolution in the first place. The adaption of "country" animals to "city" life (namely, those that have successfully domesticated), along with man, is an apt illustration.

However, we may note the emergent theme that it is not human society alone that has this property. We find the atomistic nesting—seeking stable arrangements in ever-widening numbers—in nuclear particles, molecular particles, biochemical particles, biological organ systems, and now in social systems of human entities. The elites are to be understood as the analogue of the solid-state dislocation. The dislocations in the solid state on one hand tie up energy and "cost" the society of the somewhat regular array of individuals for their maintenance. However, they stabilize the arrays by directing them or interlocking them. This is also the nature of human elites. In fact, we may finally draw the tentative conclusion that such dynamic gating, that is, ac active leadership of an elite mechanism or structure, is the foundation of catalysis—whether subatomic, chemical, physical, biological, or sociological. One illustration is the pacemaker acting in the heart as a "catalyst" for the cardiovascular system cycle.

There is a point contained in this discussion of major social significance both today and for all other times of crisis. A society of any kind of particles, in the present discussion human, which is in process of being whipped up into its potential internal modes, of diffusive and propagative nature, begins to pile up stresses in some regions. These must be relieved or "walls" will crack, a process which certainly can take place. However, there is another way to stabilize this nonlinear field. A nucleating center, an "elite" entity—in some sense—can attempt to spin the field into a self-sustaining "movement," most characteristically a vortex. This or related processes now can bleed off energy into "angular momentum" or other local internal modes. A historical movement may start. What these descriptive words cannot do, which instead requires detailed local theory, is describe the conditions under which the movement will be viable.

It is clear that some men, notably elites, intuitively understand this process or learn it in time. It is part of their ability to wire themselves up as automotive "taxis,"[1] as demagogues, as executives, i.e., more generally as capable of being or performing as any required formed or functional entity, solely by self-control of their own system in time and space. The only thing that is not clear, at any moment, is whether they have sufficient wisdom to "read" the requirements for sustained limit-cycle performance; namely, performance that will trap bystanders into their movement who are either going faster or slower than the pace of the movement.

A speech and coordination center in the cortex may have provided the unstabilizing source to put man on the road to culture. Agricultural changes (e.g., the appearance of new grains) may very well have been the immediate densifying cause for human civilization. Its sporadic development in various parts of the world, with time differences perhaps in the range of 1,000–5,000 years, points to the possibility of a spreading instability.

Once densification has occurred, interpersonal interaction then begins to take place. The patterns that develop must fit the human hungers. Viewed in statistical mechanics terms, the "atomistic" oscillator, the human, must range discretely by quantum jumps through his energy-associated oscillator patterns. An a priori cynical indifference (perhaps shown by humans to other species when they raise them) might associate an equipartition of "energy" among all of these state patterns. The energy associated with breathing is equal to that of dueling, of voiding, of eating, of sexing, of angering, of admiring, etc. (The $\frac{1}{2}kT$ per degree of freedom of statistical mechanics.) This, then, leads to the loose idea of the "temperature" of a society. Societies, thus, speculatively can be "heated" up into activity or "cooled" off. Such an idea would not be foreign to such a source as Huntington (77). (Bayesian philosophy would permit us to modify our judgment about the weightings of these states as we learned more about them physiologically. One must remember that an assignment of $\frac{1}{2}kT$ per degree of freedom was derived for mobile degrees of freedom. When these are impeded, the weightings may be different.)

Growing evidence that it is not solely idealistic economic factors or materialistic economic factors that govern the adequacy of a society leaves us with an uncomfortable feeling that the social patterning of a Communist society like the U.S.S.R., a cradle-to-the-grave highly mixed economy like Sweden, a mixed socialist-capitalist economy like England, a petit bourgeois society like France, or the richest society of all times, our own, has not succeeded in providing the basis for a colorful, satisfying life. The appeal of Wallace in the 1968 election to a significant segment of the American population (78) is a case in point. Our major point is that the quality of life has to be judged by the satisfaction of all "hungers," including those beyond the material ones. Many of us who have come from financially "poor" backgrounds know that it was not the material poverty which governed our satisfactions but the quality of interpersonal relations—of the chance to find security, to vent frustration, to find our peers. The outlook of a generation later, of American youth today, is well told in the January, 1969 issue of Fortune. We have no desire to sound either radical or reactionary, optimist or

[1] "Hey, call me a taxi!" "O.K., you're a taxi." The author finds a very succinct message in the two-line joke.

pessimist, but the quality of morality we referenced in (79) and the social problems that we most soberly alluded to in that study and that have exploded in our cities since that report do not represent a currently stable social system. It is the possibility of unstable cultural change that is well presented in Huntington.

Since the coalescing forces in a society are of a plastic or liquid droplet form rather than ideal gas or even regular solid, the question may arise as to what is its functional "purpose," i.e., what is the guiding algorithm which determines the system's action. The basic conclusion that we have to come to is that the individual atomistic system moves about and circulates internally within the grand social ensemble to satisfy its "hungers." (Satisfying its hungers means removing its internal instabilities by forming limit-cycle orbits.) In time, the individual couples and produces a similar model that grows up into the same functional pattern. The difference between individual and society is the difference in location and scale of its orbits and circulations.

> It is noteworthy that both in the society and in the individual there is no permanence of membership. There is a constant turnover in the population and its raw material. This is best seen in the physiological study of Schoenheimer (80). The detailed problem, not of concern to this section, is how the lower-ordered raw material is orderly transformed into the arrays that make up the higher-ordered systems.

These large circulations require storage depots, time lags, and dynamics so as to keep the patterned circulation going. The act of coalescing does not bring the power packs and the local oscillators or chains into immediate being. They must feed from some potential energy and work out ac conversion mechanisms. What is characteristic of the human pattern capability is that it can play internal games of some complexity. [The dog, retrieving a thrown stick time and time again, indicates that other animals have play capacity. Monkey and dolphin are further illustrations. However the coloration, the patterning complexity is not extensive. In humans, it is. The semipopular book by Berne (42) is illustrative of this capability.] Man captures the idea of role-playing.

For example, in society he catches on to the role of supplier; namely, a dynamic transform element that can speed up or retard fluxes. He must have learned this in the manipulation of his extensions: that toolmaking speeds up food acquisition (possibly a very difficult concept for him to grasp, yet one that goes back a million years); that speech speeds up[1] communications which can speed up many processes (i.e., the small energy changes of coordination and persuasion, rather than the large energy changes of mechanical force); that agriculture speeds up food acquisition even more. Finally, in society, he grasps that he can control the flux of supply of all of the "hungers"

---

[1] "Speed up" here is in the sense of the speed of encoding, decoding, and compression in the complexity of the message, not in the sense of the speed of transmission.

(note the common allusion to the "oldest" profession in the world). What he has caught on to is that there is a demand for circulating fluxes. (Get your food, get your sex, get your motor assistance, get your aggression, get your euphoria, etc. Note that it is a matter of indifference whether the individual is provided with drives or an opportunity to be driven.)

Thus, supply and demand arise. Thus, the role of society of shaping the milieu until the patterns of supply and demand fit comes into being.

What is the difference between the widely distributed sparse "liquid-droplet" population and the dense "plastic-solid" civilization? The large territorial space—particularly with the short range of exchange force—makes for very small collections of individuals. The small family molecular group finds its own orbits as it roams the land. (One can ponder on the coulomblike plus-minus nature of the sexes that binds them into a molecular array.) The phase of inefficient food-gathering uses up the easily available "fruits" of the land, so that in turn it can only support this dilute population.

> This is a possible clue as to why the human neither expands to infinity, like a gas, nor contracts to zero. Within a given ecology, there may be a free-roaming sphere which will permit sustenance without a great number of extensions, such as a closed ecological chain of domesticated grazing animals and grazing pasture, to which man can be added and which in that form really creates an oscillator chain. Mammals of the extremes of size, for example the shrew and the elephant (81), both have to spend an inordinate amount of time foraging. The choice of nearly closed orbits, forming practically a microcanonical ensemble, probably represents the lowest stressed optimal paths that the species can select consistent with its metabolic intake. To continuously expand into the unknown is likely more stressful. The most primitive selection of paths involving man likely is the oscillator chain involving the growth of berries, nuts, fruits, greens.

Conversely, when the society is condensed or localized into a region, the circulation patterns require speedier action. Storage, distribution, production become needed roles.

Must the elite role be played by the individual who sees the need or can even create the need, or may it be played by the society? This brings us up to today's practice very quickly. A priori, the answer is not clear. Furthermore, systems of both sorts have worked for various periods of time. In addition, there is a tendency noted for societies to "dissolve" after 500 years. Thus, the question is more vague than it appears.

> It is noteworthy that at this point the issue of the nature of history—of a system—begins to become pressing. Written history has most often sought to identify individual elites who controlled or represented salient foci in their epoch or else has tried to identify the "theme" of that epoch. In some senses, this can be done by looking at patterns of behavior. However, it may miss the community of the threads that may be involved, such as food, sex, and all the other "hungers." Dirtiness, disease, complacency, etc., many of the indices that might be used to measure the quality of a civilization, are not as important as the mobility in

providing channels to pattern all "hungers," e.g., its mores. In time the society opens up and flows out of its constraints, whether territorial, political, or ideological. Humanists are concerned, further, with how the new threads of change unfold. An interesting observation was made by James Webb, former director of NASA, in an afterdinner speech. He pointed out that NASA prepared the way for transportation to the moon with costs and manpower similar to what it took to build the American railroads. The remarkable thing was that it was done with such little cost in social destruction. It is thus conceivable that social change can be effected by purely political decision in peace, if carefully channeled.

However, today some sort of direct comparisons have been made for highly industrialized civilizations. (It is a great mistake to think that one or another conclusion holds necessarily for underdeveloped nations or for superdeveloped nations of the future. However, within the current epoch, comparisons may be well made.)

A useful dialectic about current technologically organized modern societies may be conducted with an article in *Fortune* (82). Its point is that both the Soviet Union and the seven socialist countries of Eastern Europe (which, apart from the U.S.S.R., represent a population of 120,000,000) are in the difficult process of reforming their economics by introducing freer market forces. "In effect," the article states, "East Europe's 'market socialism' . . . acknowledges that an economy planned and administered by the state . . . cannot deliver the goods, and that there is no substitute presently known . . . for a competitive market and a tolerably free price system." It appears that "perhaps the biggest failure was the policy of sacrificing everything to capital investment in a grand attempt to lay the material basis for socialism. Much of the investment proved to be the wrong kind, in the wrong place, at the wrong time. Thus, by 1955, East Europe found itself using more and more capital to get the same increase in output." However, to make changes, it was found that "In the communist world there were legions of vested interests. . . . Plant managers who made a facile living finagling easy quotas . . . didn't relish the idea of slugging it out in the market. Everywhere there was a great shortage of enterprising managers."

A great impediment to bringing changes is that it "bristles with technical difficulties. A market economy, like a human being, performs well only when it is not continuously forced . . . but a command economy . . . is one in which the planners assign the country a bigger job than it can do. Therefore, the bureaucrats can balance supply and demand only by force. . . . [I]n reality, supply and demand are explosively out of balance."

By trying free-market reform, the first consequence is that "all prices soar. . . . [W]hen consumer-goods prices are freed, too much money is still chasing too few goods, [and] bureaucrats . . . discover weighty reasons, both ideological and 'human' for retaining controls. . . . [A]nother source of trouble is that a command economy usually cannot even partially free its enterprises without causing unemployment.... On the face of it the situation seems tough.... The bureaucrats ... stall or even sabotage.... The more freedom for the enterprises, the less power and the fewer jobs for bureaucrats. ... [M]any economists and even some party leaders are convinced that all East European countries, as they accord their enterprises more freedom, will have to adopt some kind of worker-management councils.... Postpone ... they may, but avert it they probably cannot. On the side of the reformers are inexorable economic forces that have already demonstrated that there is no substitute for the market.... [E]nterprises badly need better managers.... The most important reason East Europe seems bound to adopt a genuine market socialism is that it has industrialized to the point where it must increase its trade, and particularly with the West...." This is the tone of the article. (We can attest in small part to the validity of the general thesis of change. Having traveled in the U.S.S.R. in the period around 1958 and having been involved in issues of concern to their Ministry of Foreign Trade, I was surprised how every journalist and economist we met raised a common question, "How is price determined in the U.S.A.?")

As entrepreneurs and intellectuals, we are not concerned with polemics for or against our own system. We live in our milieu and have some reasonable idea of how to maneuver in it. Our concern here is purely with scientific analysis.

The essential element in a free-market thesis would appear to be not so much deciding the price of commodities in the open market place but the buying and selling of all the commodities that satisfy hungers, both in long- and short-term, i.e., in more than poetic terms, "the stuff that dreams are made of."

This now is a key thesis. It suggests a little of the quasi-ergodic character of human societies (namely, that the orbital hungers may change only moderately in time as new ones emerge). However, it suggests some of the problem of regulating a society. In principle, neither the socialist thesis nor the capitalist thesis is quite right. What it takes to run an economy (a good introductory theme was expressed in a recent *New York Review* article that said that 100 men of diverse talents are needed to run and organize an underdeveloped nation) is an understanding of the hunger foci of its existing members and some technical competence to guide and regularize its under-currents. What comes out of the *Fortune* article is the suggestion that the Eastern socialist economies have not learned to do this. (In the English case, following its changes in the late 1960s through the eyes of a journal like the *New Statesman*, which is sympathetic to but critical of the labor government, has been like watching the performance of a tightrope artist.) The question is not immediately the capital cost. In a good equilibrium state of a system (or organism), the satisfactory solution is operating at that "capital" cost for the system which is a minimum for the required degree of hunger satisfaction. If such a cost is extremely high, then the system may only have a short life (that is, "a short life but a merry one" may be as optimal a solution as any other for some systems).

In speaking of the regularization of the supply of hungers by managerial elites, we can propose the following scheme for its accomplishment. We find that the principle here is the same one we have learned is the principle of organization of many other systems—notably the characteristics of a river-valley system, of the blood-flow arterial system, or of the flow of traffic. We propose it for the regularization of the flow of traffic among human hungers:

1. In a Bayesian sense, we can make an a priori estimate of a key principle, such as water flows downhill (which leads to a network of lines of steepest descent for any given topography and therefore the a priori assumption that rivers must be disposed along the direction of the lines of steepest descent) or humans have key hungers. From such a principle, we can draw some spatial field lines or space-time flux lines (in chemical engineering, the flow charts) for necessary paths.

2. The second key principle is the principle of comfortable fit. This is a principle surmised from nonlinear mechanics that energy can be regularized through synchronization by a small number fit among "quantized" (i.e., nonlinear limit-cycle) elements (12).

3. Thus, whereas water flows downhill, rivers do not. They meander about the path.[1] Thus, the traffic patterns must fit added prescriptions and constraints. (See Fig. 12-3.)

> In fact in an earlier study (83), we have developed enough hydrological theory to believe we can lay out an a priori satisfactory river system for a given landmass, including its meandering and channel shape, with some very minimum assumptions about the land. The river meanders because, in addition to carrying the water in its channel, it must also carry its bedload. Water and bedload are

---

[1] A stream of water playing on a hard surface nearly exactly flows along the lines of steepest descent. A river, subject to more complex constraints, develops a path that only grossly must follow such lines; instead it develops meanders.

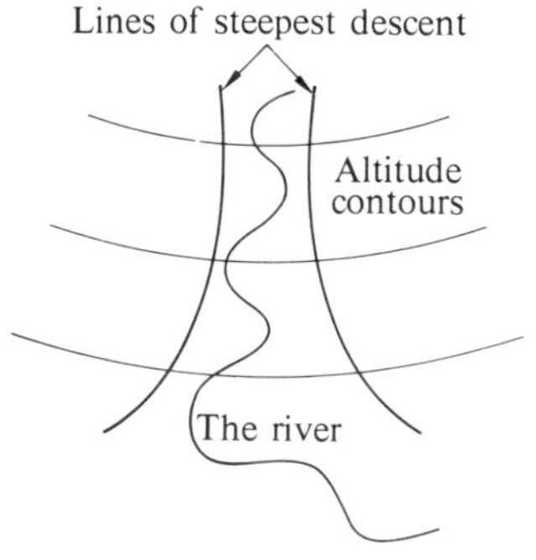

Fig. 12-3 Example of river flow among lines of steepest descent.

coupled and do not form a simple hydraulic equilibrium system. In another study (84), we have sufficiently developed the theory of the topological-anatomical design of the vascular bed so we believe that we can similarly lay out a satisfactory arterial tree.

Fairly optimal traffic patterns from A to B (e.g., home to work) are not the most direct or shortest elapsed-time pattern. One meanders with known and unknown space-time instructions, typically one that requires least command-control effort. (See Fig. 12-4.)

4. The general principle of minimizing energy expenditure is the principle of maneuvering at the lowest possible "temperature" (e.g., the philosophy of "don't make waves").

5. For complex human societies, it is not clear (in the sense that it is still very complex and that not too many people are experienced) how to draw up the initial a priori human social network, even when the hungers are known. Four common principles used are to start from and retain or modify the existing state (the conservative or liberal point of view), to start from a new preconceived state (the idealistic), or simply to satisfy known technical constraints (the "engineering" or technician approach).

The essential first step, it would seem, is to see whether the hungers are being satisfied in some reasonable way. If not, then change is required. This represents a combination of known engineering and science.

6. With some beginning of satisfying hungers, managers can move toward more desirable patterns. Such answers could be equally satisfactory to primitive or complex societies, socialist or capitalistic, etc. What then are the remaining problems?

7. The managers (that is, the regulators of supply) perform their function by being elites, i.e., by being able to tie up large amounts of people-energy. However, they are human themselves. It is not sufficient to specify that they must exist in order for the system to run. At any level, ecology, epoch, and time, if their existence is not conserved or newly created out of the matrix,

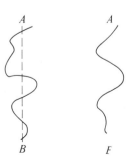

Fig. 12-4   Illustrating traffic patterns from A to B.

the system decays to a less energetic one.[1] (When the old entertainer retires and no new one arrives, the excitement in a party subsides.)

8. On the other hand, the elite attempts to run away in energy. There arise loss processes that limit his growth. (One may always think of Hitler, Stalin, Napoleon, Ghengis Khan, Alexander, and others when preparing to draw the conclusion that the individual is highly limited in capability. What he can do, in a resonancelike reinforcing situation with plastic human material, is fantastic.)

Our only historical answer to preserve the elites but not to let them run away is to supply them with a little more in their hunger demands.

> On the other hand, today the pay to top entertainers or top management has become fantastically high. We hope that this does not happen to scientists! An article on the teeny-boppers of the mid-sixties in the *New York Times Magazine*, May 28, 1967, indicates clearly that its music has become a dominant thematic hunger among adolescents. We are not concerned with the poor quality or our inability to avoid expressing "old-fashioned" parental views, but with the fact that the amount of energy tied up is so wasteful.

However, it is the regularization of the supply and demand of elites that becomes a significant social problem. They are the ones who as the command-control system perhaps make or destroy civilizations. That is, the elites become the distributed "cortical" elements that do the thinking. They shepherd the society from A to B. Often in the first instant they make the society, and in its last instant they break the society. The society's inertia, etc., governs the system's speed of response. This picture must be the rudiments of a picture of the social-economic continuum of the higher psychophysiological living system.

One may initially question whether this differs as greatly from a classical view of economics as might first appear.

Classical economics, at least in an elementary sense, deals with production, ownership, distribution, etc., of goods and services that satisfy human needs. This really does not differ so much from the previous concept except that the previous thought is based on the physiological-psychological cyclic "hungers" that form human focal patterns. This just generalizes the economic factor by giving it a more precise foundation. This foundation is in process of discovery by advertisers, service entrepreneurs, and others, even if it is not yet expressed so fully in economic texts.

Thus we can reconcile a materialistic view such as economic determinism, the doctrine that the sole factor governing history is the economic factor, with an earlier idealistic view by accepting the principle in part if it means that the summation of man's emergent hungers is the sole factor governing large-scale historical change. It is not true in toto because the hunger patterns

[1] An extensive development of very similar ideas may be found in a fascinating study by Zipf (85).

are governed also by ecological boundary conditions, which also include the many physical boundary conditions. In fact, one must see that the unfolding story of a long-term history is the changing hunger pattern response, epigenetically unfolding, including formation of elites under changing ecological conditions when starting from certain initial patterning positions, plus memory (of patterned information). Further, we have suggested that the "hungers" of the elite have a significant role in determining the course of history. The advertising man has learned that the consuming desires of the "high mobile" cast a shadow five years ahead of time of the wants of the rest of the population.[1] Somehow these desires are communicated and propagated. "Class before mass" is a rule for introducing new products.[2] (By this view, we propose the unification of quasi-ergodic theory and catalysis at every level. The systems expand into all the phase space that is available. New regions can unfreeze through the action of catalytic elites and become available for occupancy.)

For those who might maintain that the materialistic doctrine and the idealistic doctrine of economics are not reconcilable, we may examine the extreme materialistic doctrine of Marx in, say, Engels (86), a basic work on the "scientific" side of dialectical materialism. (Part 2, on political economy, is in effect a summary of some of the major principles of *Capital*.) What we propose to do is carry on a dialectic with a central Marxian thesis. In so doing, we hope to leave in resolution some view of the basic factors governing man's sociological-economic conduct. That man operates within very complex social patterns we accept as self-evident.

Political economy, Engels starts out, in the widest sense is the science of the laws governing the production and exchange of the material means of living, of subsistence in human society. Political economy is essentially a historical science; that is, it is constantly changing. Engels proceeds to outline elements in its organization and points out that it cannot be based on external laws of nature. In history, he states, the economic side is more fundamental than the political. Private property (pursuing a polemic dialectic against Dühring) emerges not through force but through the transformation of exchange from immediate producer's goods to commodities. With the changing of economic order, change in political order occurred. Historically, division of labor gradually permitted the servant of specialized social functions to become master. The emergent form of economic order became the pressing into the service of production of one people by others. Under warring conditions, which were as old as the existence of groups of communities, with such specialization enslavement then became the major

---

[1] Not all shadows lead to substance.

[2] Of course, the public relations man also has to learn and judge who the new elites are as times change.

source of labor, e.g., in Greece and Rome. In fact, it was only slavery that was able to take man from his older, barbaric animal life to his condition in the ancient (historical) world of social classes—of masses performing simple manual labor, and a few privileged persons—from which the ancient civilizations flowered. Beyond this the more technical questions of a theory of value and capital, which Engels discusses, do not concern us at present.

The key word in Engels' definition of political economy is "material." The story that Engels is sketching out and that idealist doctrine must also sketch out is a boundary-value-problem "history" of man as a distinct biological species. Else we are left with the ethological story of any other animal species. We can say that generation 1 of fruit fly begat generation 2 of fruit fly, etc.; or bee colony A leads to bee colony B, etc.; or the development of plant life in a particular region leads to the following ultimate forest with such and such prey-predator dynamic balances, etc. The emergent cultural history of man does not arise from such simple stories.

However, the boundary-value problem must include the animal inside as well as outside. As long as man was viewed idealistically as mind and spirit, the only doctrines of explanation possible were split by the dual view of animal body and spirit mind. However, if the internal milieu is considered part of the boundary-value problem, then properly human history must be that of man endowed with a complex endocrine and nervous system unfolding his history in the external milieu. If this is accepted, then economics is more broadly the laws governing the production and exchange of "material" wants of the human. However, the patterns emergent from both brain and endocrine system are involved, including man's emergent extensions. The governing internal electrochemical chains are quite "material." However, these patterns must be described partially as informational in content. These may be described "idealistically." None of these are distortions of physical reality, but statements of real issues in description.

With regard to man's history, the one reasonably certain thing that can be said is that few writers have an "explanation" for historical emergence of civilization. Marx and Engels do propose one, that it was by man's enslavement of man that he emerged from barbaric animal life. Although we cannot pursue an adequately scholarly study of this question, we can pursue it in rudimentary form. The problem is to what is modern man's culture due?[1]

What is clear is that it is the combination of the tool extensions of man

[1] Although this question is a system's start-up problem, which we have excluded from consideration in this volume, it is worth some preliminary exploration. Start-up questions about our other levels of systems science are beset by their distant and uncertain occurrence in time. The billions of years involved since the origin of nuclear particles, of the earth and sun and the solar system, of life, and of galaxies are hardly as capable of capture by touch and taste and closeup conjecture as man's growth into urban civilization a "mere" 10,000 years ago.

and his internal patterns, particularly communicational, that was required before civilization exploded.[1]

All complex animals essentially have a familial unit of interaction. Many animals have social organization. Some animals may even have a culture —a community of individuals, implied systems of communication, a division of labor, and a coordination of effort.

Dictionary definitions[2] call for more. They call for a community of individuals that show learning and transmission through the use of tools, language, and abstract thought; or for a recurring assemblage of the artifacts of their way of life that differentiates a particular group or region, independent of their presence (i.e., as it may be discovered archaeologically); or for a body of belief, form, and traits constituting a distinct tradition (i.e., recurring assemblage) of a homogeneous group. However, if we are to insist on an essentially objective view, we cannot inquire into the "beliefs" of nor establish the foundation for "traditions" of human beings, as the need for hundreds of hours of psychoanalytic investigation can testify. There is an uncertainty in the ingredients of an adequate theory of learning or the rapid reorientation of many social mores in current society. Thus it remains that, in community, degrees of homogeneity, recurrence (i.e., patterning in time), localization (i.e., patterning in space, as by settlement, or a specialized wandering migration path), task achievement (i.e., fulfillment of biological hungers), as well as objectively identifiable extensions are cultures to be identified. Although not completely proven, it is furthermore clear, ethologically, that behavioral patterns can be recognized and partially classified. (If "emotionality" hungers are not fully classifiable, heart response, adrenaline response, body attitude, etc., offer many recognizable objective complexes that can be "named"). Recognition of a division of labor to accomplish tasks and some complex of artifacts are its common traits. It is likely certain that what social scientists most expect in culture is considerable learning (i.e., as distinct from "instincts") and a complex of extensions. [Illustrative quotations are: "There are, of course, many social animals besides man, and certain insect species possess social organizations in some respects more elaborate than man's. But genetically stereotyped, instinctive behavior dominates insect societies ... while human ones are uniquely founded on culture, which individuals learn rather than inherit through their genes. . . ." (48); or "Culture, in the sense of modes of behavior learned rather than inherited by inborn instinct, is certainly no monopoly of man" (55).] Now it is clear that many colony animals—ants, bees, beavers, various crustacea—have elaborate extension structures. There are no hive-building genes. There are no mother-associating

---

[1] These are not sufficient. To what extent external factors in the ecology assisted in the explosion remains quite speculative.

[2] *Webster's Third New International Dictionary*, 1967.

genes and so on. Such structures and functional characteristics must emerge from epigenetic unfolding. This clearly must be a very complex nonlinear process. It is the elements of the process which are entrainable into pattern complexes. This must be just as true among human as among other animals. The only apparently certain difference between man and other animals is a speech and coordination center so that an extensive memorizable (imprintable) abstract verbal coding can represent and correlate many different motor actions. (Other animals do use tools.)

Thus, for example, a very specific 4–5-day estrous cycle with biochemical concomitants can be traced in rats (estrogen as a triggering hormone). No such specific endocrine concomitant has been proved or found in the human being. Nevertheless, sexual rhythm in man is quite strong and possesses a long range (87). It "cries" for chemical explanation. Endocrinologists mainly say not, arguing that human sex drive lies in the mind. Other scientists are much more willing to request that biochemists look further.

With regard to "learning," it is a mistake to consider that animals do things by "instincts." [See for example, "Learned Tradition on the Subhuman Level" in (48) or see (88).] As animal experiments show, animals deprived or restrained do not exhibit "normal" patterns of behavior, i.e., their normal patterns were learned as part of an epigenetic unfolding. Reference (12) attempts to make the point that the first nonlinear law of behavior is that patterns develop and entrain around those physical-chemical–space-time elements which are available and that the combinations which emerge are the adaptive evolutionary ones (i.e., the genetic coding produces, from its mutational content in the population, space-time elements that are adaptive and the epigenetic unfolding selects and binds them).

It is a mistake to view learning as the structure achieved by artifical "learning curves." One gets to Lysenko-type positions. You can attempt to teach cats calculus or men to bay at the moon and perhaps show some kind of learning curve. Whether adaptive selection can succeed in breeding "friendly" dogs and heavy milk-producing cows or friendly cows and heavy milk-producing dogs is a matter of an experimental process over a number of generations. Whether dogs so selected over 500,000 years may develop species that can more easily be taught to be friendly or even to talk or memorize extensively is not yet a closed subject. The limited history of human cultures, even if based on the as yet "exclusive" ability to talk, does not guarantee that other species may not in time develop cultures in the more broadly accepted sense.[1]

[1] A fundamental principle of systems' science, which "learning curves" perhaps illustrate, is that a viable atomistic entity may have internal modalities which will permit any particular type of orbital pattern to develop and form either a repeatable pattern or a more enduring structure. Systems' characteristics can emerge this way. A learning curve illustrates the course of entrainment.

We believe, for example, that it is possible to reconcile the antagonistic views held by Claude Bernard and Louis Pasteur with our own. In their day, these men were protagonists in a basic physiological battle, in which Bernard's thesis was that it was the fixity of the internal milieu which represented the essential condition for life, whereas Pasteur was the most important advocate for the germ theory of life, in which it is the coursing of the individual germ and the cell in the body that determined the course of life. These concepts were not and are not mutually exclusive but represented the emphasis of their age and their (French) physiology. In our view, now, we can reconcile these positions by regarding the body interior as an extension of the ecology available to parasites. Epidemiology then becomes the study of the interaction of species A—the parasite—within the extension of the ecological continuum furnished by host B. This once more is an ethological problem, both on this continuum level and the previous atomistic level.

12-3 ON THE START-UP OF HUMAN SOCIETY; PREHISTORY

Thus, whether viewed from an inclusive concept of culture or a more limited concept, we are up to the question raised previously. What is responsible for complex human community? We propose to view the question in our homeokinetic sense; that is, to see how the orbits and chains of social and cultural history begins with densification.

First, we will view it by means of the thesis of slavery. The problem of start-up that exists is whether Old Stone Age man's communal achievement exhibited many learned elements and was based on enslavement or not.[1] If it were, then Engels' thesis has considerable merit. We offer three popular sources to permit the arguments to be followed graphically and easily. The first is *Life's The Epic of Man* (89). (Genus Homo 500,000–700,000 B.C.; species sapiens 300,000 B.C.; Old Stone Age 500,000 B.C.–10,000 B.C.; last glacier retreat 20,000 B.C.; Middle Stone Age 20,000–8000 B.C.; New Stone Age 8000–5000 B.C.; settlements found after 8000 B.C.; soil cultivation begins 7000–6000 B.C.) The second is *Everyday Life in Ancient Times* (90). The third popular source is Van Doren Stern (91). A fourth source is Piggott (92) (Genus Homo 400,000 B.C.; species sapiens 30,000–40,000 B.C.; Old Stone Age 400,000 B.C.–10,000 B.C.; final retreat of ice 8000–9000 B.C.). A fifth source is Braidwood (52).

As viewed in *The Epic of Man* (pp. 18–27 and 30–41), as one sweeps from 500,000 B.C. to 10,000 B.C., from the Abbevillian to the Magdalenian

---

[1] One may surmise that some anthropologists would consider the archeological evidence for the existence of enslavement in Old Stone Age man's times as dubious and not worth discussing. Yet from the point of view of illustrating the difficulties of discussing the start-up of real "natural" systems, it is worthwhile to indicate what happens if you try to pursue causality associated with any particular thesis.

traditions or industries in the activity pictured for Old Stone Age man, the future of intellectual, spiritual, rational, esthetic potential is already there (burial ceremony, for example, and religiomagic activity are clearly indicated). The slavery argument would be weakened but still have some merit if enslavement were the organizing feature in the middle stage of the Old Stone Age (100,000 to 30,000 B.C., see pp. 20–24) and would have only marginal merit if it applied to the very late stages of the Old Stone Age (14,000 to 8000 B.C., see pp. 23, 33, 37, 39).

Certainly by 10,000 B.C. (see pp. 33, 34, 37, 39), as shown in the cave paintings at Lascaux, by community, esthetics, burial decoration, religiomagic, ritual likelihood, art, and ceremony, culture and the seeds of civilization existed. It no longer matters whether slavery was in existence by then or not. For, if the earlier phases had been free of slavery, man emerged from animallike "instinctive" conduct into "culture" without enslavement and through stages involving some division of labor and communal interest.

In the Middle Stone Age (after the last glaciers of 20,000 years ago, beginning about 8000 B.C., pp. 46, 48, 51, 53), tools and weapons, the extensions, became more skilled and, as shown in Star Carr in England, communal gathering and industry certainly began. Mattocks, bows and arrows, boats, sleds, skis, tents, and traps came into existence. There is no evidence that enslavement played a part in their societies. It remains still the achievement of Old Stone Age man, 300,000 to 10,000 B.C., that set the stage for Middle Stone Age man and for civilization, which represents the continuum properties of man the biological species with his particular form of culture.

With Neolithic man, 7000 to 6000 B.C., cultivation of the soil and settlement began [see page 55; also see (90, 92)]. With settlement, the age of recorded history is closer to beginning.

At that point in history when records begin, there is much more doubt that slavery has not come into existence. It would likely have been the intrapersonal conflict, not just the possibility of a people able to cope with their environment, that cooperatively created the leisure time for extensive ceremony—of priests and warriors—and the abstraction of symbolic ritual, instead of the more primitive earlier primate patterns of behavior. Thus, the possibility of leisure with man's brain brings a leisure class into existence; it forms, not "essential" physiological want, but newly created "psychological-ethological" want. Thus, we take a position different from Engels' "Anti-Dühring," as well as Dühring's.[1]

[1] The question, of course, suggests itself why civilization has not come to all men who have progressed to the stage of food production (52). One issue is likely the specific productivity achieved with the land. However, another point is that it may not be assumed that such processes take place "instantaneously." It is clear that 5,000- to 10,000-year differences in achievement in different areas have happened in the past. That such lags still exist or are only in process of disappearance currently (65) should thus be no surprise, only a measure of the slowness of the rates governing epigenetic changes in the biological system.

One could have not too much doubt that the complex of Tepe Gawra, 4000 B.C. [say Gawra XIII—see (90) pp. 24–27], in southern Mesopotamia was quite possibly built with slave labor. From that point on—Mesopotamia, Egypt, ancient Greece—there is little doubt that slavery made leisure class activities possible—art, science, philosophic thought, etc. However, this is not the issue that confronts us.

> By themselves, these issues are not so pressing. We do have an operational modern society and our modern problems. But what is pressing is to understand what the effect of density itself on a biological population is. On one hand, this is being studied without historical reference by ethologists. On the other hand, it must also be studied currently within our own society as well as historically, by anthropologists and others, on the basis of man's total history. Man and his experience still remain unique in biological annals.

Our concern has been with what turned ancient animistic man into modern civilized man. The Marxian thesis will not hold water.

The alternate, more plausible, thesis is that somewhere along the line of hominid evolution and development there evolved one or more hominids with enough behavioral patterning complexity that the seeds of man's culture—many external extensions, particularly tools, utensils, clothing, weapons; and internal extensions, particularly speech—quickly emanated from the emergent new species changes. Described in (11) as the "critical point" theory, so named by Kroeber, it postulates that the development of the capacity for acquiring culture was a sudden, all-or-none quantum leap that occurred in the phylogeny of the primates. Whether man's emergent culture is quantized or gradual is, of course, still debatable.

> One of the lines of evidence for quantization, Tax points out, is that the onset of languages must also be all-or-none. As Howells (50) states, you cannot go to the Bushman and find him speaking a language more rudimentary than yours, whereas it is not possible to teach a primate as advanced as a chimp how to speak, even when raised with humans from birth.
>
> Yet, as Tax further points out, paleontological study "has been steadily piling up evidence that makes the drawing of a sharp line between man and nonman on an anatomical basis increasingly difficult." There is no "cerebral Rubicon" in size, for example.[1]
>
> Alternatively, Warren McCulloch[2] has pointed out to us a not too well-known line of thought that up through Neanderthal man there is no evidence of left-right-handedness in tools and that speech is likely associated with a dominance and cerebral dissymmetry of one side of the brain. None of the Paleolithic tools show handedness, whereas many of the Neolithic tools do—

---

[1] The temporal lobe in the cerebral cortex is largely involved in the associative functions needed for speech. With its motor speech area, its auditory speech area, and its visual speech area, the presence of complex cortical coordinations can serve as a "cerebral Rubicon."

[2] Personal communication.

95 percent are right handed (A. Kappers).[1] If true, then speech and such a level of abstraction might not have been possible before 10,000 to 20,000 B.C.; yet "culture" in many other senses may very well have existed before.[2] It is thus conceivable that "civilization," i.e., rule under learned and transmitted formal rules of conduct, is what came into existence with speech.

The causal elements are quite indeterminate. However, the changes seem to have been associated with natural selection of genetic components during the climatic changes of the past few million years (92).

Simpson [(46), Figs. 33–36] depicts something of the quality of the changing structure of animal phyla, originating perhaps 500,000,000 years ago, waxing and waning, with some dying out.[3] The vertebrates are then shown, in more detail, with mammals starting up about 200,000,000 years ago. Hawkes [(54), Chart 1] then depicts the flowering evolution of man from beginnings 25,000,000 years ago.

We can jump now to the more detailed story of man. Let us review the issue from more specialized sources [remembering that many of the facts of classification are in controversy; Pfeiffer (166) is an excellent survey].

Hominids (manlike creatures) likely began with australopithecines 1 to 2 million years ago. More conservatively estimated as $\frac{1}{2}$ to 1 million years ago, in controversy is Leakey's dating of Zinjanthropus back to 1,800,000 B.C. "The australopithecines... would have to be considered candidates for the honor of being man's ancestor." Their remains have been associated with primitive stone tools, pebble tools in which a few flakes are removed, and possibly with bone tools. They were both tool users and makers. "They were close to crossing 'the Rubicon' between animality and humanity" [Dobzhansky, (48)].

[1] Many present day tools show handedness.

[2] Such a thesis regarding the origins of speech is, of course, very speculative. However, it might be tackled by experts possessing a deep knowledge of brain structure and anatomy. More conventional judgments are illustrated in the following sources. Whatmough [(93), p. 160] states, "The date at which human speech began may have been a hundred thousand years ago, presumably among a group of already highly selected primates, whose genetic properties made this stupendous development possible and at once obtained thereby a tremendous advantage in social cohesion and material gain." Oakley [(94), p. 126] states, "The development of speech, which is generally regarded as one of the chief attributes of man the toolmaker, of course greatly accelerated cultural evolution by facilitating the communication of ideas. The brains of the earliest tool-making *Hominidae* were probably functionally advanced enough for speech, but nevertheless speech as we know it may have been a comparatively late cultural development—an invention. The earliest mode of expression of ideas was perhaps by gesticulation, mainly of mouth and hands, accompanied by cries and grunts to attract attention. There is no means of proving when speech became general, but the rapidity of cultural change in Europe during Upper Palaeolithic times would scarcely have been possible without the medium of language." One may note that the start-up dates suggested for speech are not that remote from our surmise.

[3] Current evidence is that life is much older, dating back perhaps to 3 billion years. See for example (47).

The second grouping of hominids may be referred to as the pithe-canthropines [Howells (50)]. They may be viewed as members of the genus *Homo*, often referred to as *Homo erectus*. (Examples are *P. erectus* and *Sinanthropus pekinensis*. There may be additional related species.) They may be viewed as the lowbrow wing of the advance toward mankind.[1] They seem to have appeared during the extensive glaciation of the Pleistocene epoch in its mid-period, perhaps 300,000 to 500,000 B.C. *Sinanthropus* was found with rough chopper and cutting tools. The contemporaneous Chellean tradition already had produced well-formed flaked stone tools and possibly three stone bola-type weapons. The extensive climatic changes, in the alteration of glacial and interglacial periods, likely represented powerful inputs that drove this group geographically and influenced the evolution of its cultural elements of toolmaking, clothing, food gathering, and possibly the use of fire. This is described, to some extent, by Hawkes (54).

A third group that appeared has been referred to as a group tending toward Homo sapiens, e.g., the "neanderthaloids close to Homo sapiens"— Hawkes. (Examples are the Ehringsdorf, Swanscombe, Saccopastore, Fontéchevade, and Steinheim findings.) They existed during the last Riss and Würm glaciation and interglacial period of 100,000 to 250,000 years ago, and they likely then disappeared. In the main they preceded, but were contemporaneous with in part, the Neanderthalers. They made use of stone implements, at least of a primitive kind.

The fourth group that appeared was the species *Homo neanderthalensis* (*Homo sapiens neanderthalensis*). It existed in the period 40,000 to 150,000 years ago and disappeared during the last glaciation. Neanderthal man lived chiefly in cave shelters and was well acquainted with the use of fire. He was a meat-eater and probably dressed animal skins for clothing. He made beautiful stone tools (of the Mousterian tradition)—flint balls, perforators, disks, scrapers, knives. He used pigments and probably practiced ceremonial burial. Doubtfully, he may have been absorbed in small part into modern populations, but more certainly he is regarded as an extinct fossil man.

The fifth group that appeared "suddenly" was Cro-Magnon man (*Homo sapiens sapiens*). He lived in the period 40,000 to 20,000 B.C. (In Europe,

---

[1] We cannot resist the quotation from a book review by A. L. Todd in the *New York Times* Book Review Section of Jan. 14, 1968 entitled, "One Summer in Dayton," of L. Sprague de Camp's *The Great Monkey Trial*. He states:

"Careful reading of this dissection of the great monkey trial can help some of us understand the thought processes and behavior patterns of the narrow-minded, no matter what their level of education. Time and again, when reading the words of Bryan and his following, I thought of the simple-minded drivel published today....

"If deCamp's book contains a contemporary message, it is that we have not yet won our rest from the long struggle against the low-domes. They have changed colors and battle cries, but they are still here, afraid to permit every man to read, think, speak, and act in freedom...."

one finds evidence that Cro-Magnon and Neanderthal men were contemporaneous for a time, the one "suddenly" appearing and the other
"suddenly" disappearing.[1]) Although culturally well-developed, possessing
both diversified stone tools and beautiful bone implements, he too is considered a fossil man. Cro-Magnon man may have represented a number of
possibly distinct varieties (possibly only different social groups). Such
esthetic elements (internal extensions) as burial ceremony and decoration and
human and animal modeling are found.

With imperfect knowledge of his precise beginnings, finally modern
Homo sapiens sapiens came into existence. His basic racial (i.e., ethnic stock)
distinctions already existed by 35,000 B.C.[2] His precise evolution and origin
are a matter of controversy. From his beginnings, he was a wanderer and
colonizer, as were many earlier men.

Some measures of his cultural achievements are suggested by the
following findings: (in external extensions) art objects were created perhaps
by 12,000 B.C.; the bow and arrow were in use in the period after 10,000 B.C.;
while there was still food-gathering, that is, hunting, fishing, picking and
digging of plants, at 10,000 B.C., plant domestication and breeding of stock
were well established in 7000 to 5000 B.C.; the use of polished-stone and
ground-edge tools was established around 10,000 B.C.; making of pottery
began around 7000 B.C.; and written language has been found in 3500 B.C.
This brings the story of man up to modern recorded history. However, we are
mainly concerned with beginnings.

First, in power aspect, it appears that the structure of most primate
societies depends on the dominant male elders with considerably more
significance to be attached to the closer and more continuous intimacy
between the females and young than in most other mammalian species
(Hawkes). [It is clear that the next decade will show an increased background
developed by ethologists. To cite an interesting comparative ethological
illustration, Eisenberg (95).]

Australopithecus[3] of 250,000 years ago lived in bands of some size,
likely with some kind of family life, possessing primitive tools, a hearth fire,
the ability to kill animals for food by hunting parties, and the ability to
gather berries (Hawkes). These hominids, certainly for all practical pur-

---

[1] The classification and identification of Neanderthals and their relationships to Homo sapiens
is sufficiently confused that any consensus of opinion is not possible today. The author is simply
offering a very nominal view of the groups of early men.

[2] "The latter half of the Upper Pleistocine beginning around 40,000 years ago witnessed developments of prime importance to the future history of man. It saw in the first place the final emergence
of modern man" (Clark). "Enough is known to enable us to see Late Pleistocene and post-
Pleistocene times, say from 40,000 to 8000 B.C., as the main formative age for the races of man"
(Hawkes).

[3] The australopithecine fossils are dated commonly in the range 1,000,000 to 250,000 B.C.

poses, have crossed the line to human social organization, if not also to human culture.

Pithecanthropus or his contemporaries with tools and fire have not yet furnished more information about the origin of culture.

Certainly the Mousterian tradition in toolmaking as well as the other known traditions of Neanderthal man indicate a considerable culture. See, for example, Binford (53).

Thus, in some crude sense, one may say that 250,000 years ago, whether at the end of the primitive line of australopithecines, the beginning Homo life of pithecanthropines, the pre-sapien neanderthaloids, or even more certainly somewhat later, by 50,000 years ago, genetically, species had emerged that were capable of performing with the content of human culture.

More classically, communication by the use of language has been considered important in distinguishing human and animal cultures. (According to Howells, Herskovitz defines language as a system of arbitrary vocal symbols by which members of a social group cooperate and interact.) Commonly, the belief is held that speech and language were quite ancient developments. However, it is not yet possible to directly trace language (in particular, written language) back further than 4000 B.C. today, nor are the technical considerations that some sources have thus far been giving to the problem [see, for example, Pei (96), Hall (97), or (11)] of sufficient scientific adequacy to cast any real light on the earlier problem. However, the next decades of ethology should reveal more about "language" structure in other species (11).[1] The language test of culture is not directly available to us at present, and we cannot be certain whether it is or is not salient.

The stress in culture, historically (since Tylor's definition in 1871), has been on the characteristics of behavior of human beings that are acquired from other human beings. Here again, it is ethology which will have to tell the fuller story in time.

Thus, it is clear that what is most often involved in all thinking about man, as compared to other animals, is the acquisition of "extensions" that obviously were not genetically coded for and thus had to be "acquired," i.e., that are epigenetic.

The stress is generally placed on the brain as the organizer of behavior. Yet, objectively, what we have to go on are the appearance of tools and their evolution from 1 to 2 million years ago, the appearance of fire use and weapon use from one-half to one-quarter million years ago, and the appearance of

---

[1] In humans, vocal sounds in language can be identified with the number of positions that mouth opening, tongue stop, and breath pump can produce in combination.

Although the teaching of a sign language to a chimp (88) opens a new line of inquiry into animal learning, it is still dubious that transmissibility and language capability have been demonstrated. Only more extensive research will clarify these issues.

art and ritual burial 100,000 to 50,000 years ago. Further, what we have to go on are severe climatic changes.

What emerged from these climatic changes of the past 1 million years was an extensive adaption of complex mammalian species. One can only surmise that such driving ecological cues provided "trying" times. Adaptive behavioral patterns that could fit the milieu found their places.

The problems to be solved were not "intellectual." In the case of the primates, it is fairer to assume that those that could adapt "tools," i.e., stones, into their life patterns and could remember and transmit such information were advantaged and selectively bred by their probing and changing ecology.

We prefer to think of the problem as neural-endocrinological rather than neural. [See, for example, Christian in (24) as an apt introduction, or Beach in (98).] As example of the pervasiveness of the point of view that we hold with, Ford and Beach, *Patterns of Sexual Behavior*, state succinctly, "primates are particularly susceptible to the effect of experience...." It is certainly true that we find it easier to demonstrate and understand the effect of experience in primates. Thus, it is not Homo sapiens' character that we are first concerned with but densification, environmental characteristics, emulation, retention, and communication.[1]

Various animal species form social aggregates. These, we believe, can more plausibly be traced to a neurological-endocrine origin than just a neurological. Primates emulate (other primates) extensively. (Other species do too, but it is quite easy to recognize in primates.) Offhand, this may be considered neurological. However, we propose another thesis—as part of the earlier exchange force concept. The emulation is used to fit as extensions of the internal rhythms, i.e., of the internal dramas.

It is the "syndromes" of anger and frustration, of "empathetic" interaction (e.g., the compelling repetition of you do what I do, and I do what you do, etc.), of a most extensive endocrine response patterning[2]—certainly with neurological concomitance—that lead to the selection of new paths and emulation that will bind the denser human populations together.

However, it may have taken the selective changes of intimate cave dwelling, exposure to the driving input of cold during some stages of hominid development, wandering and exploring and being forced to colonize, and other such alternations of environmental patterns to forge and select systems of emulation.

[1] An interesting reference for primate characteristics of emulation and diffusion is Kawamura (99).

[2] It has been common to regard the central nervous system as existing for behavior. Yet it has been difficult for neural researchers to implicate processes for much longer than a tenth of a second to a few seconds. We have hypothesized that the endocrine and neuroendocrine responses would be salient in longer-term behavior. It is only recently—as was mentioned, late 1969—that longer-term periodic processes are being discovered. Here once again we are trying to stretch the evidence and story out to a new frontier of thought about the brain [see for example (167)].

The ingredient of retention (which, to repeat again, may have been more neurochemical than neuroelectric) and finally, with its appearance, the extensive emulation possible by the ultimate ingredient of an abstract "verbal" communications system may very well have been the extra needed elements that put across the system of dependence upon extensions.

Possibly the mouth-tongue-vocal cord structure with many positions, the walking gait freeing two appendages, the opposed thumb and digital manipulability, the freer-running sex system, the extended brain size, a complex coordination center that made speech possible, and the availability of sufficiently lively "motor" systems provided a sufficiently developed endocrine system with the opportunity to manipulate the total organism into a more complexly patterned behavior.

One test of this thesis is the nature of human behavior. One can conceive that if behavior were determined neurally there could be quite rapid directed learning as our extensions are developed and codified. Yet this does not seem to be true.[1] (An extensive probing of this thesis would properly be the subject of a significant exploration of systems engineering. However, it is outside our present scope, except for a few passing comments.)

> *Australopithecus* and early hominids were carnivorous predators, possibly cannibals. [See (48–50, 52, 54–56). Monkeys and apes are vegetarians.] We remain nearly so.[2] We cannot plan our mutual living in rational fashion. Instead we depend on strong rhythmic alternation of peace, love, tranquility, accomplishment, and consideration with war, hate, anger, frustration, and unkindness to conduct our affairs. Beneath every man's surface, it is the logic of his internal and formed chemical rhythms that governs him, both at the split-second level, the life level, the level of the social era, the level of the whole age of civilization, and the level of man as an emerging species. Yet, verbally, he can produce most beautiful messages with his central nervous system.
>
> Timely discussion of man's aggressiveness may be found in Freeman's contribution to (100) and Holloway's to (101).
>
> These qualities of behavior are not being stressed for moral reasons. It is only a recognition, as a hypothesis, that the system is run by an abstract chemistry that even produces its "emotionality" and all other binding. Recognizing this, we propose the problem to man. Can he "govern" himself (in the specific physical technical sense meant in this study of what creates and maintains viable limit-cycle societies)? Can he understand the sources of his instability and begin to manipulate his environment into a pattern that suits his kind? This is the obvious purported aim of all humanistic societies, including our own.
>
> Here, as part of systems science, we are concerned with the question of whether man—the "sapient" civilized species—forms an active persistent

[1] From the neural system, one visualizes a simple logic emerging something like the computational logic of switch networks. Within a limited number of trials, one might expect rapid convergence to a unique conclusion. Instead, one seems to find complex patterns of behavior, as if the coloration of many slowly etching chemical streams had patterned the nervous system's response.

[2] Not being able to distinguish eating your fellow man from eating with your fellow man is like the schizophrenic lack of distinction between symbols and their meaning. Gilbert and Sullivan's travesty on the inability to distinguish pilots and pirates beautifully caricatures this.

ethological atom as part of the biological continuum. If neurally run, one might expect a gain in logical organization of behavior from experience and transmission of that experience for an increased gain in logical organization, until some higher disorganizing noise level due to random inputs is reached. If endocrine-run, satisfying more nearly internal biochemical algorithms, the "logical" organization of behavior, i.e., its concordance with external affairs, might be low. Technically, the latter might seem to be the case. It is clear that man's extensions—both internal and external—develop in technical complexity. Our engineering structures and our theatrical dramas testify to this. Yet we have no guarantee that the quality of our civilization has changed in 5,000 years if viewed from the respective stations of "have" and "have not" individuals, groups, or nations. What is of concern to man, as a species if not to individual men, is whether he has any manipulative ability to control his long-term destiny.

Toward this end, we must see that our ethological-anthropological-ecological-economic entity, the "city" or "city-state," forms an ecological complex. Human, virus, bacterium, insect, rat, bird, domestic animal (cat, dog, pig, sheep, cow), and higher plant (berry, nut, grass, cereal, bush, tree) are thrown into intimate association. A social patterning of behavior emerges. It is the directed problem that we see for man to make it emerge in a way that is "satisfying" for all.

However, we must recognize the internal forces that shape men, the flux of internal juices and their patterning that make men irrational, mad, frustrated, irritated, etc. If they cannot regulate or train for these, they remain the same beasts from which they claim to have developed.

We can now briefly view civilization (as it is known today) before the written word of 4000 B.C. and thus before we have much detail about its elite structure. (We will skip the older traces of culture defined by Neanderthal man, and start with the colonies known to be of modern man, *Homo sapiens sapiens.*)

CHILDE (56)

Perhaps 10,000 to 8000 B.C., some societies appeared in the Near East that cultivated plants and bred domestic animals and thus had a food-producing economy. This differed from the savage food-gathering, beast-of-prey existence that had prevailed earlier, albeit with tools and weapons and even with "houses," trade, rituals, and art. By 3000 B.C., in the rich river valleys of the Nile, Tigris-Euphrates, and the Indus, some riverside villages were transformed into cities. Farmers produced a surplus of cultivated foodstuffs and supported a new city population of specialized craftsmen, merchants, priests, officials, and clerks. Writing was a byproduct.

These New Stone Age barbarians of the Near East[1] were possibly aided by the "spontaneous" growth of wheat and barley ancestors and the existence of wild sheep, goats, cattle, and pigs. Ecologically, hemmed in by the desert and mountains, surrounded by bodies of water (Caspian Sea, Black Sea,

---

[1] A chronology of developments in the Near East may be found in Mellaart (58). Early European settlements are illustrated by Piggott (59). The complete period 10,000 B.C. to 3000 B.C., of course, covers a transition from the Neolithic to the Early Bronze Age.

Mediterranean Sea, and the Persian Gulf), and confined to water areas, a cultivation and domestication cycle might have been formed. Extensions of utensils (pottery), housing, and even more specialized tools came into existence.

All of this was possible, perhaps with or without a well-marked, specialized elite structure, although with a division of labor. It simply may have fitted the primate structure that was man's, that is, beyond the family structure of apes. The "elite" elements, the focal centers, may have been precipitated by climatic-ecological factors, although not immediately; or they may have stemmed from the behavioral elements of man—his neural-endocrine structure. In any case, we must stress that conditions of biological instability may have come into existence at that time. A dilute continuumlike ecology could not persist. It began to create a spatial vortex. Man began to "mill" about. The spatial-geographic distribution at this "start" is quite striking [see (52), p. 125]. Limit cycles and patterns of behavior began to emerge. The constellations of civilization were forming.

For those who might consider the process of pattern formation mystical, we can offer the following hypothetical picture: Animals stabilize endocrinologically by day-night cues, by yearly cues, and by driven cues. [An older description may be found in Beach (98). Current work is widely fragmented. For example, the circadian researchers attempt to make a strong case for the near-daily system. We believe that a much broader endocrine base exists, which, however, includes the circadian cycle, and we are attempting to create such a foundation. See, for example, (76) and Richter in (87).] If driven by weather extremes, a very drastic selection process takes place. There is selection for food-gathering, for mobility, for shelter seeking. This problem faced all species every day, every year, and a great number of times during the few million years of the Pleistocene era. Driven to roam, to adapt to changing climatic conditions and to entrapping enclaves, various kinds of species, not surprisingly, achieved dominance in different areas. It is not impossible that selections for food-gathering and its extensional augmentation (by tools and weapons), for covering and its extensional augmentation (by mobility, by skin covering), and for communications were forced from these changes.

An example of an early known settlement is Jericho I in the Jordan Valley, dated about 7000 B.C. The inhabitants of this village—perhaps 8 acres—hunted and collected, grew crops, watered them by use of a spring, and grazed sheep and goats on irrigated meadowland. They used no ground-stone axes (the work of New Stone Age men) and did not fire pottery. [A much more detailed and more recent breakdown of the early settlements and their use of pottery and tools is given in Braidwood (52) and Mellaart (58). For example, (58) indicates that the earliest Natufian culture arrived in Palestine and Jordan in about 10,000 B.C. Braidwood (52) suggests 8000 B.C.]

At Jarmo, in Kurdistan, about 5000 B.C., farmers cultivated cereals that clearly originated as wild grasses of the region and bred cows, sheep, and goats. They used axes of ground stone and made unbaked clay figures. They had no pottery but used stone and perhaps even wooden vessels, even as at Jericho.

Other settlements, of later vintage, were found associated with perennial springs for domestic and wild animals and with irrigation. They also exhibited: the practice of spinning and weaving; stone and pottery ware; fishing; linen clothing; huts in regular rows with streets; fertilization from the ashes of burnt scrub brushwood and from manure; defense against human foes; methods for cultivating the soil; storage of grain; baking; brewing ("By 3000 B.C., ... intoxicants had become necessities to most societies in Europe and in Asia. ..."). Some settlements were intermittently nomadic. The New Stone Age men lived in small communities, perhaps 1 to 10 acres, perhaps a handful of households and up, with 25 to 35 households not uncommon.

These spatial aggregates formed social organisms in which members cooperated for collective tasks. Their "public works" were all collectively built and communal. A division of labor of male-female existed, likely with no other division such as industrial specialization. Each Neolithic household would prepare its own needs. Each village, as territorial community, was probably self-sufficient. This differentiates Neolithic barbarianism from civilization and the higher barbarisms of the metal ages. It is likely "that a Neolithic economy offers no material inducement to the peasant to produce more than he needs to support himself and his family and provide for the next harvest."

The Epic of Man (89) aptly depicts the quality of Neolithic man in these formative stages for the nonexpert.

HOWELLS (50)

There was a spread of these Neolithic farming villages from their center in the Near East to Europe, reaching South Germany before 4000 B.C., and to Africa.

MONTAGU (51)

The Natufians (Mount Carmel caves) of 7000 to 5000 B.C. present early evidence of plant domestication and agricultural implements (sickles). They had no domestic animals other than the hunting dog and no pottery, though they made stone vessels. [Also see (58) and (52).]

By about 4000 B.C., there were well-developed Neolithic communities in lower and upper Egypt, such as the Tasians, the Badarians, the Merimdeans, and the Faiyumis. The first evidence of a loom, for example, is dated as 4400 B.C. at Badari. The Badarians learned to hammer copper, though not to melt or smelt it. They were followed by the Amratians, whose appearance marks predynastic Egypt in about 4000 B.C. They made use of alphabetlike signs, but not as writing. The Gerzeans were successors to the Badarians. They used metals and cultivated olives. They used the rudiments of a script based on older Stone Age hunting signs. They played a game like checkers. For the first time, a king made his appearance. They may also have invented the calendar. Thence on, there is a steady development in civilization. The overgrown villages became towns with officials, copper and metals were used, the potter's wheel was invented, and writing history began.

There is no evidence of the manufacture of weapons as instruments of offense or defense against other men until the Neolithic. When some groups began to cultivate herds of livestock, raiding by other groups followed. In order to protect property against marauders, defenses were erected, and new and more weapons manufactured. Sometimes a marauder was captured, leading to the discovery that instead of killing the enemy one could enslave him. Thus was slavery born, in the Neolithic of what is now the Middle East.

Thus our previous dialectic with the Marxian view was valid. One must realize that the issues being argued are speculative. Nevertheless, they are properly focused.

It appears reasonable that the elements of culture began earlier than man's development of a fixed civilization. It appears that it is the production of extensions of man—both internal and external—that arose out of the necessity posed by driving climatic changes and by the appearance of an "imaginative" brain, i.e., one that could produce many new internal memorizable and communicable configurations. It is the highly, detailedly manipulatable motor systems and the highly unstable nervous and endocrine systems which produce enough patterned "coloration" to permit the creation of extensions to burgeon. For a million years, this was in process of genetic selection and epigenetic adaption. Certainly, the selection pressure was toward furbearing animals, such as the mammals, and burrowing animals. Practically all of these animals found adaptive solutions. One species, man, found solutions by the use of extensions. Toolmaking, clothing making, hunting, artifacts, and speech communications were the main results of this rich nervous-endocrine system. It is conceivable that speech communications may very well have been the last and most recent acquisition.

"Suddenly" a new dimension emerged—the cultivation of the land. Toolmaking and hunting had long since developed to a rather sophisticated level. It is possible that the combination of tools and newly appearing grasses after the last ice age, 12,000 to 15,000 years ago, and the fertile location led more easily to cultivation of crops. One must remember that the populations were very sparse. Cultivation tended to fix man's locus and, more significantly, gave him a productivity measure more near unity. A man could more nearly produce his own needs without a wide grazing and hunting range. See, for example, Schaller (19) for a picture of the field life of another primate; Hallet (102) for a somewhat lurid but interesting quick view of recent colonial Africa; Junod (64) for an older classic piece of anthropological study; or see a picture of a barbaric culture as reconstructed from its "last man," Kroeber's *Ishi* (103).

This Neolithic culture, starting perhaps 10,000 B.C., represented a viable species capable of living with tools for hunting, fishing, grazing, and cultivating, with arts and artifacts, with clothing and shelter, and with magic and ritual. Its extent may be noted in (58) and (52). This quite plausibly represented a stable (or quasi-stable) animal culture, likely as advanced or more so than any other animal culture (e.g., bee, ant, beaver). The great advance that had already taken place— over 1 to 2 million years—was the use of a complex of adaptive extensions. The older life was essentially a communal life.

The pace had already quickened, for after only 10,000 years or at most 20,000 years of communal living, cultivation was "mastered"; namely, by about 3000 B.C. "The high yield of about 30 bushels of cereal to $\frac{4}{5}$ths of a bushel sown revolutionized man's diet." At this point, a large urban society[1] could be supported. There were surplus goods. The stuff of life could be produced in less than full time so that leisure came into existence. At this point, modern economics was born.

However, note all of the correlated consequences of this step. The imagination of the settlers was freed. They built extensions more lavish than their immediate needs. They had to insure the continuity of such surplus. Thus magic and invocation. They had to keep track of such surplus. Thus records and other abstractions such as mathematics. A proliferation of the division of labor was required for the complex of storage and distribution.

The abstraction of value had to arise, and a more complex community of effort came into existence. However, it was the imagination of the nomads outside of the settlement that was also excited. Thus, attack, defense, weaponry, and slavery came into existence. Out of the complex of all of these achievements modern civilization was born!

However, as many primitive cultures have shown, it was not necessary that the path toward civilization be uniquely selected. Thus detailed explanations are required why these Neolithic cultures in the Near East were sufficiently unstable that they had to coalesce toward both civilization and urbanization.[2]

One should note that primitive man's culture was essentially stable enough to be ergodic (namely, it is often considered that studies of prehistoric primitive culture and more current primitive cultures will reveal about the same range of characteristics). Further, the characteristics of such primitive sparse populations are like those of many other ecological interacting populations; that is, if they are in "equilibrium" with other species in prey-predator relation, they tend to cycle up and down in limit-cycle fashion. Such population oscillations are shown by Kerner (72). They tend to involve the time domain covering the daily period, the yearly seasonal period, and the period of maturation from birth to growth.

This does not account for the walled cities and urban structures depicted in ancient Tepe Gawra and Ninevah in northern Mesopotamia or the Babylonian and Sumerian cities in southern Mesopotamia. [See (90), pp. 24–31 or (89) or (58).] Discussing the coming of civilization, (89) states, "Man's creative fever first struck in the area loosely known as the Middle East. . . . A formative period, from 4500 to 2900 B.C., witnessed the great transition from simpler societies to civilization. Then came the full flowering of Sumer's civilization during the Early Dynastic period, down to 2400 B.C. . . . At the beginning of the Early Dynastic period, Sumer was a land of separate, self-sufficient and politically autonomous city-states. . . . At each site a walled and fortified city, dominated by one or more monumental temples, rose high above the flat plain. . . . Since the welfare of the

---

[1] In our view, civil society would begin with the act of living together by abstract rules of conduct (requiring codification by language)—whereas an urban society would follow on the basis of additional fixing of the living site of an appreciably sized community. A large urban society requires considerable mastery of ecological factors.

[2] Here we have an example of the start-up problem in systems theory. Clearly, the elements are unstable in the environment, but the path that may be taken toward coalescence is not clear. In this present volume, we do not pretend to know the answer to start-up questions.

Sumerian city depended utterly on the good will of the gods, the temples played a vital role in city life. Over and about their religious function . . . , they also served as focal centers for the city's social and economic activity."

It is possible, by a thought that might be described as the "crossroads" concept, that it was those Neolithic settlements that were in exposed positions at crossroads which had to be traversed, say for water, that led to the condensation toward a compact protective community. This is pure conjecture. This and the large gain in food productivity may have been the causes.

It was not solely the end of the glacial period that led to the explosion of urban settlements. As described in (52), men lived for quite a few thousand years on the hilly flanks of rain-watered grassland in the nuclear Near East before cultivation and farming villages began. It was a few thousand years more until towns began.

However, one may reexamine Calhoun's section on "The Social Use of Space" in (24) or Freeman's section on "Human Aggression in Anthropological Perspective" in (100) to see how flimsy are the scientific bases we now have to work with.

Montagu continues, pointing out that occupational differences come into being with human society, the division of labor between sexes being the first. The Neolithic revolution was probably the creation of women. Raiding activities in the late Neolithic were certainly the work of men.

In every nonliterate society, there is specialization. There may be medicine men, priests, council elders, skilled artists, stoneworkers, woodworkers, metalworkers, educators, engineers, and so forth. As society advances, there is occupational differentiation and institutionalization. However, industry had its beginnings and came into being with human culture and the early flint tools.

All this had been developed by 3000 B.C. in Mesopotamia, Egypt, and the Indus Valley. Then in the Fertile Crescent, in the alluvial plains between the Tigris and Euphrates, "the great second revolution in the history of humanity took place, the Urban Revolution, the growth of the city and of the city-state." Here both urbanization and dehumanization of man began with the control of human beings as commodities as well as the growth of the good and just side of civilization. "It is not the justness of a society's weights and measures which is the proper mark of its civilization, but its kindness." The problem that has always faced men in community is to live lives of sentiment.

In the line of studies we have been engaged in [(79, 104, 73, 12, 76) represents or references the work] and in the effort to "sell" our research studies—which we view as the directed activity of science for society—we have begun to appreciate and marvel at the large irrational component in man. We have arrived at one catchphrase. "There exists no persuasion of man by man, only demagogic reinforcement. You can only convince people who have no opinion." As scientists, we have wondered as to its basis. It is only recently, in collaboration with endocrinology colleagues, that we believe we are approaching a most useful second round of hypothesis [see (167)].

In the complex mammal, the guiding algorithm is perhaps dominantly a shaping of neuroendocrines that play—through the sympathetic system—on the endocrine spectrum to create patterns of behavior, particularly long-term ones.

Consider the latter thought. It seems clear to us that the internal drives, the guide algorithms—since it dawned on us (79) that man does not operate through rationality[1]—are drives that shape the endocrine responses that then manipulate the motor systems. It was quite "clever" of man to discover extensions. However, as expressed in (12), there is no external metric to man's behavior. Physical external extension, internal dramatic extension, and the reality are mixed. A person, a thing, an idea, a hate, a pain, etc., all become mixed and equatable, under some circumstance.

It cannot be "rational" communications processes operating internally that can be laid out by "strict" logic. It must be blind biochemical, including of course neuroendocrine, processes.

Thus each new development in man's culture, by epigenetic unfolding, results in a shift in character of the community of living individuals. The same is really true for all species. However, in humans, with the extensive external and internal dramatizations possible, the changes loom so much larger. It is the new orbits, the patterns, the foci, the emergence of so many new elites that mark this "second revolution" in man's behavioral patterns.

To summarize what we have gleaned from such sketchy reading in prehistory, we now believe that as soon as man became formally civilized (i.e., as soon as he had to regularize into detailed civil patterns all of the consequences of a hard-won high-yield–long-storage food source produced by cultivation) he enslaved his fellow man and began to treat him as goods (i.e., as soon as he began to "own" things, he conceived that he could own men). We believe that this point in history is the beginning of the social quantization that represents man's modern culture. The reason for our long excursion was to reach such a point which would mark the end of the start-up period for man's current cultural life system. (It is not always possible to discuss or even localize the conditions under which a system becomes unstable. We do not propose that our discussion is correct, only plausible. It would take much greater scholarship than we possess, particularly in this area, to establish the exact thesis.)

Its major characteristic can be described as follows: If an animal works on a short "daily" time constant (say, animal A kills animal B, immediately animal C takes the kill away), an internal pattern of neural-endocrine systems' behavior emerges. The same can be said for animals gathering and storing food for a seasonal period during the year. However, if high food

---

[1] We make no especial claim for having discovered that man acts irrationally. However, in (79) we undertook the task of pointing out that planning and planning documents are most often based on the assumption that men act rationally, whereas this seemed to us contrary to the facts. In our particular case, the conclusion was derived from observing managements for two decades.

yield is possible by considerable effort with longer storage, then it would seem that a greater neural-endocrine behavioral complex likely has to emerge. This coupled with an extensive memory (characteristic of mammalian species, not only man); and the complex competition of "two brains"—the older limbic system, the so-called "visceral" or "smell" brain; and the neocortical system, with its complex visual capacity and integrative centers that permit coordinated speech (i.e., with a complex communications system) —possibly permits a much greater internal complex pattern to emerge from the endocrine system in long selection and development time.

It is our thesis that the complex biological system is unstable in the following way. If any local organ system is "heated up," as by temperature or by other chemical demands, it gets the enriched service of the blood flow and other humoral agents with the assistance of the nervous system. (The system is not free of dissipation of signal. However, the dissipative components in the cycle are not too clear to us yet. Generally they are regarded as nervous inhibition of one part of the communication net by a neighboring part.) The net effect is that problems or concepts that can be held on for long time and can be involved in excitation of an extensive chemical flooding (by "reverberation" as it were, although the current trend is to believe, perhaps dubiously, that the storage may be by holography)[1] create large-scale storms. They "mark" the system structurally (function is transformed, in time, into form). It is these poorly damped internal storms that make up the extensive endocrine and biochemical coloring in human behavior. However, much of the reason for this evolution is the effect of density upon the closely packed internal cellular colonies. A viable guide algorithm must be that the internal system which can call out its need most vociferously gets the service.

We are now proposing that the social continuum of mankind also evolved along this path. Much of the social instability that arose to form discrete societies may very well have arisen from a density effect, a consequence of enforced "huddling" for food supply.

The groups involved may simply have been larger in Neolithic times than in Paleolithic times. For example, in (100), Lorenz comments, "Animals can be made to behave like men and massacre the fellow members of their own species. If one crowds a dozen roe deer into a pen in a zoo, the most gory massacre is the result." The point is not that this is characteristic of all species but that some species do not huddle; instead they show antagonistic behavior. The social form depends on the nature of the interpersonal binding systems in different animals. However, it is not "rational," i.e., the product of higher cortical processes of a computational electrical nature. It is a chemical

---

[1] Still more recently the formal identification of oscillator patterns in the nervous system associated with behavior and memory has begun (i.e., the earlier references to Pribram and John).

"visceral" interaction with the nervous system.[1] Thus, its objects—things, men, ideas—can be confused. Now we project this scheme to the longer time constant that emerges for "civilization," i.e., for such numbers as up to 500 years, that spans a considerable number of generations. It is the interaction within and between generations which provides an atomistic background to the "ecology" and "ethology" which concerns mankind.[2] One must note that with the Neolithic revolution we find enduring records and enduring monuments. Elites of that time held and planned such ideas for more than a decade. Life could be made to revolve around these monuments. They "captured" human "imagination." Their building could extend over many generations.

All of these changes, we conjecture, may have come into existence as the result of an excess of produce obtained by considerable effort. After perhaps 5,000 years of cultivation, the critical parameter to produce the new orbital patterns was the jump in food productivity. Thus, it is not only the human that may be capable of this kind of social revolution. One can conceive that a selective process may make it possible for some other species to follow a similar path or at least "mimic" such social complexity. (e.g., whereas one of Aesop's fables pointed out that "clothes make the man," in this case we might say "rationalizing food makes man civil.")

12-4   AN INTRODUCTION TO SOCIOECONOMIC THEORY

Having provided a sketchy outline and some source material on prehistory, we begin the dialectics of socioeconomics, lightly, with one master of economic thought. We have chosen a source to seek to illuminate the question of differentiation of function, the division of labor in society. Durkheim, for example (105), starts with a discussion about corporations as producing groups.[3] He points out that "they appear as soon as there are trades, which

[1] We are not literally asserting its visceral origin in the individual. The play more likely occurs at the pituitary-hypothalamic axis. However, it involves the visceral systems.

[2] It is not the case for all species and in all circumstances that the time constant of 500 years is relevant. The biologist does experiments involving generations of bacteria in a day. The period is somewhat longer for hydra, and fruit flies, and rats. Sometimes the social-ecological milieu can present such accentuated stresses that even rats and humans will change their character over fewer generations than might be commonly expected. Nevertheless, 500 years is an approximate measure of the time constant for major human societies.

[3] As Braidwood (52) discusses and (89) illustrates, an ensemble of tools that can be identified as being used together in a cultural group are referred to archeologically as an "industry." They are evidence of the use of extensions in an organized fashion for "economic" purposes. Although fairly developed industries can even be traced back to 100,000 B.C., their complexity in the pre-agricultural age of Middle Stone Age man of 8000 B.C. [see, for example, (89) pp. 48, 51, and 53] was certainly considerable. Agriculture then fixed and urbanized the population. This represented the period 8000–4000 B.C. "Corporations" were feasible, in Durkheim's sense, after the gain in productivity of the land, namely, after 4000–3000 B.C.

means as soon as industry ceases being purely agricultural," that they were unknown in Greece because trades, being looked down upon, were carried on mainly by strangers and thus not part of the legal organization, and that they are dated conventionally from the earliest times of the Republic. He says further, "If from the origin of the city up to the zenith of the Empire, from the dawn of Christian societies up to modern times, they have been necessary, it is because they answer durable and profound needs." He points to the closely connected parallel between the family and corporation and states that the corporation has been heir to the family. He goes on:

> As long as industry is exclusively agricultural, it has, in the family and in the village, which is itself only a sort of great family, its immediate organ, and it needs no other. ... Economic activity, having no consequences outside the family is sufficiently regulated by the family, and the family itself thus serves as occupational group. [The] case is no longer the same once trades exist ... customers are necessary [Division of effort, specialization, storage, all abstract transformational concepts come into existence]. ... [G]oing outside the house ... is necessary. ... [R]elations with competitors, fighting, ... coming to an understanding [are necessary.] ... [I]n addition, trades demand cities. ... [C]ities have always been formed ... from the ranks of immigrants ... who have left their native homes. ... A new form of activity was thus constituted ... burst from the old familial form. In order not to remain ... unorganized [it was] necessary to create a new form which would be fitting to it. ... This is the origin of the corporation. ... [J]ust as the family has elaborated domestic ethics and law, the corporation is now the source of occupational ethics and law.

He points out, however, that under the Roman emperors, conditions of obligatory employment specified by the state degenerated into an intolerable servitude which ruined the corporate institution. (One might also say that under the Nazis, the image of the corporate institution again became tarnished.) The rebirth in the eleventh and twelfth centuries of the corporation clearly united people of the same occupation by strong interpersonal bonds. Reciprocal obligations and rights came into existence between master and workman, personally and regarding the quality of the work. As within all political societies, groups sharing common interests will associate, bind, subordinate particular interests to a general interest, and thus enter into moral activity.

> Society is not alone in its interest in the formation of special groups to regulate their own activity; developing within them what otherwise would become anarchic; but the individual, on his part, finds joy in it, for anarchy is painful to him. He also suffers from pain and disorder produced whenever inter-individual relations are not

submitted to some regulatory influence. It is not good for man to live with the threat of war. . . . [T]his sensation of general hostility [and] the tension it necessitates . . . are difficult states when they are chronic. If we love war, we also love the joys of peace. . . . Common life is attractive as well as coercive. . . . That is when individuals who are found to have common interests associate, it is not only to defend these interests, it is to associate, . . . to have the pleasure of communing . . . which is to say, finally, to lead the same moral life together. . . . Domestic morality is not otherwise formed. . . .

Having briefly discussed the corporation in the Middle Ages, Durkheim proceeds to a more modern state. "The organization of trades and of commerce seems . . . to have been the primitive organization of the European bourgeoisie," that is, essentially of the inhabitants of cities, he concludes. The commune, an aggregate of trades bodies (in which, he says, the trades bodies were communes on a small scale for the very reason that they had been the model on which the communal institution developed), was the cornerstone of our societies. As a combination of corporations, it has served as foundation for all the political systems which issued from the communal movement.

Solidarity was without inconvenience as long as the trades had a communal character. "While, as originally, merchants and workers had only the inhabitants of the city or its immediate environs for customers, which means as long as the market was principally local, the bodies of trades, with their municipal organization, answered all needs. But it was no longer the same once great industry was born. As it had nothing especially urban about it, it could not adapt itself to a system which had not been made for it."

Before continuing, it is worthwhile to make a number of comments. We purposely made the jump from the slow-paced anthropological background, attempting to let history rush to its climax of the beginnings of recorded civilization, then quickly to an early master of the twentieth century to emphasize that we were dealing with the same problem in both cases. This problem is the appearance and stabilization of man's social structure. It is interesting to find concurrence for the belief in the fundamental nature of interpersonal relations as determining the basis for internal institutions in "political" societies. It is interesting to find out how the structure of social institutions arises as the humans wrap themselves up into local space-time patterns that fit their growing local problems. It is interesting to speculate that it is, practically, stability questions associated with population density itself which are the source.

In archaic societies, the small group must have been sufficiently stable to get the young grown up and to provide sufficient contact for mating and reproduction to support a population. Such sparse densities must have been good enough for the first $\frac{1}{2}$ to 2 million years of man's life. Man's social diluteness and his domination by climatic conditions such as the ice ages is indicated even by the current belief, which may modify in time, that man did not come to the Western Hemisphere until 12,000 to 40,000 years ago. It is only recently (June, 1967) that evidence for

35,000 to 40,000 years ago is beginning to have some solidity as an estimate for man's presence in the Americas. Even in Asia, Europe, and Africa, the remnants of ancient man are still rarely discovered.[1] Thus, one may infer that the possibility for self-propagative increase in density was quite limited in the past few hundred thousand years.

With the end of the last ice ages, man finally was in a position to make his play. (Whether it was the end of the ice ages or the beginning of an ability to talk is still moot.)[2] Agricultural clustering appears likely possibly during the past 15,000 years in the transition from Paleolithic to Neolithic and, more certainly with some evidence for man's selection process at work, within the past 10,000 years. The family group, the small settlement, local supply, division of labor, and morality—as a regularization of local modes of conduct—had all likely come into existence as well as the more nomadic wanderings of earlier years.

The thesis which we can begin to see, in common probably with many others, is that when the density begins to approach various critical levels certain kinds of specialized properties and elite characteristics arise. Individual members of the species who develop a slightly specialized skill are more immediately copied. Why? Durkheim suggests the answer. To copy from a stranger has no significant content. It is the act of communing, the concomitant of something associated with interpersonal relationship that has meaning. Interpersonal relationships are fostered by denser crowding, by "togetherness." The growth of such density begins to develop a "crystallite" structure; that is, a limited range over which a certain regularity can exist. (Having found that a family unit can persist, it is not possible to fill the earth space by an indefinite geographic extension of a regular array of family units filling all the space. Disordering relations arise. The system becomes unstable. Instead a finite unit—a community, a "crystallite" structure—arises.) To illustrate, a linearly extended population distributed along a river can void freely into the river on the basis that the stream purifies itself. However, as the linear density increases, one arrives finally at a density such that every so often something has to be done. The waste must be collected and a disposal plant—per some larger unit population—must be supplied. The same is true with regard to food and other things. Remember man has many needs and hungers. Yet these must all be provided for even though density impedes easy process exchange. In physics, the idea is encompassed in suggesting a sphere of influence, a cell, or a capture cross-section, a variety of ideas that suggest that different phenomena have different ranges. This extends to the other "hungers." There are densities for parks, movies, brothels, competitive sports activities, even research activities, public dances, places for young people to get together, supermarkets, and living clusters. These change with density and with the concentration of time available to the members of the species for other than the more primitive physiological hungers.

An essential factor seems to be how closed the local society is and whether at the existing density the ecology permits adequate interpersonal relations. Crowding, mild hunger, and dirt are not fatal if various "dis-ease" states do not arise. On the other hand, in open societies, although in the short run there is no or negligible transfer, in the long run there is sufficient density that the transfer

---

[1] An interesting example of Asian prehistory is coming to light in the Niah River caves of the Malaysian state of Sarawak.

[2] Whereas conventional wisdom, which may likely be right, would date man's ability to some much more remote past, we have put forth the thought that speech might only date back to Neolithic, or not much further back than the earliest Mesolithic phase.

creates new appetites. The problem, as we can agree with Durkheim, is how to create new patterned institutions that fit the needs to "ease" the system from its status of "dis-ease." The general method is to pattern them like more primitive institutions, like the focal centers, like the family, etc. The persons who do the patterning are the elites. They grasp the problem. They see a focal center. They drag the system of people around by active "transactions" in the "marketplace" (12).

Conversely, all this newly created internal momentum heats up the system. Its "temperature" rises. If you raise the social temperature of particular densities of population, perhaps it can be cooled off by expansion to lower density.

One may think about this for relevance to our urban problems. One type of problem worth considering is this: How de we retain high levels of productivity in a locale without extensive international interaction; without requiring exceptionally high speed, short reaction time, transportation and communication; but provided under conditions of somewhat relieved density.

What the past has shown, Durkheim concludes in his preface, is that the framework of the occupational group must have relation with the framework of economic life. The corporation, in order not to die, has had to become as wide as the market, both national and international. Corporations of the future will require a greater complexity of attributes by reason of increased growth. There is every reason to suppose that they will become essential bases of political organization.[1] They must fix themselves in proportion to the development of economic life.

One must avoid the trap associated with ideas of a number of our great revolutionary innovators of the past 100 years—in particular Darwin, Marx, Freud, Einstein—of trying to extend them beyond the contexts for which their constructs are useful. The nature of the universe was defined by Einstein. The character of man as a biological species was defined by Darwin. The nature of interpersonal relations was defined by Freud. Man in society was most nearly defined by Marx. Nevertheless, attempting a more detailed science from their rudimentary theses for "clinical" purposes requires much more illumination than dogma. In examining human social structure, Durkheim here has put forth the corporate structure as a next one beyond the commune or storekeeper community. This certainly seems historically valid. As a structure that determines business "morality," the corporation has certainly taken over that role. However, one should not draw the conclusion that Durkheim is leading to the fascist concept of a state capitalism in which the corporation governs all state and political morality.[2]

In his introduction, Durkheim points out that Adam Smith was the first to attempt a theory of the division of labor. Although it seems obvious that modern industry is moving toward concentration of power—both

[1] Although his writing may be viewed as heretical by many academic economists, the same theme is ably treated in Galbraith's new book, *The New Industrial State* (107).

[2] A more extended exposition of the nature of the modern corporation is under development by A. A. Berle. See five of his books, *The Modern Corporation and Private Property*, *The 20th Century Capitalistic Revolution*, *Power Without Property*, *The American Economic Republic*, and the latest, *Power*.

mechanical and monetary—with infinite specialization, Adam Smith and John Stuart Mill still hoped that agriculture might remain a small-scale industry. However, even this is no longer possible.

Economists see division of effort as the supreme rule of human societies and the condition of their progress. This, however, applies as well to biological systems. Although it suggests two moral directions for the development of man —one, to be complete and sufficient unto himself; two, to be only a part of the whole—the speculation will not do. Instead, the functions of the division of labor are to be examined, i.e., the social needs it satisfies (Book 1); its causes will be sought (Book 2); and its pathological forms will be examined (Book 3).

In examining the function of the division of labor, he concludes, as a new light on the thesis, that the division of labor does not operate for the economic service it renders but in the "moral" effect that it produces, its true function being to create in two or more people a feeling of "solidarity," to cause coherence among friends. He then uses the sexual division to illustrate the source of conjugal solidarity.[1]

> It is hard to avoid identifying the community between this thought and our own expressed ideas. "Economics" deals with all the human hungers. What binds a society together are interpersonal forces; in Durkheim's terms, what creates a division and specialization of effort and labor. We are parallel in thought, not coincident. We would agree with him that division of labor is common in biological problems and that it likely arises from the internal breakdown of the biological system into quantized chains and organs of more than one kind. Internal wants thus exist both in parallel and in series. It is the oscillator systems, common to the species, each shared by communications links, which then create the interpersonal forces. Thus, we have cause and effect interchanged; i.e., it is both the actual common cause and the feeling of community that leads to a division of labor.
>
> We find it interesting to note the pervasiveness of the potential concept of "exchange forces." In physics, exchange forces are viewed as arising from a process peculiar to quantum mechanics with no classical analogue. After such potential "forces" as gravitational, coulomb electrostatic, direct mechanical change of momentum (collision), and polarization magnetic have been considered, exchange forces are viewed, in descriptive form, as "forces" that arise because indistinguishable particles "exchange" position in a probabilistic sense between two systems (of atoms). It is such rapid positional exchanges that are said to give rise to this nonclassical force.
>
> We will continue to stress that such "exchanges" represent a binding "force" in all kinds of phenomena and, in fact, have a "classical" explanation. The thought has been discussed for two types of phenomena—transport coefficients in molecular physics and interpersonal bonding between people. We note now that it begins to extend beyond pairs of individuals—closest neighbor forces —to the larger colony of social organization.

[1] We would not deny, for those who would oppose this thesis, that in the division of labor there are feelings of both competition and empathy. Such polar feelings illuminate the specific complex content of man's internal dramas. The forces tend to be more multipolar than homopolar.

Less technically, people may just enjoy doing things together, in community, because it will serve a common purpose. When they work together to get a difficult and dirty job done, they feel a shared bond when finished.

Pursuing the example afforded by marriage, Durkheim concludes that the most remarkable effect of the division of labor is not to increase the output of functions that are divided but to provide them solidity and stability.

That is, in our terms (12) the synchronous orbits become more certainly fixed rather than more orbits being proliferated.

Following Comte, he stresses that division of labor is not a purely economic phenomena but an essential condition of social life assessed as the immense totality of human endeavor. "It is thus the continuous repartition of different human endeavors which especially constitutes social solidarity and which becomes the elementary cause of the extension and growing complication of the social organism." (With which we would agree—if it were understood that critical density and ecology are sufficient to begin the chain reaction of continuous repartition among the elementary hungers.)

Durkheim's path then takes him through the development of the forms of law, which do not concern us at present. His discussion of crime and punishment, however, remains quite biological and elementarily familial and the concept of the common conscience looms strongly. (This resembles Freud's superego image and the image ideal which we borrowed from Freud.) He pursues this thought first in the religious doctrine that "everything social is religious," then in application to political, economic, and scientific functions as they free themselves from the religious. As society develops, there is an evolution of the common conscience from its foundation in a mechanical similarity in the biological entities that furnish the basis for the solidarity to a freer organic solidarity. "Individualism, free thought, dates neither from our time, nor from 1789, nor ... Reformation, ... scholasticism, [or] the decline of Graeco-Latin polytheism.... It is a phenomenon which begins in no certain past ... but ... without cessation all through history...." The role of the division of labor fills the former role of the common conscience, a role more important than ordinarily assigned to it by economists.

We can begin to describe this in our terms; how the great variety of personal orbital patterns, which had been less bound in lower-density societies now, through the action of keystone elites, begin to fall into different classes of patterns. These patterns are cooperatively or complementarily interwoven. In simpler primate societies, it is often only one older and stronger male leader who, as elite, sets the basic orbital pattern.

We must make the issue much clearer. It represents, in our opinion, an extension of the concept of exchange forces. In the human, for example, we do not have coulomb forces, gravitation forces, etc., that make up the usual simple pattern of physical attractive and repulsive forces. Yet human beings are attracted or repelled or join. However, so do complex molecules chemically. What is common is that both have many oscillator elements. It is not sufficient to recognize that molecules and human beings have translational degrees of freedom, that they move about in space. This does not bind them, it only helps make them viable by making contact likely. It guarantees that they will interact under suitable conditions. What binds them in the case of simple molecules is a few direct bonds. In the complex molecule, as in the complex human, it is a greater array of bonding points that determine the matter. One might easily think that 2^{20} is more like the number of contacts to consider for humans. Yet the question of whether molecular complex A will bond to molecular complex B or whether A is a catalyst for B and C is more likely to be decided by less than a handful of emergent rules. We cannot illustrate with generality. However, we can refer to the Hume–Rothery rules for alloy formation as an illustration (or Watson–Crick coding for biological replication). Alloys can be formed if the added atomic type does not differ by more than a few percent in volume from the base molecule. The mysteries of most basic biochemical binding, however, still remain unfathomed.

The exchange rules for humans are whether two individuals that meet "empathize" or come under the same "conscience." However, we propose to use an idea that is common for nuclei, atoms, molecules, cells, humans, and societies. "Emphathizing" nuclei are hard to define. (The physicist has a modest panoply of binding forces, and he must attempt to make do with these.) However, what is more nearly common is the "exchange" idea that what makes two individuals bind is the exchange of body image. For a fleeting moment, persons look at each other's external images and actions and fit their own view of the internal image and command-control system that leads to such action. If it fits, by complementarity or congruence, there is empathy—there is an interpersonal force. Two young people meeting—we older fogies can remember—catch such a fleeting glimpse. They begin to talk, to share experiences, to note likes and dislikes. Are they the same or different? As the picture emerges, there are a variety of bonds that form, some based on attraction and some on repulsion, that is, some on likenesses and some on differences. What emerges is a complementary pair. These can bond. This is the most rudimentary form of the division of labor in human aggregates, of a cooperative structure of two or more pieces which no longer perform with chemical patterns but with complementary patterns of internal and external actions.[1] With no bonding, there is no division of labor.

The exchange forces are thus a bonding system that arises between two somewhat alike particles, whose internal actions can tolerate deformation or penetration resulting from the other particle's internal action and which by "suitable" identification of that action (that is, by some "magic" which detects the dimensionality of each other's force fields)[2] can exchange or share small fluxes or energetic signals. This small flux is a "communications" flux. The exchange, which involves imperfect fitting at the order of the relaxational time constant of the small flux, not of the gross system, provides the transport coefficients that govern the fluxes that flow between the systems.

However, this would not bind them together. It would provide viscous transfer, energy transfer, etc. The binding must also involve a direct force. The

---

[1] By the complex nature of the bonds, yet aware of the range of bonding found in molecules, we can only propose descriptive mechanisms for these forces in primitive terms.

[2] Many societies have some specific identification for these personal "force fields."

most plausible one remaining must be sexual or at least erotic or pleasurable interest. Thus, the regularization of the family group, the male-female-child configuration, is the only one that can create a bound unit, not necessarily permanently, but at least until its central function is discharged.

Socialization or the socialized continuum then begins when the density is high enough that the forces involved in the transport coefficients become appreciable.[1] At this point, the energetic elites begin to give a form and sweep to the internal circulations. A stabilized social pattern may now emerge. The nasty nagging question that, of course, remains is what determines the history of the elites.

It is clear that they are not dynamic entities. They sweep into crevices and form governing interfaces. Their population depends on population density and ecological complexity. However, the transformations they influence are relaxational processes. They are diffusive not propagative, yet their presence gives form and color to the social continuum.

At this moment, we have no better answer than to suggest that their number is approximately determinate. We infer this from some knowledge of what makes dislocations in the solid state. One can expect so many managerial types, or intellectual types, or entertainment types in such and such a society and at such and such a stage. Which individuals and what specific actions are not clear, although the general color of their actions may be. It is a most illuminating experience to have grown up alongside of such "elite" entities. Whereas their specific talents or suitability may be apparent in retrospect, one seldom can see it in advance. Yet when they appear with the kernel of an idea and the suited drive to push it across, the holy fire of public acceptance sweeping them into critical position must always impress the observer. Since their effectiveness is indeterminate, they slowly change the form of their societies in a diffusely emanative fashion. This is contained in the catchphrase that in 500 years they generally succeed first in building up the society from an earlier one and then generally spoiling it. See, for example, McClelland (106). This suggests that a somewhat common type of unfolding takes place: a general growth, a dynamic steady state, and then a decline and fall of the civilization. Obviously, all the attendant complexities of the background ecology, the level of density, and technological complexity with which the system begins are salient factors. However, Durkheim may have been a good first choice to start economic history with.

Durkheim then discusses organic solidarity in the forms that social organizations may take, in which horde and clan are possible prehistorical or conjectural types without such solidarity. Then, territorial conscriptions arise rather than the family aggregates which remain still segmented. A slow transformation takes place, "a leveling analogous to . . . liquid masses put into communication." Since territorial divisions are less basically grounded, they

---

[1] In the plastic-elastic state of solid material, we have indicated that the abnormally high growth of viscosity at the end of the liquid state already begins to represent form. Our asphalt roads attest to this—until the day gets too hot! In the fluid, the presence of viscous drag also "conditions" the fluid for the action of various auxiliary factors to become amplified and sweep the field into the form of turbulent eddies. These two cases both illustrate "formed" societies.

lose their significance, and occupational organization comes into being. The village-city concentrates the industry necessary to supply the country.

We find this "systems" discussion much more preferable than any loose biological concept of a territorial imperative in man.

Occupational organization had to adapt to what existed before. "The same law holds of biological development." He traces this through colonies and the segmental organization of organs in animals.

Thus, gradually there is a place created for the individual in society as the civilization grows.

In summary of his first part, Durkheim points out that social life, therefore, comes from a double source, the likeness of consciences (i.e., our superegos) and the division of social labor (in our opinion, the consequence of interpersonal forces). In the first instance, man operates similarly whether in small group or in a collective. In the second instance, he develops an individual patterned appearance. Rules for like behavior and divided functions arise. The rules of collective practice make up the "morality." The rules of occupational morality become as imperative as any others. "Consequently, even where society relies most completely upon the division of labor, it does not become a jumble of juxtaposed atoms, between which it can establish only external, transient contacts. Rather the members are united by ties which extend deeper and far beyond the short moments during which the exchange is made." Altruism is not simply an agreeable ornament to social life but its fundamental base. It is the accommodation of conflicting motions between components which would otherwise always be in extremely high stressful relation to each other.

In a second part, Durkheim relates the causes for the division of labor to the pleasure-pain principle.[1] However, there is no relation between the variation in happiness and the advances of the division of labor. "The division of labor develops, therefore, as there are more individuals in contact to be able to act and react upon one another." (That is, as density increases, as we had surmised.) He points to condensation of societies through population concentration, from the nomad, to agriculture, to the city. The added factor in increasing density is the increase in modes of communication and transportation. "The division of labor varies . . . with the volume and density of societies. . . ."

A number of secondary factors, such as heredity, that influence the

---

[1] In our view, we have placed the problem of behavior on the patterns of oscillators, which now come to rest in exchange forces, complementary patterning, and division of effort. We have put the overall drive in the anxiety-euphoria (perhaps better, anxiety-complacence) behavioral oscillation that we suspect is a major behavioral modality of the human brain. We surmise that it operates with a 2 to 4-week period.

division of labor are discussed. The final section of abnormal forms of the division of labor do not concern us.

In summary, Durkheim was useful to us to point out the social validity of our organic model of how individuals work. The biological system with a human brain leads to interpersonal forces, which with sexual attraction lead to primitive social organization and a division of motor effort. It should be pointed out that we defined the living system as one which eats to move so that it can continue to move to eat and intermittently couple and reproduce its own kind (12). Particular instability characteristics of the human nervous system and social density were what made the human behavioral patterning so potentially complex. This is not characteristic only of large brain size. Some animals—dog, porpoise, monkey—show patterning complexity; others do not at normal ecological densities. (Patterning complexity will also increase in other animal species with density increase.) Among human beings, density itself has been responsible for increasing the complexity of the division of effort. The foundation for the interpersonal forces and the resulting division of effort is exchange forces in which body images are exchanged and compared. The division of effort, corollary to density increase, increases the extensions of man—his tools, weapons, vehicles, machines, communications, and structures. The social organization changes from familial to territorial to communal to occupational. Structures (the city, its commerce and market-place, the agricultural country, its trade, the corporation) come into existence. Elites came into existence to regularize the flux of motor effort that satisfies the human hungers. The city becomes the crossroads for purveying the content for all hungers.

Beyond the spectral response of the individual [see (76) for example], there is an extensive spectral response of the individual in society. We cannot pretend any adequate experimental foundation. We cannot say with precision what the data would indicate in the U.S.S.R., Rome, Athens, or Jericho. However, in a most tentative way, we can base our comments on a "seat of the pants" intuition of what may be some elements in the spectral responses to be found in human societies.

We can thus continue to structure internal society a little further. It is clear that the occupational foci, specifically the degrees of freedom that are associated in oscillatory and limit-cycle fashion with the occupational foci, are not at equilibrium (i.e., equipartition of energy) in a statistical mechanical sense. Thus, at best we may be able to discuss near-equilibrium equations of change. The other oscillators (feeding, anger, anxiety, etc.) may be more nearly at equilibrium. Their relaxation time constant is more nearly of the order of 30 days, whereas the relaxation time constant of the individual occupational focus is of the order of 20 years. Thus, psychological equipartition does not encompass the social changes, although it may encompass social change for the individual.

If we accept our most provisional estimate (104) of a 200-year "time constant" for society's changes,[1] then we can make some crude comments. Society acts like a thixotropic material, a very yielding plastic-elastic solid, a material in its softening range, a supercooled liquid, a high polymeric gel, or "silly putty." In our discussion of the equations of change of molecular phenomena, we found the following two classes of systems: (a) those whose internal energies were comparable to their external translational energies and (b) those whose internal energies were large compared to their translational energies. For processes that were not too rapid, phenomena in the first class of systems could be described by nearly homogeneous phenomenological changes associated with a coefficient of shear viscosity, thermal conductivity, simple diffusivity, and wavelike propagative elastic changes. The second class of systems could be considered to be "abnormal" liquids (abnormal in a variety of ways) or plastic-elastic materials. The basic property associated with these materials seems to be that the coefficient of bulk viscosity was large compared to shear viscosity (formally, this just means, essentially, that the internal degrees of freedom absorbed large energy compared to the translational).

In the first class, the scale of processes which are not too fast or too grainy to be considered as continua (here, social continua) depends on the mean free path being very small compared to the size scale of the social enclosure and the propagative wavelength of the society.

For translating the parameters which we have used (2) to qualify the continuum limitations of hydrodynamics quantitatively into social terms, we will have to take some educated physiological guesses. The mean free path of an individual man is likely the "Lebensraum," in which he must have near-space territorial control. He may crawl into bed with a loved one, but he will bristle at enemies closer perhaps than 10 ft. Thus, 10 ft is an a priori estimate of his mean free path (100 ft is certainly too large).[2] We are impressed by the ideas of R. Gregory on the relation between the advantage offered by vision, as a detector of enemy motion, remotely, and the development of a complex nervous system. See for example (109). In the present context, the mean free path for human beings is likely related to the immediate field of sharp detailed optical images and the degree to which the optical field is filled. Using our physical criteria for continuum systems (2), social continua begin at societies a thousandfold larger in scale. Thus, 10,000 ft or 2 miles is not a bad estimate for the size of an elementary social continuum. (At 100 ft, a small village or commune has the variability that one would associate with molecular

---

[1] We cannot speak with any kind of precision here. The 200 years may be a three-time-constant "range," so that 70 years is more nearly the true effective time constant. Or it may be the real relaxation time itself. The former may be more attractive if viewed in the context of significant social change at 100-year intervals.

[2] Social distance in subhumans and humans is a topic that has been discussed by Hall (108).

aggregates, as in Knudsen flow. The size of early Neolithic communities, of the order of a few acres, represented diameters of the order of 300 to 600 ft.)

The more difficult phenomenon to estimate is the time-dependent process. The second temporal criterion (2) is really the ratio of relaxation time to the time of interest. The relaxation time—for interpersonal relations—is likely more of the order of 30 days, the postulated anxiety-complacency period or the period keyed by the menstrual cycle rather than 1 day or 300 days. We know that 1-day behavior is powerfully polarized and keyed both by night-day variation and by an autonomous rhythm. However, intense behavioral rhythms can be locked at the 7-day (i.e., near-1-week) scale. The socialized calendar week is its reflection. For example, as codified in the Hebrew religion in sitting a week in mourning for dead, it is infinitely difficult to hold an intense feeling of grief for more than a week. The gross scales of feeling good, feeling bad, holding a fixed state such as a grudge, getting whipped up for an election, even enjoying oneself highly, etc., can be held for somewhat less than a month. In a year, many such episodes have taken place. Thus, 30 days is the a priori guess. In accordance with statistical mechanical results, continuum properties begin at tenfold longer periods, namely 1 year. The year is the social differential of time; that is, it is a time long compared to individual relaxation processes and short compared to the continuum unfolding in society.

The issue becomes important in the second class of phenomena whose internal energies are large compared to their translational energies. We have stated that the time constant of the society is of the order of 100 years, (e.g., perhaps 70 years, perhaps 200 years). We now find that the atomistic time constant is of the order of 30 days. Thus, there is adequate time for the social processes to move toward equilibrium. However, they do not. Why not? Because their bulk viscosity is too large. They could only move toward equilibrium for periods long compared to 100 years. The equilibrium we are discussing is an equipartition of energy among all of the degrees of freedom. The problem is a high "bulk viscosity" in human societies. It limits the rapid relaxation of energy among internal degrees of freedom.

What represents the bulk viscosity in the human? Provisionally, we submit that it is the action of the slowly releasing neural-endocrine systems (commonly it would be considered the "brain" and its memory) that holds certain internal-external patterned systems with long time delays. Man "remembers." His internal networks are "adaptive," but only at a very slow time scale.[1] If someone hits you, you may not hit him back for 20 years. In a Freudian view, somehow you tie up a large amount of energy among some rich circulating, reverberating, oscillating, or field-stored repressions.

Governed likely at the high-frequency end by the adrenals, the highly

---

[1] To cite an illustration for the atomistic 30-day time scale, brain recoveries from disturbances commonly begin with this sort of time scale, as does growth of new vascular networks upon change in condition of body activity. Also see (87).

energetic internal degrees of freedom whip up a storm. They create spin, like whirling dervishes or vortices. Their constellation cannot relax quickly, although they can exchange quickly with neighbors. Why? They do not have the translational relaxation range to carry them far. As chemical concentrations, they have specific limited diffusional paths. Thus they lock up and become local internal elite structures. In a year or two, they can control a large amount of energy. (The effects may show up, for example, in the individual's GI tract, as ulcers.)

This complex process results in temporary frozen-out or newly appearing degrees of freedom. They look temporarily like structures. They likely make themselves evident in endocrine systems responses. (We have mentioned physical examples before—the flecks in mixing polymeric solutions, phenomena around the critical point, crystallitelike structures appearing in liquids especially near freezing, the entire nature of ions, catalysts, and precipitates appearing in solution.) It is a growth of structural form out of function, the more apt generalization that Hegel suggested and Marx somewhat distorted of the transformation of quality out of quantity. It requires generally many degrees of freedom (as in polymers or the complex form of large molecules, with the possibility of using the extensive arrangement for coding information. One may note that it only takes two states, heads and tails, for a beginning of all kinds of logical games. With six chess pieces, the game of chess is beyond current computational complexity).

It is desirable to have some sort of physical measure for such non-equilibrium states. Obviously, one static classification is the states of matter—solid, liquid, gaseous. Another dynamic possibility is "temperature." It would not have the same full meaning as for near-equilibrium degrees of freedom, but it would not be far off. [See for example (3).] To illustrate some of its disequilibrium aspects, two states would be "equitemperature," even though not at the same temperature, if some very energetic local state (say, revolution in the street) could be carried off into another reachable state (say, a few weeks' outing in the country); or, somewhat similar, some high-temperatured people can easily be picked up and handled because, although their internal activities represent high temperature, they have such an adiabatic envelope and can switch or be switched to other high-temperature states without communicating any heat. The latter is commonly illustrated by adolescents.

Equations of change for the social continuum are thus almost reachable (not necessarily by a few weeks' study but perhaps a few years' study). Is there any merit to their present discovery? Our tentative answer is No. Their existence is enough for us to try to think through boundary-value problems, the histories of societies. What we cannot do too well is describe the changing frozen-out chunks. We should be able to do statistical boundary-value problems; that is, how does an existing society change its degrees of freedom?

It is not yet clear how to write the history of the changing elites. They are superficial outcomes of the state of society. However, they then change their particular social histories. It is not yet clear how to manage this in theory. On one hand, the easy thing would seem to be to write a two-class theory of the populace and the elites. This is like a theory of electrical conduction, not in metals with their free electrons, but in semiconductors.

We have thus laid down a very primitive view of social ensembles—the potential ground for "economic" organization ("economic"—the summation of goods and services that satisfy human hungers)—and seeds for thought on which economic theory might be planted.

To see how the human atom fits within his society, we can postulate the approximate connectivity scale shown in Fig. 12-5.

One must first note that the individual is the weld between his grandparents and his grandchildren. Second, his temporal range of concern is at the order of 100 to 120 years (i.e., longer than his life). In toto, the society is so welded over a 200-year range, in which the ideas developed through the grandparent's generation are transmitted over perhaps a 25-year gate to the individual through the parents and in turn to the grandchildren through the children. It is this large degree of redundancy and binding that tends to give society such a high degree of continuity, even though the individuals are atomistic. The social fracturing, i.e., the boundary-value problems, must arise from epochal divisions other than the life and death of the individual.

12-5   INTRODUCING ECONOMICS

As far as classic economic history is concerned, Taylor (110) and Schumpeter (111) are apt sources. The historical line of economics is commonly listed as:

> The physiocrats (Quesnay)
> The sentimental school (Adam Smith)

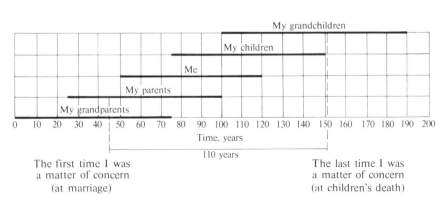

Fig. 12-5   A man's connectivity within his society.

English utilitarianism (Benthanism)
Malthus
Ricardo
John Stuart Mill
Marx
Victorian neoclassics (Jevons)
Marshall
Twentieth-century ideologies (communism, fascism, socialism, capital-
  ism, mixed economies)
New economics (Chamberlain, Keynes, Galbraith)

We will not discuss that historical content. (Nor will one sentence summaries of each school help the reader. He should consult these sources.)

Another cross-section of economics (other than the history of its ideas) was cursorily chosen from a textbook, Whittaker (112), to serve as an introductory ideological history of economic thought. We will briefly consider only his first three subjects.

SOCIAL EVOLUTION AND THOUGHT

It appears that man at all times lived a social life even in the most rudimentary cultures. (For example, Malinowski is referenced as a source for information about this.) Of the theories that are centered on such organizing concepts as the social contract, the family, or the horde, it is the central significance of the family grouping that seems to hold merit.

Reviewing the period of written history, the family grouping was authoritarian in ancient Egypt. It involved a greater individual diversity in Greek and Roman civilization with commerce and industry. In feudalism, it involved religious submission. In the Reformation and the modern national state, it involved mercantilism. In the modern individualistic state, it has involved laissez faire. Now there is a decline of individualism (!?), and there is state regulation.

> The essential element seems to be that after establishing a physiologically satisfactory unit, the intercoursing family unit, then ecological factors governed population density accretion under a variety of schemes. Group memory retained technological advances, whereas war and conflict—i.e., social mismatch—degraded the political systems. System patterns grew that seemed to fit each ecology. No "logical" evolution existed, only an "emergent" evolution. Thus it is only the topology of certain critical relations that counts. Organization for food, for its distribution, for deferred "payment" (work now, get paid later), for control and satiation of needs, and for relief of anxieties and supply of complacencies are some of these relations.

> As far as the last thought about decline of individualism and an increased state regulation, we would have taken issue in 1940, in 1960, and now. Although the industrial states have become more the subject of manipulation by "managerial" elites, nevertheless they have remained laissez faire. The ideological thrust of our

societies has become the vector sum of force of various special-pleader power groups. Thus it is simply not the individual who acts in a laissez-faire mode but his power group. In this respect, the reference shows an overly conservative bias.

WEALTH AND HUMAN INDUSTRY

Satisfaction of human wants is the kernel of economic study. Wealth is the sum of those things that conduce to this satisfaction. However, the scope of wants and wealth requires amplification. Primitive people seem to have simple wants, and the collecting of deferred wealth is not so important. Labor with incentive is not important to them (Malinowski, Firth). On the other hand, civilization exhibits education in wants. Money, as a means of exchange for deferred labor (wages), begins to have meaning. Greek, Roman, and Christian thinkers became concerned with the ethics and morality of behavior,[1] i.e., what should guide a man's behavior, and in contrast, what does guide a man's behavior, such as the pleasure principle (i.e., the "ought" and "is" of behavior). Their ideology began with first favoring the more primitive occupations— agriculture and then mining rather than manufacturing or commerce. However, with excess wealth, this did not serve people. Acquisition of part of the community took place and was more important. "In most ancient countries, labor was held in some disdain." Then too, Christian precepts opted for the suitability of simple labor. However, those in authority did not really accept this view.

> Note in primitive man that the familial group must exist for survival and in fact did exist. Increased density creates new elites as structural elements freezing out of the social mix. Their sphere of influence—represented by their "force of attraction"—polarizes their "regularized" (i.e., crystallite) domains. Identical occupational effort by all cannot coexist. The system is unstable. One effort succeeds better than another. "Wealth," i.e., deferred payment or ownership, comes into existence. More and more is produced. The system shortens the path of effort. Productivity increases. It is the contrasts that suggest the way within the human brain. The system is always stabilized by changing contingencies in the milieu, i.e., it is only ac stabilized, not dc stabilized. A fully static canvas essentially represents sensory deprivation, which can lead to anxiety and, in the extreme, to terror. It is the patterns that appear against the night-day alternation that govern. It is from man's waking "dreams," his internal dramas, that the "is" and "ought" and the new tastes and fancies arise.

With the Protestant Reformation, there simultaneously arose modern capitalism. The morality of labor underwent change. Such matters became matters for extensive written debate. By the sixteenth and seventeenth centuries, such moral matters (God, wealth, the state, happiness) were

---

[1] This is not meant to imply that more primitive men were not concerned with ethics and morality. However, from this point on in history, we are more familiar with a nearly unbroken written tradition.

institutionalized matters of discussion and in drastic process of change as mercantilism (the management of the flux of material wealth, of money, treasure, and goods) became of consequence.

> Following upon the crowding density of societies, the European map was beginning to become filled and sketched out. A period of 500 years had passed in isolated feudal form. Mobility of the Italian cities, their contact with the East, and the Crusades—a clarion call that had made the older social elites mobile—removed the elite inhibition and began the interdiffusion between isolated societies that then required a new elite, the more mobile merchant. The new human wants were both internal—new religious goals, new morality, new entertainment—and external—new goods, etc. The tangible changes always seem to be most easily identified as of "things." However, the internal revolution was just as great.

Among the intellectual elites, the discussion proceeded—as with the Greeks before them—on the goals of human endeavor and specifically on the relation of health and happiness. As modern times approached, the relation of wealth and the whole of life was dropped. The later economists have in general abandoned the examination of psychological (and physiological) factors on which their predecessors had grounded economics.[1]

> One cannot be satisfied with this thesis. It seems clear that "intellectuals" have been concerned with functional pattern (internal and external behavior) and formed things (materials and possession) for a long time. The abstractions of magico-religious significance existed with Neanderthal man in Paleolithic times 100,000 to 200,000 years ago, with considerable likelihood. Quite certainly they existed in modern Homo sapiens in Upper Paleolithic times 10,000 to 20,000 years ago. The first identification of specific moral codes can be identified 5,000 years ago. See, for example, Hawkes' *Prehistory* (54). Since they influenced life and death, the growth of agricultural and husbanded or domesticated animals were the earliest matters discussed in these codes. At the end of prehistory, i.e., at the dawn of history, the conduct of man in society already appears. The extent of its concern in early history—Egyptian, Mesopotamian, early and later Greek, Roman, etc.—is only limited by the extent of the documentation that outlines our familiarity with their everyday concerns. The extensive religious apparatus of the Egyptians hardly precludes their concern with the "ought" and "is" of conduct, with its internally governed and externally ungoverned aspects. Indeed, it would emerge as our thesis—as we expressed it in describing human behavior—that it is only the external patterns that seem to have no external metric, whereas the internal patterns have very specific "psychological-physiological" metric.

[1] The salient (nonexhaustive) chain of names involved in this transition extends from Adam Smith, Hume, Hobbes, Bentham, Shaftesbury, Hutcheson, Locke, Helvetius, Quesnay, Malthus, Ricardo, Senior, John Stuart Mill, Jevons, Edgeworth, Pareto, Lauderdale, Cournot, MacLeod, Müller, List, Carlyle, Ruskin, Thoreau, Rae, Veblen, Marshall, Davenport, Cassel, Dickinson, to Keynes. The list is offered for the would-be philosophic, who want to fill out their background.

From our physical-scientific point of view, we must confess to a considerable inability to come to grips with historical economics. What strikes us is that its dynamics are missing. Thus we are forced to the very pedestrian task of sending the student to the initiating literature. Perhaps one of our readers can perform the dynamic synthesis.

Thus, the thesis of an explosion during the Reformation—as the precursor to the eighteenth-century Enlightenment—is simply a convenient academic fiction to describe the overt patterned act of "writing." It is more to the point, likely, that printing had become cheap enough. At this point, then, the detailed splitting of hairs, the endless extension of ritualized repetitious debate until all corners of the verbal argument are sketched in, might have been expected to take place.[1]

Thus, one social characteristic of the human system in groups is that the "accidental" arriving at a newly practiced arrangement—the hula hoop, travel to Samarkand, perfume, immolation, genocide, atoms—will be explored and transferred. The human mind is a nonselective carrier. Its sensory system slices anything in the image field, stores it in the memory space, and transmits it. At some time, almost any such memory will be put on the agenda of history. Whether it will become a major historical factor is another matter.

How things go on the agenda of history we have described in an earlier report on technological forecasting (104). The emergence of a new patterned arrangement in the social system is related to the characteristic that an overwhelming number of the necessary genetic mutations are already in existence in the biological species. (This thought does not mean that new mutations are not forming all the time but that the genetic "bins" are quite full.) The changing milieu selects what can be dominant for the era. This scheme is genetic selection.

In the same way, we propose the generalization that an overwhelming number of idea fragments—however weird—exist in the composite memory of society. (This thought does not mean that new ones are not forming.) The changing ecological-social milieu selects what can be dominant for the era. This scheme is epigenetic selection.

If the generalization about epigenetic selection be true, it leads to a next consequence. The elite does not select the era; it is the era that selects the elite. Thus, the intellectually stated aims of education as training for society have been a little fallacious, and it is perhaps society which has unconsciously trained better than its leaders have recognized.

Those who are intellectuals and moralists (not a value judgment—just a statement about some people who talk a lot and think about "ought" behavior) would like to define the aims and ends of education. Yet, in a viable society, what is needed is accomodative preparation for all kinds of uncertain and emergent futures. What the needed social elites of tomorrow are may not be completely determined. Thus, a mix is required, and society has to have a way of dealing with the problems of the nonelites. Not all elites see the necessity for such a mix and for education for such a mix. Thus, optimally viable societies do not exist at all times, and in fact many societies go under.[2] Actually this may be the reason that one finds elites involved in the first phase of building up societies and in the last phases of tearing down societies.

In the large, this is not exclusively a human problem. It is only one problem in the development of atom-continuum-atom organization. There is no a priori

---

[1] We have no intention of insulting any of our readers. We hasten quickly to point out that our guilt is of no lesser order of magnitude than our colleagues!

[2] A very pointed discussion has begun in this era on whether the continued existence of our complex American urban society is threatened, i.e., has it become unstable? We mention this because the terms of reference are moving closer to those of concern to us in GSS—those involving conditions for social stability and boundary-value problems. References (113, 114, 115) are useful starting points for consideration.

guarantee that atomistic or continuum societies, at any level, will persist in viability. They—the universe itself—may be running down or up. Whether any system has a survival-emergent evolution will depend on whether its elements have survival factors that meet the emergent gross "ecology." One may suppose that what we are seeking in this book, over and over again, are clues or ways to assure sustained limit cycles. This must always be a delicate balance between degradative and explosive behavior.[1] In the present instance, we are discussing the verbal limit-cycle patterns that have become concerned with managing goods and wordly conduct. The need for verbal or abstract identification of how to maintain the social system has arisen in crowded societies following upon man's development of technically proficient extensions in the preparation, manufacture, storage, and distribution of the "foodstuffs" for human wants.

ECONOMIC INDIVIDUALISM

From the near-individualism that is found emphasized in dilute (low-density) societies to the need for regulation of some economic functions of society, there has been a nonmonotonic development of ideas from Greek times to the present. The philosophic basis for a doctrine of laissez faire was that in a free interchange among individuals (in a dilute society), the interests of all of the individuals will be served. This doctrine was maintained up through the eighteenth century, for example, in the physiocrats, in Adam Smith, and the classical and neoclassical economists descended from Smith. Some later economists were more pessimistic after the end of the eighteenth century, i.e., Ricardo, Malthus, John Stuart Mill, Sismondi, and the Hegelians.

The author (112) appears to have oversimplified history here. The individual and the state—as the spokesman for society—have been in considerable conflict and not only in the past 200 years. It is a somewhat pedestrian economic history that can propose such a weak summary. Even at that, if attention is placed only on the past 200 years, it appears clear that the problem of advanced societies has been that the economic liberalism-expansionist development of a moderate-temperate-zone–agricultural-mercantile society, where intranationalist or intra-societal interests were served by loyalty and cooperation in the division of labor, have actually worked; whereas the intermittent periods of exploitation of the individual's labor that lead to the evils of body slavery and wage slavery and, more recently, the economic-ecological poverty of have-not societies have created social problems that a laissez faire morality and ethic would not cooperatively solve solely on the basis of the individual. Thus in the advanced states, whether by capitalism or socialism, a very complex social regulation has so far emerged as necessary. It is clear to any open-minded individual that we are also watching the emergent evolution of an individual-welfare-concerned state. To what extremes it will go and how many nonviable "sports" it will show certainly cannot be told now. An extensive application of the working concept does not go back further than Bismarck, and it has only been in post-Great Depression times that the

---

[1] Many readers may have gotten the impression that limit cycles are very rugged affairs that will invariably emerge, given energy sources and degradative processes. This is likely true. However, the state of affairs that emerges may not be the one desired. To achieve this most often requires a delicate balance.

seriousness and reality of the problem has become apparent.[1] One may go one step further and point out that it is the disparity of have and have-not societies and segments that will likely represent the next emergent phase of economic doctrines. Thus far, only a few international catastrophes that were aggravated by dense society have emerged in history. There have been recurring famines in a few areas. There was the Great Plague and, more recently, the large-scale killing capacity of modern warfare. We still have to taste and look forward to a more widespread mass starvation, mass poisoning by air and water, major earthquake affecting a large segment of western civilization, unchecked disease, and an international holocaust of war.

In the light of what we have learned in a review of history and prehistory, it would seem that you cannot "plan" society by blueprints. The chemical nature of behavior confuses things, people, ideas, etc. Thus, all that can be done for planning is to change parameters that are involved in the cyclic patterns and note the direction of motion of the patterns.

Some may submit that this is a nonprogram. It is not. It is an approach, a methodology for acquiring experience about nonlinear systems' operation. We believe that both we and various kinds of social planners have experience in observing nonlinear operating systems. However, it is another matter to gradually discover the "laws" that govern their pattern formation. It is the intent of this first document to set the stage for such later systems' "engineering." At present, we are attempting to discover the problem areas.

For contrast to (112), another source for the ideas of economics is Catlin (116).

We believe this very brief excursion into the history and ideology of economic thought reveals an important weakness. One might say that economic thinkers have not been sufficiently culture-oriented but rather have written as specialists from the point of view of their own culture. (This, we believe, is part of the sin known as Hegel's glorification of the Prussian state.)

Thus, the elements of culture are more to be found in such anthropological writings as Mead (9). For those who want an apt introduction to our own culture, we might propose (62), (117), and (107). Of course, there is change brewing in modern economic theory. An apt sampler of its flavor is Shackle (118). This may be compared with Catlin.

It may be useful to the reader to set up the background of economic thought by a better introduction to its sources and ideas. In approaching the subject of economics with a little closer detail, we must be guided by the thesis of Thornton Wilder's play; man (as well as all other species) escapes socially only by the skin of his teeth. What this means is that his social behavior and its patterns are driven into orbits and modalities by the endocrines, using the nervous system as gating networks. These patterns change with the changing "ecological" milieu, in the large, the local, and the internal. The endocrines

---

[1] As (113, 114) discuss, there was concern in the late 1960s with whether the liberal reformism of the welfare-oriented state will be capable of dealing with the highly stressed modern urban technologically advanced society.

are subject to the vicissitudinal winds of the milieu. (Long-term viability of the operating chains of the species is thereby not assured.)[1]

It is this which creates the very fragile dynamics of society and which therefore creates the scientific difficulty. At present, we could not describe such boundary-value problems (as the neurally gated endocrine response) by physical or psychological means. It will require a behavior-chemical (i.e., a motor chemical with nervous-system gating) theory. Yet, it is out of this subtle complex that the patterns of civilization emerge. If this were to be true, it would be our thesis of homeokinesis with a vengeance!

It would be remiss of us not to point to a remarkable effort to connect physiology and economics by Noyes (119). It is unfortunate that the author may not have succeeded in establishing a full complement of useful theses. (We would suggest, for example, that his theses came into more complete view by adding (12) and (76) on the physiological side. However, we profit by his undertaking the beginning of the economic side.) Nevertheless, the work is scholarly and provocative at every point and helps furnish any reader some preliminary background toward any further hypotheses. In any case, we must characterize the work as monumental, even if not definitive. (We can only admire its boldness and wish that we had thought of it first.)[2]

Second, the material produced by such investigators as Calhoun and Christian, for which (24) is a good starting source, is another significant guideline.

Third, Wynne-Edwards' article (33) is a very important article for our GSS theses. In it, he basically states our homeokinetic theses, applying them not only to physiology and individual behavior, but to society and population. (The thought of its applicability to society has obviously appealed to us earlier but with no authority other than our own opinion. In our first review paper on the biological system (26), we stated, "With regard to the property of relaxation oscillation, in a hierarchy of systems the general type of nonlinear limit cycle continues to repeat—in the individual behavioral level, the group level, and the culture and civilization level. One might say that the mechanics and organization of external behavior mimics the internal biological behavior." Now we would simply add, "Naturally, they both represent similar ecologies!" Further, we have Wynne-Edwards for added authority.)

The ecological background is exemplified by Odum in (75) and (120) and references to be found therein. This is a fourth background line.

---

[1] Suppose there are species adapted to the tundra, to the taiga, or to the grasslands. An "intelligent" animal, run by his nervous system, might adapt his life to another zone if conditions changed in the zone to which he was acclimatized. However, if biochemically run, more likely most of his species would disappear. Perhaps some marginal numbers would adapt.

[2] Zipf (85) though not as bold as Noyes, is a source worth scanning as a rather interdisciplinary attempt at society.

Fifth, an introduction to the economic thinking of the past 100 years would also furnish the interested reader with a suitable background. Batson (121) provides an extensive, briefly annotated bibliography in economics before the current era. It represents, quite well, Western thoughts on economics. Without doubt, the (brief) line of economic thinkers was Adam Smith, Malthus, Ricardo, J. S. Mill, Marx, Marshall, and Keynes. The poles behind modern thinking are Marx and Marshall. A short sampler of readable books from the era reported is (122–144). (These books are available in a variety of editions. Generally, a late edition is referenced. Often there is a more recent reprint.)

One might add another thirty references for a more-rounded list, or one might use minimally four references such as Marshall, Fisher, Schumpeter, and Knight. [A few more modern references to complete the current economic story are (145–149). For an overview, see (168)].

The content of economics, as viewed by Batson, is contained in the following topics:

Scope, methods, and principles of economics (i.e., its theory)
The nature of capital and income
The theory of the productive process
The theory of population
The theory of productive variation
The theory of incentive
The theory of risk
Division of labor
International trade
Value and distribution
The concept of value
Determination of value
Supply
Demand
Competition and monopoly
Theory of differential prices
Functional distribution of wealth
Wages
Interest
Profits
Rent
Personal distribution of wealth
Fluctuations
Public finance
Public revenue
Particular taxes (income, etc.)

Public expenditure
Public indebtedness

One may compare these topics with the contents of a favored current textbook, Samuelson's *Economics* (150). It appears that there is no significant change. Samuelson covers essentially:

Control problem of economic society—its limited resources
Theory of price
Supply and demand
Total product and income
Saving, consuming, investing
Theory of money, and banking
Competition and monopoly
Theory of costs
Theory of production

What does this content indicate? Basically, it assumes the existence of a system that has form and function, a human society. It assumes that there are limited potential sources, "wealth," which can be acquired in a number of ways and their rate-limited fluxes; that there are humans who use their motor systems in work; that there are other material goods; that these are engaged in cyclic processes so that human wants are satisfied; that various fluxes and potentials have value measures; that a formed and functional civil organization, the government, enters into these processes to satisfy other functional requirements. In all, it assumes a very busy system.

Add to these other descriptions, some of the sociological ideas of Pareto (151), Max Weber (152, 153), Murdock (154), C. Wright Mills (155), and Drucker (156). Such material is quite representative of the content of modern economic-sociological thinking about the nature of civilization, of the "forces" in the marketplace regarding the commodities of interest to man, and of the social structure and its elements.

What are difficult to find are the dynamics of the system and its causal elements. (The field is overwhelmed by a barrage of words.)

[An excellent introduction to current sociological perspectives, particularly as viewed in the United States, written for the nonspecialist is Parsons (157).][1]

To cite a few illustrations: regarding population, Batson states, "Much that has been written on population in recent years will be of little interest to

---

[1] The more complete historical foundations of sociology are assembled, in spirit much like "economic perspectives" have been assembled in this book, by Parsons et al. in (158). The difference is that they provide brief "samplers" of the literature of sociology. This requires nearly 1,500 pages. We have taken the liberty of only hinting at the ideas of a real working economics. This is admittedly inadequate.

the ordinary student of economics.... American ... discussions of the biological and ethnological aspects ... have brought to light a large quantity of useful data, but very little in the way of general economic theory." Yet it has awaited Wynne-Edwards as ethologist to begin real questions of population dynamics (which is why we included the Yule and Stevenson references; refer also to their discussion).

As an example of what kind of material is missing, there is a large gap between Keynes' *The Economic Consequences of Peace* (1919), and the recent satire, L. Lewin's *Report from Iron Mountain on the Possibility and Desirability of Peace* (1967), on the relation of civilization, economics, and war.

As an additional example of missing material, there is the question of the real nature of human "wants." The literature mentioned does not cover the sources of human wants. Neither Noyes nor Pareto, for example, adequately derives the dynamic source of wants.

What we have done in (12) is to propose a tentative list of human "hungers," modalities of operation. It is these that lie at the source of human wants. Food and sex and water and oxygen and voiding represent operating modalities. However, sensory excitation, sensory cessation, aggressiveness, motor activity, euphoria or complacency, anger, fear, and greed also represent them. There are elites who have come into existence who are becoming expert in the manipulation, creation, and development of wants. It is, of course, likely that serious attention to these wants cannot be provided until other more autonomic hungers are provided. The oxygen-deprived, water-deprived, motion-deprived, food-deprived, sensory-limited animal has limited capacity available. However, as soon as leisure time exists, the internal instabilities can drive the system. Then, new hungers (new paths) can be created. (In *The Theory of the Leisure Class*, Veblin was groundbreaking in illustrating the creation of wants for a particular upward-mobile social group.)

The creation of a new want, hunger, or modality is not a one-step or one-shot process. By sequencing, gradually the system can be shifted into the new orbit. At the present, it is difficult to tell how extensive and exotic these shifts can be. We can take the example of Richardson's description of the creation of war moods (159). We can review in our minds the whole real and fictional history of carefully planned sexual campaigns that are quite capable of leading all from the most innocent to the most sophisticated to the most perverse practice. (The sexual revolution that has taken place in the past few years is quite clear in any group ranging in age today from 10 to 30 years old.) We can take the very real issue of sensory interference exemplified by drugs or smoking as a very real danger to health and note that the problem cannot be prevented by words or logical argument. We can take the demagogery of a Hitler or Stalin to see how people can be led to the most perverse practice for a generation or more. It is the dynamics of these processes that do not appear in the economics books. (At most, they appear to want to manage the already existent and make some vague estimate of the latent. To cite examples, one may examine (160), a book on modern methods of forecasting, or Gross (161), a book which has become a popular guide for managing organizations, for the degree to which systems dynamics is currently described.)

Noyes elegantly describes the issue in a section titled, "An Untoward Incident." Out of his window, he sees a farmer plowing with horses. The farmer stops to talk to someone. The whole process stops—the plow stops, the horses stop. What is missing then is the human's motivation (we prefer to point to the term "motor-actuation," the "automotive" nature of man) to initiate or continue the process.

It is this process of rearranging things in the milieu and of arranging new orbits that represents the creation of new wants.

What we human beings seem to have is a few periods of high influence-ability, not necessarily a full imprintability but certainly periods of high impressionability, during which new orbits are more easily formed. These periods are childhood and, especially, adolescence. It is clear that the appeals are endocrine (behaviorochemical) not electrical-rational.

Further, in similar vein, there are similar periods in the mature adult that might be described as "childlike" or "adolescentlike" periods in his mature life-orbit development, in which he is similarly impressionable with regard to new wants and orbits. The new homemaker is like that. The nouveau riche are like that. (The jet set are also like that.) The divorced person is like that. The enterprising business man "sees" new ideas, ideas that he might not have accepted a few years earlier. This is certainly true today, when successful businesses of all sizes have to turn over much of their product lines within a decade [see, for example, (160), p. 318].

As a final example of the creation of wants, one may page through F. L. Wright's *The Natural House* (162). Although Wright is very difficult to understand and read, yet a few things can certainly be gleaned. He put glass into the modern consciousness as a material for living (see pp. 51–54). He, at least in part, put the slab, particularly as cantilevered tensile load-bearing element, functionally into modern building. He put the stressed skin into modern building. He put slab heating (so-called radiant heating) into modern building.

All these illustrations represent the striking out by "imagination" into new orbits on the part of elites. The potential ideas are not new. They sit around in various forms waiting for discovery. Finally, the selective age occurs, and the idea goes on the agenda. The latent instability is there. The idea captures the imagination of some elites. It then takes its social course and finally explodes.

The second characteristic, beyond the invention of new paths—wants—is the fixing of orbits, of traditions. After having acquired an orbit, the human animal is often prepared to accept the orbit "forever" (or until it no longer fits the system's acceptable pattern of endocrine excitation).

For greater technical detail of how a modern economic state is run or at least influenced, a suitable reference is the National Bureau of Economic Research Conferences (163).

Summarizing this section, the "hungers" of the atomistic human are connected to its internal "quantized" limit-cycle oscillators. Their successful patterning into a melodic line of inhibiting and releasing these oscillating chains from inhibition makes for a viable human system that eats and moves about so that it can eat and move about and periodically couple and reproduce its own structure. This atomism is thus not essentially different from other levels.

Interpersonal force systems must exist that stabilize the system into larger constellations (else the cooperative effort of self-replication and nurture of the self-replicated system would not come off). The topology of the human nervous system is sufficiently complex that a rich behavioral patterning emerges, rich enough to show appreciable evolution in time scales faster than most other biological species. However, other species may have been quite capable of evolving and adapting right alongside of man.

The larger constellations of family [i.e., the nuclear family seems to exist in all cultures ever studied, see (154)] and larger kinship groupings seem to develop to increase the group stability. The temporally and spatially extensive characteristics of such systems, ranging through an ecological field, represent that species' culture.

Major orientation of the human to food supply is essential, in that man and other biological species are only slow-moving entities that essentially can only move on a two-dimensional surface. Clustering (59) with regard to fixed geographic centers occurred when agricultural productivity passed a significant yield threshold. Human culture now began to take on its complex character, particularly when man developed many extensions of his physiological apparatus. The cultural continuum thus now resembles a complex, as in the solid state, of nested levels that seem to stabilize in some larger unit.

SUMMARY

1. The physical base whence a fairly bound continuum (that is, in a liquid, plastic, or solidlike state), here a social continuum, arises is sufficient density that the interparticulate forces act within near-bound cell structures. The specific transport coefficient that probably tends to measure the state of viscoelasticity of the social continuum is its bulk viscosity.

2. The characteristic phenomena associated with the atomistic entity, man (who acts much like a free electron in that continuum), who is involved in the social continuum are the flux of materials and energies that move in and out of his system and his characteristic behavioral patterns (eating, drinking, defecating, breathing, moving, talking, anger, anxiety, euphoria and complacency).

3. The essential character of socioeconomic history of the human species is that it is immersed in an ecology and has a history and memory of things past.

4. The characteristic cellular organization against which the social history of man plays itself out is the family unit of organization.

5. The individual himself shows the following behavioral phases: involvement in the mother-infant constellation; the peer-peer play constellation; the adolescent-pairing-with-the-opposite-sex constellation; the fixing of an adult orbit of career and a tie with a person of the opposite sex.

6. The major binding force between individuals is an analogue to the exchange force in quantum mechanics. The individual carries an extensive internal body image. This copenetrates his fellow human being. The individual judges the complementarity or congruence of the other person's actions with his projected body image. Empathy or bonding takes place. This internal body image scan begins to take place at high frequency (near 10 Hz).

7. A bonding force having been established between individuals (but one that has no potential; that is, the bonding force is not one that depends on distance but on the qualities of exchange), then condensation takes place among groups greater than two. Here the boundary interaction with the ecology emerges as significant.

8. These ecological factors are the quality of energy, water, food, temperature range, density of population, etc. The epigenetic factors are the state of transmittable memory, the patterns of interpersonal relations, the patterns of motor activity, the emotional patterns, the intellectual and creative patterns, etc.

9. In the case of the human biological species, its dynamic history is not simply a monotonic unfolding of social behavior. Instead it shows aperiodic change. It evolves. Although the content of the patterns of behavior does not change that much, the externals—of organization, of extensions—change.

10. The evolution seems to stem from the capability to create extensions, both internal and external. Internally, man develops role playing. He wires himself together—first as a controller, then as a habituated regulator—into the playing of roles. He manipulates the environment in these roles. The particularly energetic individuals—elites—polarize the society. They are the proximal creators of the aperiodic changes in pattern. Thus there emerges a differentiation of social function, a division of labor among individuals.

11. Although the elites are "creative," it is not that difficult to forecast the directions that their creativity can take. It runs through a limited repertoire—of arts, sciences, and techniques. Particularly effective are the creation of new wants.

12. In the socioeconomic system, Western philosophy would have it that the free market, in which all the stuff that dreams are made of is freely bartered, is an essential ingredient for a viable social continuum. However, it is useful to review Aristotle for a broader view of the foundations of such thought.

13. What elites learn is not only how and what hungers to regularize the flux of but also how to orient the society along the comfortable paths toward these hungers.

14. History then emerges as the unfolding dynamic pattern of newly emergent hungers being organized for and satisfied.

15. Whereas Marx and Engels have viewed, in their doctrine of historical and economic determinism, that it was man's enslavement of man that made the advances of civilization available for a privileged few and that the story of history unfolds only through the autonomous evolution of the economic factor, man's material wants, we propose another view of man's culturization.

Man's beginning history can be noted in *Australopithecus* of 300,000 years ago. Family life, tools, and the hearth fire all attest to his human culture. However, the spectacular starts of civilization begin with Neolithic man of 10,000 years ago. Speech and cultivation of agricultural foodstuffs with higher yield than its accidental gathering were its concomitants. It appears the active exploitation of his coordination center that simultaneously permitted speech and the common abstraction of a number of internal modalities made the broad exploitation of extensions—the mark of civilization—possible.

16. The development of economics among men begins with a division of effort within the family, within the village with the beginning of agriculture, within the corporation when man's industry is no longer purely agricultural, according to Durkheim. The foundation for the common association is the same as for the individual's interpersonal forces of attraction. The common life of interests and association, of communing, is the condition of life.

17. The statistical mechanics of social systems obviously involve different time and scale constants than for molecular systems. Yet we can show that the same measures for the conditions of the continuum hold.

In a physical system, a continuum begins at a scale a thousandfold larger than the mean free path of motional relaxations in a field. The mean free path of a human, his "Lebensraum," is about 10 ft (the range of excellent visual detail). Thus, it is estimated that a social continuum begins at 2 miles, the size of a large-sized community.

In a physical system, a continuum begins at a time scale 10 times its relaxation time. The relaxation time—for interpersonal relations—is of the order of 30 days (a postulated anxiety-complacency cycle or a governing menstrual cycle). Thus, it is estimated that a social continuum begins at the scale of 1 year.

18. However, the social continuum is nearly plastic-elastic. It thus has a large bulk viscosity. Its full relaxation of internal modes has a scale of the order of 100 years, not because of external momentum exchanges, but because of internal momentum exchanges. "Times" and "ways" can be held onto for such lengths of time. The relation between the atomistic scale of the individual and the social scale of his social unit is shown in Fig. 12-6. This "plate-to-plate" welding or binding is similar, in its temporal character, to the spatial binding between crystal planes in a crystalline solid. The weldment only penetrates a few layers, but it is sufficient to make a continuum structure (i.e., the 30-day episodes in his life are bound in the individual for a lifetime of 70 years; the individual's availability in body and memory tends to bind his ancestry and progeny for a period of 200 years).

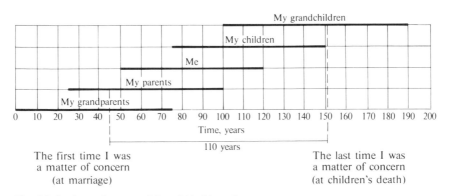

The first time I was
a matter of concern
(at marriage)

The last time I was
a matter of concern
(at children's death)

Fig. 12-6 A man's connectivity within his society.

19. The details of the economic system concern how costly or scarce resources are manipulated by the species, as extensions, to satisfy its hungers. Its history of change emerges in the longer time scale.

QUESTIONS

1. Take one or more cultures that you are familiar with. In terms of the number of generations it lasted, *a.* lay out a description of its start-up phase (indicate by maps changing each generation or two for a few hundred years before its start-up how the population assembled, what they were before, what the apparent drives were, etc.) so that the boundary conditions that led to the start-up are clear. Carefully trace, by generations, the changes taking place in salient human institutions and with regard to salient human hungers and modalities. Do this for each economic class, indicating their size, etc. Similarly *b.* lay out a description of its life phase. Similarly, *c.* lay out a description of its deterioration phase.

2. Assuming that you have some cultural descriptions in mind as requested in Question 1, what are the biological-ecological-sociological factors which determined the changing stability leading to each phase?

3. Discuss your own culture, in the terms of the references that apply to your own age, as to how the future of that culture might unfold and modify its operating conditions with changes in the point of view of its elite structure (i.e., if "election" or "evolution" or "revolution" brought about these changes). Pay careful attention to the difference in point of view of (155), (164), and the writings of A. A. Berle.

4. Attempt to cast light on the origin of speech within hominid development.

BIBLIOGRAPHY

1. Iberall, A.: "Phenomenological Basis for a Generalized Elastic Theory of the Solid State for Application to Structural Design," 1954, unpublished study.
2. Iberall, A.: Contributions Toward Solutions of the Equations of Hydrodynamics, Part A., The Continuum Limitations of Fluid Mechanics, *Contractors Rep. Off. Nav. Res., Wash., D.C.,* Contract No. Nonr 34-5(00), December, 1963.
3. Herzfeld, K., and T. Litovitz: "Absorption and Dispersion of Ultrasonic Waves," Academic Press, Inc., New York, 1959.
4. Tolman, R.: "The Principles of Statistical Mechanics," Oxford University Press, New York, 1938.
5. Hirschfelder, J., C. Curtiss, and B. Bird: "Molecular Theory of Gases and Liquids," John Wiley & Sons, Inc., New York, 1964.
6. Alfrey, T.: "Mechanical Behavior of High Polymers," Interscience Publishers, Inc., New York, 1948.
7. McGill, T. (ed.): "Readings in Animal Behavior," Holt, Rinehart and Winston, Inc., New York, 1965.
8. Tobach, E. (ed.): Conference on Experimental Approaches to the Study of Emotional Behavior, *Ann. N.Y. Acad. Sci.,* vol. 159, p. 621, 1969.
9. Mead, M. (ed.): "Cultural Patterns and Technical Change," New American Library of World Literature, Inc., New York, 1955.
10. Arensberg, C., and A. Niehoff: "Introducing Social Change," Aldine Publishing Company, Chicago, 1964.
11. Tax, S. (ed.): "Horizons of Anthropology," Aldine Publishing Company, Chicago, 1964.
12. Iberall, A., and W. McCulloch: 1967 Behavioral Model of Man, His Chains Revealed, *Currents in Mod. Biol.,* vol. 1, p. 337, 1968.

13. Perry, H., and M. Gawel (ed.): "The Collected Works of Harry Stack Sullivan," volumes 1 and 2, W. W. Norton & Company, Inc., New York, 1953.

14. Harlow, M., and H. Harlow: Affection in Primates, *Discovery*, January, 1966.

15. Goodall, J.: *Nat. Geographic*, vol. 124, 1963; vol. 138, p. 802, 1965.
————: "My Friends the Wild Chimpanzees," National Geographic Society, Washington, D.C., 1969.

16. Gesell, A., and F. Ilg: "Infant and Child in the Culture of Today," Harper & Row, Publishers, Incorporated, New York, 1943.
———— and ————: "Child from Five to Ten," Harper & Row, Publishers, Incorporated, New York, 1946.
———— and ————: "Years from Ten to Sixteen," Harper & Row, Publishers, Incorporated, New York, 1956.

17. Smith, F.: "Explanations of Human Behaviour," Constable & Co., Ltd., London, 1960.

18. DeVore, I.: "Primate Behavior. Field Studies of Monkeys and Apes," Holt, Reinhart and Winston, Inc., New York, 1965.

19. Schaller, G.: "The Mountain Gorilla," University of Chicago Press, Chicago, 1963.

20. Schrier, A., H. Harlow, and F. Stollnitz: "Behavior of Nonhuman Primates," volumes 1 and 2, Academic Press, Inc., New York, 1965.

21. Thorpe, W., and O. Zangwill: "Current Problems in Animal Behavior," Cambridge University Press, New York, 1963.

22. Thorpe, W.: "Learning and Instinct in Animals," Harvard University Press, Cambridge, Mass., 1963.

23. Berelson, B., and G. Steiner: "Human Behavior," Harcourt, Brace & World, Inc., New York, 1964.

24. Mayer, W., and R. Van Gelder: "Physiological Mammalogy," vol. 1, Academic Press, Inc., New York, 1963.

25. Piaget, J.: "The Language and Thought of the Child," George Routledge & Sons, Ltd., London, 1926.
————: "The Origins of Intelligence in Children," International Universities Press, Inc., New York, 1952.
————: "The Construction of Reality in the Child," Basic Books, Inc., Publishers, New York, 1954.
————: "Six Psychological Studies," Random House, Inc., New York, 1967.

26. Iberall, A., and S. Cardon: Control in Biological Systems—A Physical Review, *Ann. N.Y. Acad. Sci.*, vol. 117, pp. 445–515, 1964.
———— and ————: Regulation and Control in Biological Systems, *Proc. of IFAC Tokyo Symposium on Engin. for Control System Design*, 1965.

27. Willus, F., and T. Keys: "Classics in Cardiology," vol. 1, Dover Publications, Inc., New York, 1961.

28. Scotch, N., and H. Geiger: *J. Chronic Diseases*, vol. 16, p. 1183, 1963.

29. Editorial on hypertension: *Am. J. Public Health*, vol. 53, 1963.

30. Walker, A.: *Am. Heart J.*, vol. 68, p. 581, 1964.

31. Ratcliff, H., and R. Snyder: *Circulation*, vol. 26, p. 1352, 1962.

32. Christian, J., and D. Davis: *Science*, vol. 146, p. 1550, 1964.

33. Wynne-Edward, V.: *Science*, vol. 147, p. 1543, 1965.

34. Calhoun, J.: *Sci. Am.*, vol. 206, p. 139, 1962.

35. Dunham, H.: "Community, Schizophrenia and Epidemiological Analysis," Wayne State University Press, Detroit, Mich., 1965.

36. Strole, L.: "Mental Health in the Metropolis," vol. 1, McGraw-Hill Book Company, New York, 1962.

37. Henry, J.: Culture, Personality, and Evolution, *Amer. Anthropol.*, vol. 61, p. 221, 1959.
38. Stott, D.: Cultural and Natural Checks on Population Growth, in A. Montagu (ed.), "Culture and the Evolution of Man," Oxford University Press, New York, 1962.
39. Pauling, L., and E. Wilson: "Introduction to Quantum Mechanics," McGraw-Hill Book Company, New York, 1935.
40. Pauling, L.: "The Nature of the Chemical Bond," Cornell University Press, Ithaca, N.Y., 1960.
41. Luria, A.: "Higher Cortical Functions in Man," Basic Books, Inc., Publishers, New York, 1966.
42. Berne, E.: "Games People Play," Grove Press, Inc., New York, 1964.
43. White, L.: "The Science of Culture," Farrar, Straus & Co., New York, 1949.
44. Steward, J.: "Theory of Cultural Change," University of Illinois Press, Urbana, Ill., 1955.
45. Sahlins, M., and E. Service: "Evolution and Culture," University of Michigan Press, Ann Arbor, Mich., 1960.
46. Simpson, G.: "Life of the Past," Yale University Press, New Haven, Conn., 1953.
47. Newell, N.: Crises in the History of Life, *Sci. Am.*, vol. 208, no. 2, p. 76, 1963.
48. Dobzhansky, T.: "Mankind Evolving," Yale University Press, New Haven, Conn., 1962.
49. Leakey, L.: "Adam's Ancestors," Harper & Row, Publishers, Incorporated, New York, 1960.
50. Howells, W.: "Back of History," Doubleday & Company, Inc., New York, 1963.
51. Montagu, A.: "Man: His First Two Million Years," Columbia University Press, New York, 1969.
52. Braidwood, R.: "Prehistoric Man," Chicago Natural History Museum, Chicago, 1964.
53. Binford, S., and L. Binford: Stone Tools and Human Behavior, *Sci. Am.*, vol. 220, no. 4, p. 70, 1969.
54. Hawkes, J.: "History of Mankind," vol. 1, part 1: "Prehistory," New American Library, Inc., New York, 1965.
55. Clark, G.: "The Stone Age Hunters," McGraw-Hill Book Company, New York, 1967.
56. Childe, A.: "What Happened in History," (1954 ed.), Penguin Books, Inc., Baltimore, Md., 1964.
57. Linton, R.: "The Tree of Culture," Alfred A. Knopf, Inc., New York, 1957.
58. Mellaart, J.: "Eastern Civilizations of the Near East," McGraw-Hill Book Company, New York, 1965.
59. Piggott, S.: "Ancient Europe," Aldine Publishing Company, Chicago, 1965.
60. Wooley, L.: "History of Mankind, vol. 1, part 2: The Beginnings of Civilization," New American Library, Inc., New York, 1965.
61. Pareti, L., P. Brezzi, and L. Petech: "History of Mankind," vol. 2. "The Ancient World," Harper & Row, Publishers, Incorporated, New York, 1965.
62. Mead, M., and R. Bunzel: "The Golden Age of American Anthropology," George Braziller, Inc., New York, 1960.
63. Mead, M.: "New Lives for Old, Cultural Transformation—Manus 1923–1953," New American Library, Inc., New York, 1956.
64. Junod, H.: "The Life of a South African Tribe," 2 vols., rev. 1927 ed., University Books, Inc., New York, 1962.
65. "Vanishing Peoples of the Earth," National Geographic Society, Washington, D.C., 1968.
66. Lewis, O.: "The Children of Sanchez," Random House, Inc., New York, 1961.
    ———: "Five Families," John Wiley & Sons, Inc., New York, 1962.
    ———: "La Vida," Random House, Inc., New York, 1966.
67. Dunn, N.: "Up the Junction," J. B. Lippincott Company, Philadelphia, Pa., 1967.
68. Terkel, S.: "Division Street: America," Random House, Inc., New York, 1967.
69. Friedenberg, E.: *N.Y. Rev.*, vol. 8, p. 11, May 18, 1967.

70. Fried, A., and R. Elman (eds.): "Charles Booth's London," Pantheon Books, Inc., New York, 1969.
71. Soboul, A.: "Parisian Sans-Culottes in the French Revolution 1793–94," Oxford University Press, New York, 1964 (part of 1958 thesis).
72. Rashevsky, N. (ed.): Mathematical Theories of Biological Phenomena, *Ann. N.Y. Acad. Sci.*, vol. 96, p. 895, 1962.
73. Waddington, C. (ed.): "Towards a Theoretical Biology, 2. Sketches," Aldine Publishing Company, Chicago, 1969.
74. Grabstein, C.: "The Strategy of Life," W. H. Freeman and Company, San Francisco, 1964.
75. Odum, E.: The Strategy of Ecosystem Development, *Science*, vol. 164, no. 3877, p. 263, April 18, 1969.
76. Iberall, A., and S. Cardon: "Hierarchical Regulation in the Complex Biological Organism," IEEE SSC Conference, Philadelphia, Pa., October, 1969.
77. Huntington, E.: "Mainsprings of Civilization," 1965 ed., New American Library, Inc., New York, 1964.
78. Hardwick, E.: *N.Y. Review*, vol. 11, p. 3, November 7, 1968.
79. Iberall, A.: "Advanced Technological Planning for Interdisciplinary Physical Research," for Army Res. Off., Final Report AD No. 467-051-L, June, 1965.
80. Schoenheimer, R.: "The Dynamics of Body Constituents," Harvard University Press, Cambridge, Mass., 1942.
81. Bartels, H.: Some Aspects of Circulatory and Respiratory Functions in Mammals, *Am. Heart J.*, vol. 72, no. 1, p. 1, 1966.
82. Burck, G.: East Europe's Struggle for Economic Freedom, *Fortune*, vol. 75, p. 125, May, 1967.
83. Iberall, A., and S. Cardon: "A Study of the Physical Description of the Hydrology of a Large Land Mass Pertinent to Water Supply and Pollution Control," HEW, Contract No. Saph-78640, four quarterly reports, 1961–1962.
84. Iberall, A.: Anatomy and Steady Flow Characteristics of the Arterial System with an Introduction to Its Pulsatile Characteristics, *Math. Biosci.*, vol. 1, no. 3, p. 375, 1967.
85. Zipf, G.: "Human Behavior and the Principle of Last Effort," Addison-Wesley Press, Inc., Cambridge, Mass., 1949.
86. Engels, F.: "Anti-Dühring," New York International Publishers, New York, 1939.
87. Michael, R.: "Endocrinology and Human Behavior," Oxford University Press, New York, 1968.
88. Gardner, R., and B. Gardner: Teaching Sign Language to a Chimpanzee, *Science*, vol. 165, no. 3894, p. 664, 1969.
89. Barnett, L., and *Life's* Editors: "The Epic of Man," Time-Life Books, New York, 1961.
90. "Everyday Life in Ancient Times," National Geographic Society, Washington, D.C., 1961.
91. Van Doren Stern, P.: "Prehistoric Europe," W. W. Norton & Company, Inc., New York, 1969.
92. Piggott, S. (ed.): "The Dawn of Civilization," McGraw-Hill Book Company, New York, 1961.
93. Whatmough, J.: "Language, a Modern Synthesis," New American Library of World Literature, Inc., New York, 1956.
94. Oakley, K.: "Man The Tool-Maker," University of Chicago Press, Chicago, 1961.
95. Eisenberg, J.: A Comparative Study in Rodent Ethology with Emphasis on Evaluation of Social Behavior, *Proc. U.S. Nat. Museum*, vol. 122, no. 3597, p. 1, 1967.
96. Pei, M.: "The Story of Language," 1949 ed., New American Library, Inc., New York, 1960.
97. Hall, E.: "The Silent Language," Fawcett World Library, New York, 1959.
98. Beach, F.: "Hormones and Behavior," Paul B. Hoeber, Inc., New York, 1948.

99. Kawamura, A.: The Process of Subculture Propagation, in C. Southwick (ed.), "Primate Social Behavior," D. Van Nostrand Company, Inc., New York, 1962.

100. Carthy, J., and F. Ebling (eds.): "The Natural History of Aggression," Academic Press, Inc., New York, 1964.

101. Fried, M., M. Harris, and R. Murphy (eds.): "War," The Natural History Press, New York, 1968.

102. Hallett, J.: "Congo Kitabu," Fawcett World Library, New York, 1967.

103. Kroeber, T.: "Ishi," University of California Press, Berkeley, Calif., 1964.

104. Iberall, A.: "Technical Forecasting, A Physicist's View," for Army Research Off., AD #635-811, July, 1966.

105. Durkheim, E.: "The Division of Labor in Society," The Free Press, Glencoe, Ill., 1947.

106. McClelland, D.: "The Achieving Society," D. Van Nostrand Company, Inc., New York, 1961.

107. Galbraith, J.: "The New Industrial State," Houghton Mifflin Company, Boston, Mass., 1967.

108. Hall, E.: "The Hidden Dimension," Doubleday & Company, Inc., Garden City, N.Y., 1966.

109. Gregory, R.: "Eye and Brain," McGraw-Hill Book Company, New York, 1966.

110. Taylor, O.: "A History of Economic Thought," McGraw-Hill Book Company, New York, 1960.

111. Schumpeter, J.: "A History of Economic Analysis," Oxford University Press, New York, 1954.

112. Whittaker, E.: "Economic Analysis," John Wiley & Sons., Inc., New York, 1956.

113. Moore, B.: Revolution in America?, N.Y. Review, vol. 12, no. 2, p. 6, January 30, 1969.

114. Kirchheimer, O.: Confining Conditions and Revolutionary Breakthroughs, Amer. Political Sci. Rev., vol. 59, no. 4, December, 1965.

115. Jacobs, J.: "The Death and Life of Great American Cities," Random House, Inc., New York, 1961.

116. Catlin, W.: "The Progress of Economics," Twayne Publishers, Inc., New York, 1962.

117. Lerner, M.: "America As A Civilization," Simon & Schuster, Inc., New York, 1957.

118. Shackle, G.: "A Scheme of Economic Thought," Cambridge University Press, New York, 1965.

119. Noyes, C.: "Economic Man," vols. 1 and 2, Columbia University Press, New York, 1948.

120. Odum, E.: "Ecology," Holt, Rinehart and Winston, Inc., New York, 1966.

121. Batson, H.: "A Select Bibliography of Modern Economic Theory 1870–1929," George Routledge & Sons, Ltd., London, 1930.

122. Newcomb, S.: "Principles of Political Economy," Augustus M. Kelley, Publishers, New York, 1886.

123. Jevons, W.: "The Theory of Political Economy," Augustus M. Kelley, Publishers, New York, 1871.

124. Keynes, J.: "The Scope and Method of Political Economy," Augustus M. Kelley, Publishers, New York, 1917.
———: "The Economic Consequences of Peace," Augustus M. Kelley, Publishers, New York, 1919.

125. Marshall, A.: "Principles of Economics," The Macmillan Company, New York, 1920.

126. Cournot, A.: "Researches Into the Mathematical Principles of the Theory of Wealth," Augustus M. Kelley, Publishers, New York, 1897.

127. Fisher, I.: "The Nature of Capital and Income," Augustus M. Kelley, Publishers, New York, 1906.
———: "Elementary Principles of Economics," Augustus M. Kelley, Publishers, New York, 1912.

128. Pigou, A.: "The Economics of Welfare," St. Martin's Press, Inc., New York, 1932.
129. Sidgwick, H.: "The Principles of Political Economy," Kraus Reprint Corporation, New York, 1968.
130. Taussig, F.: "Principles of Economics," vols. 1 and 2, The Macmillan Company, New York, 1961.
131. Schumpeter, J.: "Ten Great Economists: From Marx to Keynes," Oxford University Press, New York, 1951.
132. Yule, G.: The Growth of Population ..., *J. Roy. Stat. Soc.*, vol. 88, p. 1, 1925.
133. Stevenson, T.: The Laws Governing Population, *J. Roy. Stat. Soc.*, vol. 88, p. 63, 1925.
134. Böhm-Bawerk, E.: "Capital and Interest," 3 vols, Macmillan & Co. Ltd., London, 1890; also see his "Gesammelte Schriften," F. X. Weiss (ed.), 2 vols., 1924–1926.
135. Knight, F.: "Risk, Uncertainty, and Profit," Harper & Row, Publishers, Incorporated, New York, 1965.
136. Clark, J.: "The Distribution of Wealth," Augustus M. Kelley, Publishers, New York, 1899.
137. Cannan, E.: "History of the Theories of Production and Distribution," Augustus M. Kelley, Publishers, New York, 1924.
138. Hansen, A.: "Business Cycles and National Income," W. W. Norton & Company, Inc., New York, 1964.
139. Kondratieff, N.: The Static and the Dynamic Views of Economics, *Quart. J. Econ.*, vol. 39, p. 575, 1925.
140. Mitchell, W.: "Business Cycles: The Problem and Its Setting," Burt Franklin, New York, 1913.
141. Veblen, T.: "The Theory of the Business Enterprise," Augustus M. Kelley, Publishers, New York, 1904.
142. Webb, S. (ed.): "Seasonal Trades," Augustus M. Kelley, Publishers, New York, 1912.
143. Seligman, E.: "Studies in Public Finance," The Macmillan Company, New York, 1925.
144. Dalton et al.: The National Debt, *Econ. J.*, vol. 35, p. 351, 1925.
145. Keynes, J.: "The General Theory of Employment, Interest, and Money," Harcourt, Brace & World, Inc., New York, 1936.
146. Ackley, G.: "Macro Economic Theory," vols. 1 and 2, The Macmillan Company, New York, 1961.
147. Hicks, J.: "Value and Capital; An Inquiry Into Some Fundamental Principles of Economic Theory," Oxford University Press, New York, 1968.
148. Klein, L.: "Keynesian Revolution," The Macmillan Company, New York, 1968.
149. Patinkin, D.: "Money, Interest, and Prices," Harper & Row, Publishers, Incorporated, New York, 1968.
150. Samuelson, P.: "Economics," McGraw-Hill Book Company, New York, 1967.
151. Pareto, V.: "A Treatise on General Sociology," vols. 1 and 2, 1935 ed., Dover Publications, Inc., New York, 1963.
152. Weber, M.: "The Theory of Social and Economic Organization," The Free Press, Glencoe, Illinois, 1947.
153. Gerth, H., and C. Mills: "From Max Weber," Oxford University Press, New York, 1958.
154. Murdock, G.: "Social Structure," The Free Press, New York, 1965.
155. Mills, C.: "The Power Elite," Oxford University Press, New York, 1959.
156. Drucker, P.: "The New Society," 1950 ed., Harper & Row, Publishers, Incorporated, New York, 1962.
157. Parsons, T. (ed.): "American Sociology," Basic Books, Inc., Publishers, New York, 1968.
158. Parsons, T., E. Shils, K. Naegele, and J. Pitts: "Theories of Society," vols. I and II, Free Press of Glencoe, Inc., New York, 1961.
159. Richardson, L.: "Arms and Insecurity," Boxwood Press, Pittsburgh, Pa., 1960.
——————: War Moods, *Psychometrika*, vol. 13, no. 147, p. 197, 1948.

160. Bursk, E., and J. Chapman: "New Decision-Making Tools for Managers," 1963 Harvard ed., New American Library, New York, 1965.
161. Gross, B.: "The Managing of Organizations," vols. 1 and 2, The Free Press, New York, 1964.
162. Wright, F.: "The Natural House," Bramhall House, Inc., New York, 1954.
163. "Conference on Research in Income and Wealth," vols. 1–29, National Bureau of Economic Research, Princeton, N.J., 1938–1966.
164. Lundberg, F.: "The Rich and the Super-Rich," Lyle Stuart, Inc., New York, 1968.
165. Friedenberg, E.: *N.Y. Rev.*, vol. 15, p. 13, September 3, 1970.
166. Pfeiffer, J.: "The Emergence of Man," Harper & Row, Publishers, Incorporated, New York, 1970.
167. Bloch, E., et al.: Introduction to a Biological Systems Science, *NASA Contractors Rep. CR-1720*, 1971.
168. Heilbroner, R.: "The Worldly Philosophers," Simon & Schuster, Inc., New York, 1967.

13

The Tribe—Settlement—City—State—Nation; Ethnological Atomism

Man, bound into ensemble, spreads over the face of the earth. However, once again, the indefinite extension of such a continuum, here the cultural continuum, cannot continue. Stability criteria tend to determine its operative atomistic size.

(Let us review what the ensemble consists of. It consists first of nuclear families that are bound together into a group—perhaps first a horde, then a tribe, then, since Neolithic times, a settlement. There may be a cluster of such settlements (1, 2) that make up a regional galaxy, e.g., northern or southern Mesopotamia or Egypt. If the cultural continuum continued, there would be a distribution of such galaxies that spread indefinitely over the territories of the earth. In the main, ecological pressures would determine their focal centers.)

As the settlements become more extensive in space, the velocity of propagation of those fluxes that are used for transport and exchange of materials and for communication are too slow to permit anything but limited size. The species must continue to roam over its food-gathering range but also in an area which it can defend itself. (One must note that communicative propagation itself starts first with mass fluxes—of person-to-person contact through speech; of augmentation with higher-speed horses; finally, with wire and wireless transmission. It is amusing to note that it was really e-m communication—line of sight observation—from which communications started.[1] However, this was simply not well coordinated in early cultures. It could have been, by signal fires, mirror substances, etc. It may very well have been better coordinated acoustically, by drums, etc.) It is very possible that the dimensional scale of a culture—its atomistic fracturing or quantization— depended on that size domain in which reinforcement or a resonancelike property could apply, i.e., messages or materials getting back from the frontiers in time to be of utility.[2]

> If so, we may then suggest another basic chord in GSS; namely, that it is internal propagative wave systems, modes derived from some foundational coupled phenomena, that help to quantize a continuum by nonlinear emanative processes—by coherent generation, pulse by pulse—thereby forming the fundamental atomistic particle at any level.[3]

The function of the communications between settlements was likely to facilitate mass exchanges—of goods and of men (as objects, e.g., slaves or sexual objects).

It is conceivable that the size of the atomistic entity that emerges from the cultural continuum is governed by a mean-free-path relaxation in space and time, similar to the scale suggested by our previous estimate [also see (5) and (6)]. Thus in primitive cultures, it might fracture at the space domain that a man could comfortably walk in short time. It finally might extend as far as a man could comfortably walk in a day or ride on a horse in a day. The day-night interval has always been a powerful cue for biological species. Thus, the culture could spread from hundreds of feet, to miles, to tens of miles. Out of such primitive stuff, without going into the detailed interactions, we can see the possibility of atomistic properties emerging in human cultures. As such

[1] Gregory (3, 4) has in fact proposed that the brain itself emerged from the possibility of early distant warning against enemies afforded by light and dark signals and the preferred evolutionary value of light detection centers.

[2] The dynamics of attempting to build an empire too fast, too large, and with inadequate substance to hold it together for very long was never better revealed than in Alexander's march and conquest that extended to the "ends of the earth." Its excitement is beautifully depicted in a recent National Geographic Society volume (7).

[3] Although we first chanced enunciating this theme in 1965, we note that by 1969 (e.g., the September, 1969 International Biophysics Congress) it has become coin of the realm, at least at the frontiers of biophysics, at the origin of life, in cellular processes, in the ganglia of the central nervous sytem, in organs, and in the total organism. Here we try to extend the concept to the social organism.

smaller entities grow, like the dislocation domains, they tend to run into each other. They grow within a limited domain around some nucleating center. However, a cluster of such domains (i.e., crystallite domains with a common culture) may finally stabilize as a larger atomistic element. As in the solid state, such macroscopic structures may extend and nest until the grain is formed. Thus structures of tribes, settlements, cities, states, and nations, with some relative cultural homogeneity within them, grow out. These are the ethnological atoms. (We are not stressing their absolute homogeneity. Just as in solids, various homogeneous intrusions may enter the solid field and by diffusion and convection ultimately make the field quite locally inhomogeneous but globally homogeneous, e.g., polycrystalline, polygot, social melting pot, etc.)

An excellent introduction to sociocultural systems is Sorokin (8).

It appears clear that the relationship among such units, ranging from tribes to nations, is viewed in political science as an n-body problem, as a balance of forces among them. [See for example Morgenthau (9), Aron (10), or any issue of *Foreign Affairs*. As a typical illustration, Marshall Shulman's article, " 'Europe' versus 'Detente'?," vol. 45, p. 395, April, 1967, states, "Turning to an evaluation of 'peaceful coexistence' in Soviet policy . . . in practice, peaceful coexistence has meant an increasingly active diplomatic effort in the theater most likely to be decisive for the balance of power— Europe."] Thus it follows that the tribe-city-state-nation is regarded as an atomistic element.

The history of such atomic cultures is viewed in a number of different ways. Some, characteristic of the historians' view of history, may be examined in prototype in the Greek historians, in Barnes (11), or Toynbee (12). This view is predominantly descriptive of human activity—of what is said and done and written about. In time, it has added the dimensions of social and cultural description and the history of what men thought about. It is still not oriented very strongly toward an anthropological description.

There is very little that we care to add at this point. The problem will be to point out what may constitute a history of such atomistic elements. The trouble is that they are not isolated, stable atomistic elements, certainly not in modern times when communications speed is global. Thus we wish to be guided by our physical experience for their description.

For the past, with isolated communities, one might first sketch the geographic-ecological background. In physical terms, we would be concerned with "insertion" characteristics; namely a description of what the operating characteristics would be upon the insertion of one mobile element into a collection of other elements (this, for example, is the philosophy of molecular orbitals, etc.). We could imagine what the quality of life would be to a visitor of various classes in that society in that particular era. Second, we might sketch the physiological characteristics of the (unstable) human being and some information about his previous history. The problem, as in the n-body orbital cases, is that we can only proceed by perturbation analysis. Namely, if the system is in a given configuration, we can describe its motion and stability, if the perturbations stay within bounds. Thus, history likely must be written from perturbation epoch to perturbation epoch. The perturbations must be with regard to fundamental changes in the patterning of cultural behavior, not in changes in the elites. Too much history is concerned with the elites, conquest, and the flow of material things. In a physically and ethologically founded history, we would not be too concerned with the detailed naming of individual elites. In this regard, though static descriptions, Mead (13), Murdock (14), and Sorokin (8) attempt to shed a little light on cultural elements and changes. An excellent introduction to the dynamics of such systems was written by Sorokin (15). Also see Huntington (16).

By the time we come to modern history, i.e., with the birth of modern nations, it is practically continuum history, although with a high noise level. The quasi-ergodic hypothesis can be used, but the fluctuations will be high.

SUMMARY

1. The cultural ensemble cannot grow in size indefinitely. Thus the social continuum quantizes, likely through mass and communications flux limitations, into spatial settlements. The atomistic humans have coalesced into these settlements because of internal forces. Internally they have the fluidity of liquid droplets. Now they are attached to the land which feeds their hungers. Within their stable range, the exchange of fluxes—of attractions and repulsions between individuals and with their boundaries—creates the n-body problem between these social entities that develop into tribes, cities, nations.

QUESTIONS

1. Attempt to identify the atomistic human settlements that make up a considerable large chunk of history. (The task is monumental, unless done lightly. For example, 10,000 years of history might be broken down in 30-year chunks, i.e., three averaging sights per century. This makes 300 segments of history. There possibly have been 50 major settlements per epoch. This makes perhaps 15,000 entities to identify.) Indicate the changing complexion of the world settlements by changing still maps. Identify new growths or the influx of new significant populations.

2. Attempt to write a description of the quality of life of a modal person (or an "average" person from the most numerous class) at various epochs in ancient history, e.g., for various of the early population centers 7000 to 3500 B.C. or 3500 B.C. to 0.

3. If you think you understand the question, "Throughout his past 10,000 years of history, how did man live?" attempt a similar treatment for how some other major species lived during its moderately unchanging genetic past, e.g., describe the quality of life for some other primate species, or dog, bee, ant, bird, etc., for grosser intervals of time, such as 500 years, 5,000 years, 50,000 years, or 500,000 years.

4. Attempt to write a brief sketchy 2,000-year history of man (8000 to 6000, 6000 to 4000, 4000 to 2000, 2000 to 0 B.C., 0 to 2000) from a physical-physiological-ethological-ecological point of view, i.e., specify the boundary conditions of what went before and the status of things at the start of the period, write the unfolding history, and end with a summary of where things were left at the end of the period.

BIBLIOGRAPHY

1. Piggott, S.: "Ancient Europe," Aldine Publishing Company, Chicago, Ill., 1965.
2. Mellaart, J.: "Eastern Civilizations of the Near East," McGraw-Hill Book Company, New York, 1965.
3. Waddington, C. (ed.): "Towards a Theoretical Biology, 2. Sketches," Aldine Publishing Company, Chicago, 1969.
4. Gregory, R.: "Eye and Brain," McGraw-Hill Book Company, New York, 1966.
5. Mayer, W., and R. Van Gelder: "Physiological Mammalogy," vol. 1, Academic Press, Inc., New York, 1963.
6. Richardson, L.: "Arms and Insecurity," Boxwood Press, Pittsburgh, Pa., 1960.
    ———: War Moods, *Psychometrika*, vol. 13, no. 147, p. 197, 1948.
7. The World of Alexander, in "Greece and Rome, Builders of our World," National Geographic Society, Washington, D.C., 1968.
8. Sorokin, P.: "Modern Historical and Social Philosophies," Dover Publications, Inc., New York, 1963.
9. Morgenthau, H.: "Politics Among Nations," Alfred A. Knopf, Inc., New York, 1961.
10. Aron, R.: "Peace and War," Doubleday & Company, Inc., New York, 1966.
11. Barnes, H.: "A History of Historical Writing," 1938 ed., Dover Publications, Inc., New York, 1963.
12. Toynbee, A.: "A Study of History," vols. 1–12, Oxford University Press, New York, 1962–1964.
13. Mead, M. (ed.): "Cultural Patterns and Technical Change," New American Library of World Literature, Inc., New York, 1955.
14. Murdock, G.: "Social Structure," The Free Press, New York, 1965.
15. Sorokin, P.: "Social and Cultural Dynamics," 4 vols., Bedminster Press, Totowa, N.J., 1937–1941.
16. Huntington, E.: "Mainsprings of Civilization," 1965 ed., New American Library, New York, 1964.

14

Civilization; The History
of Nations as Social Continuum

Finally, as territories on the two-dimensional earth fill up, atomistic nations divide up the globe, and population density increases, we begin to have one world, a continuum of grainlike cultures, with political subdivisions. Man, as a cultural composite, now makes up the temporally historical and spatially distributed continuum of civilization, in which the composite history of nations and cultures past and present make up a continuum. Because of the time constants (the relaxation age of civilizations is 500 years; the relaxation age of major "moral" principles, i.e., major moral imperatives, is likely of the order of 2,000 years) and the changing space constants (because of the change of communications speed and mass transportation flux speeds, the physical scale is about 2 to 10 miles for the city, 100 to 500 miles for the state-nation, 1,000 to 5,000 miles for the supernation), the number experience is limited.

Our statistical mechanics should hold for such a system. However, we must grasp the considerable limitations. Translational mobility is not like that of atoms (which have a relaxation time constant of the order of 10^{-10} sec).

Instead, there is communicational mobility—a phone call can be made around the world in 5 minutes, but only within limited time periods during the day and for limited spatial zones (there may not be a phone). There is personal mobility—a trip to any region in the world can be made within 15 hours, but with all sorts of excluded space and time domains. There is mobility for man's extensions—heavy goods can be delivered within 3 months, but again with excluded domains.[1] In these excluded domains, it is possible for all of the relaxation times to drop back to those for primitive Neolithic man. Thus, neither our extensions nor our memory elements are yet so widely available to all parts of the phase space. Our civilization shows large shifting patches of frozen-out degrees of freedom, e.g., the ghetto. This is a consequence of the very large "bulk viscosity" tied up in the actions of the individual and of the interpersonal ensemble. Yet, in the limited regions of phase space that are available to the highly mobile degrees of freedom, the quasi-ergodic hypothesis would be quite applicable. However, neither the haves nor the have-nots in the societies of men or nations are yet ready to grasp the discrepancy.

Assuming that social history is the dynamics of the continuum of civilization, what is needed is such a theory of social history. Even though the nation may involve its own detailed individualistic story of past events, in the continuum of such interactional elements the individual events must lose their significance. What must remain are the gross equilibrium and gross equations of change of the entire system of nations. Thus must be written the dynamic history of civilization.

In point of fact, as is illustrated in such a source as Russell's *A History of Western Philosophy* (1), histories popularly tend to introduce the Near Eastern origins of civilization (per written history) of 3500 B.C. to 2000 B.C., mention the ancient Greek (Minoan, Mycenaean) civilization, touch inadequately on the Hebrew Biblical traditions from about 1700 B.C., then quickly jump more comfortably to 1000 B.C. and to the classic Greeks, starting say from Thales (600 B.C.), or Herodotus (500 B.C.).

It is useful to have such sources as a sampler (2–10).

Mellaart (6) and Piggott (7) are excellent starting points. Perhaps 30,000 to 40,000 years ago during one of the last glacial periods, the hunting population of Homo sapiens in Europe might have been a few thousand.[2] Human shelters in Europe date back to perhaps 40,000 B.C.[3] In 11,000 B.C.,

---

[1] Purely personal and trivial but indicative, the author has spent a frustrating 6 months attempting to get a crated painting delivered by a major shipping firm from London to Philadelphia, dock to dock.

[2] The Mousterian culture of Neanderthal man was rapidly replaced by men of modern subspecies as the Upper Paleolithic culture between 30,000 to 40,000 years ago (Würm II/III period) all over Europe (11).

[3] Man's earliest known constructed dwellings, structures built by nomadic hunters about 300,000 years ago, have been uncovered in recent years near Nice [see (12)].

a small migratory reindeer-hunting population existed. By 8000 B.C., the last glacial phase was over, and Europe became temperate. The dog was domesticated. The red deer became a principal source of protein food. The population of Britain might have been perhaps 10,000 at that time. A simplified source, Mellaart (Figs. 1, 2, 26, 41), depicts the Near East communities from about 9000 B.C. to 5000 B.C. Piggott (Figs. 12, 20, 26) depicts some measure of the distribution of agricultural communities in Europe in about 5000 B.C., 4000 B.C., and 3000 B.C. In a loose way, the densities depicted are sufficient to regard them as a near-"ideal-gas" continuum (or slightly condensed vapor continuum in which the settlements are liquid or supercooled liquid droplets) of Neolithic settlements, not much different from animal populations ecologically distributed in some region. There is nothing, Piggott states, in these remains to suggest any structures within the village settlement that can be interpreted as temples or communal shrines forming a focal point in the layout. Suddenly, in Europe, by 2500 B.C. at Stonehenge[1] and by 2000 B.C. during the Minoan period in Crete, ritual building—temples and palaces— begins.

In Linton (5), the origin and spread of early civilization is given in the following order:[2]

Tigris-Euphrates Valley	c. 4500 B.C.
Nile Valley	c. 4300 B.C.
Indus Valley	c. 4000 B.C.
Crete	c. 2800 B.C.
Huang Ho Valley (China)	c. 2200 B.C.

Between 4500 and 3500 B.C., city life and Sumerian culture emerged. (Pages 298–312 provide an excellent description. About 3500 B.C., there likely was a great flood, which was the basis for the biblical myths.)

At roughly 2000 B.C., the Sumerians succumbed to a Semitic conquest, and the two groups fused. (Abraham, for example, probably represented one of these Semitic Akkadians, living in Ur—a Sumerian city—in about 1700 B.C.)

The Sumerian-Akkadian culture spread into the south of Mesopotamia. There, it was taken over by the Babylonians, at their height about 1700 B.C. (the age of Hammurabi, its best known ruler).

Lower Egypt was united under a single ruler by 4500 B.C., and it conquered Upper Egypt about 250 years later. By 4000 B.C., the two regions were again independent. Little is then known of Lower Egyptian culture until the

[1] In the past few years, a number of somewhat speculative studies, from an archeological point of view, on ancient Britain have appeared. Thom (13) presents an interesting study of calendrical knowledge in the British Isles in about 2000 B.C.

[2] More recently Dilmun has been discovered and proposed as a bridge, a trade link (3000– 2000 B.C.) between the Sumerian and Indus civilizations.

Dynastic period, 3300 B.C. In Upper Egypt on the other hand, a rich culture, the Badarian, flourished. In this 700 years, the foundations of later Egyptian civilization were laid (pp. 402–424).

Although there were occasional contacts between Egypt and Mesopotamia as early as 3500 B.C., little real interaction occurred until Egypt became a military power in Asia (1500 B.C.). Thus Egypt and Mesopotamia, though first centers of city life, seem to have grown up independently and simultaneously.

The Indus Valley civilization, as a Neolithic culture, seems also to have produced a distinctive cultural configuration, cut off by geography and climate. However, reconstruction of Indian history (such as racial history) is still largely conjectural. Yet, there is abundant evidence of southwest Asiatic affiliations. The civilization goes back at least to 3300 B.C., with a considerable deterioration in culture, population, and cities by about 2500 B.C. Either barbaric invasion earlier or the classically dated formulation of an Aryan invasion in 1500 B.C. ended the era of the earlier population and then turned India toward its more mystic philosophic bent (pp. 467–485).

The fourth birth of cities took place along the Huang Ho. (The first European culture, of Crete, was not a city culture.) This center became ancestral to the great Chinese civilization and had absolutely no connection to Europe.

At present, there is a gap of about 200,000 years between early man in that region of Northern China and the appearance of Neolithic culture in 3000 to 3500 B.C. The Yang Shao culture developed in about 3000 to 2500 B.C. From this, near the Yellow River, there arose the Lung Shan culture— perhaps 2300 to 2000 B.C. Both these cultures showed highly complex development levels without metal. From the latter culture arose the Shang Dynasty (1766 to 1122 B.C.) of about 1800 B.C., when China drew abreast of Western civilization. It flowered, recognizably Chinese, by 1550 B.C. (The South China history, other than by penetration from the north, still remains a blank, pp. 520–536.)

It is not remiss to point to Linton as one source for the interesting statement, "According to the Chinese historian, Mencius, Chinese history moves in 500-year repetitive cycles." (It has been our provisional estimate that this is the magnitude of time constant for most other civilizations.[1])

Cretan civilization was derived from the southwestern Asiatic Neolithic center, more influenced by Egypt than by Asia. The first Cretan settlement may have been about 5000 B.C. Contact with predynastic Egypt seems to have been established by about 4000 B.C. Not agriculture, but the production of olive oil and pottery and the conduct of seafaring and commerce were their likely fortes. Copper, the "metal of Cyprus," was in trade by 3000 B.C. The

---

[1] The 500-year time constant of a civilization should not be confused with the 70 to 200-year time constant for major social and technological ideas referred to earlier.

establishment at Knossos was the center for the Minoan priest-kings. In about 1500 B.C., Knossos fell, and Crete became culturally insignificant. Probably Indo-European invaders in the period 1700 to 1500 B.C. conquered the Greek peninsula, and possibly the Myceneans overthrew the Cretans. The tribes invading the Greek peninsula were of diverse culture and "apparently they followed the familiar pattern of conquest, each noble family . . . setting up a separate principality. They concentrated in their hands the economic surplus of the conquered communities and employed it first in the building of great fortified holds from which they dominated the surrounding countryside. . . ."[1]

> It is clear that the books cited, particularly Linton, bring the beginnings of civilized man into elegant view. We could do no more than to whet the reader's appetite to own and read Linton or to see the pages of Life's *Epic of Man* often turned within his home.

What we have done is very briefly to have alluded to the growth of civilization from its beginning in Neolithic times, 7000 to 5000 B.C., well into historic times of 1500 B.C. The nominal termination of 1500 B.C. was a pause at the beginning of the Western cultural tradition; namely, both the Biblical tradition and Mycenean pre-Homeric tradition have their roots here. [The next 2,000 years may be traced in (9).]

From their origin, agricultural settlements had flowered, spreading from their southwestern Asian center on the hills surrounding the Tigris-Euphrates River Valley and also springing up independently in other river valleys. A small number of city-based centers grew up. Their possible foci were abstract religious symbols. Walls, towers, symbols, slavery, class distinction, and division of effort marked their endeavors.

All the cultural configurations were not equally developed at any point in that time period. There was some intercourse, but certainly no free flow of culture.

Turning to a remarkable feature of human cultures, their detailed codification of rules of conduct, Barnes, for example, lists five great stages in the evolution of jurisprudence:

The Code of Hammurabi (1700 B.C.)
The Justinian Code
The English common law
The Code Napoleon
The German Hohenzollern Imperial Code

(Surprising—at least to us—is the omission of the Mosaic Code and the American Constitution.[2])

---

[1] At the time of writing, a frame of reference for Atlantis and its relation to Greece is in process of emerging from myth to testable hypothesis.

[2] We commend to our readers, as an introduction to the exciting story of its origins, S. Padover, *To Secure These Blessings*, Washington Square Press, New York, 1962.

Yet, there is a note in the University of Pennsylvania Museum that states there are now three earlier codes known [see also (8)]:

The Code of Ur—Nammu of Ur (King of Sumer and Akkad), 2100 B.C.
The Code of the city of Eshnunnu (Akkadian)
The Code of Lipit-Ishtar (Sumerian), 1870 B.C.

Invasion, enslavement, the walling of cities, law, feudalism, empire—civilization, its rise, flowering, and decline and fall, had come into existence.

Viewed from our remote perspective, the period 1500 B.C. to 2000 A.D. (just as remote as the originating age 5000 B.C. to 1500 B.C.) exhibits only a limited number of new elements. Of course, a major element that had not emerged yet was the monotheistic doctrines, which can be dated as:

Hebraic (Moses—1300 to 1200 B.C.)
Buddhist (Buddha—560 B.C.)
Christian (Christ)
Islamic (Mohammed—570 A.D.)

The sharpened intellectual foundations of western thought had not been developed by the Greeks. Modern science had not been created. [See, for example, Taton (14).] The full range of the political development of nations had not yet unfolded.

From this time on, we can begin to name elites in increasing number —Hammurabi, Sargon, Rameses, Midas, Nebuchadnezzar, Tut, Nefret-ity, Moses, Homer, Xerxes, Cyrus, Darius, Pericles, Alexander, Aristotle, Socrates. One question that arises is whether they were essential to the story of mankind?

Our provisional answer is no. By 1500 B.C., we had the full range of human interactions, including full ranges of religious belief and practice, intellectual belief and thought, political belief and practice, economic belief and practice, sociological belief and practice, esthetic invention, technological invention, and "Edisonian" innovation. Countless elites had come into existence, made their contribution, and passed away. The extra drama of large-sized "public" figures did not represent that much change in the rhythms of history.

There is no point to us continuing the story of history. The references cited can do it infinitely better. Thus the dawn of Western history is the end of our excursion into human history. The social continuum of the nations of the world had fully begun.

With regard to the dynamic picture of man, most of us are much influenced—probably prejudiced—by the historian who forces a literary description onto the pages of history, barely having time to crowd in those "colorful" individuals—elites—whom he feels represent the story and thread of history. It is as if the high-frequency oral response or manual-motor response were the only indicators of behavior. The same complaint voiced differently is contained in the goals of the archaeologist.

History is not the description of what particular individuals did or of general changes in things. It is the description of the dynamic quality of life in social groups, i.e., the structure of the patterns of life (15–17).

The only quality still missing in the early history (and even current history) is the full ergodic quality of a freely exchanging society. For this we must return to our thesis. The biosystem has too high a bulk viscosity. Thus it ties up tremendous internal energy which is not easily released by contact. Why? Because, basically, behavior has a chemical base. It is the entwined endocrine and enzyme signaling that is involved. The nervous system simply gates the response. (It shapes the high-frequency response and is shaped chemically in the low-frequency domain.)

With the neocortical structure in the nervous system, it makes use of extensive visual signaling. (Two aphorisms are illustrative: "Out of sight, out of mind!" and the current aircraft pilot's rule, "See and be seen.") The foundation for so much of behavioral exchange is the exchange system set up by visual signals. This makes up the context of the highest-frequency inter-changes. However, its influence on the endocrine chain patterns is somewhat slow. In this step are the slow energy exchange bound and the limitation to near-neighbor exchanges, i.e., to a short-range force between people. (People affect each other mainly by close contact; memory trace is not as significant, except for frozen-out "repressions.")

Organ and gland response is the order of the day, not central nervous system rationality. Thus what emerges in civilization are the "emotional" rhythms of war, of conflict, of new patterns of changing social behavior, i.e., all of the concomitants of civilization.

As Ansoff, a planning specialist for Lockheed, points out in (18), a 1954 Brookings Institution study showed that few of the 100 largest United States corporations which have grown have stuck to their traditional products and methods in the period 1909 to 1948. Of the 100 largest in 1909, only one-third were among the 100 largest in 1948; about one-half of the new entries to the group in 1919 were left in 1948; less than one-half of the new entries in 1929 were left in 1948, etc. A majority of the giants of yesteryear have dropped in a short span of time.

With regard to the nature of individual elites, Mills' book (19) is still an excellent picture [just as (3) depicts quite vividly the kinds of unknown elites of yesteryear]. Recent attempts to do better, by pointing up the characteristics of today's "ruling class," do not quite come off [see Heilbroner's review of two recent books, Domhoff's Who Rules America and Rose's The Power Structure (20)]. A more recent entry, Lundberg's The Rich and Super-Rich (21), certainly deserves study. Revising his earlier book, he enters into spirited controversy with Mills. Reviews of this book have ranged from the assertions of its triviality to statements of its major importance. Truth, we believe, lies between him and Wright. The reader will do well to familiarize himself with these sociological poles.

Essentially, elites are "temporary" precipitated structures. They cannot last historically. However, they polarize human effort. (Their range of influence is a few generations at most as they attempt to pass power and prestige to their subsequent issue.)

> Many of us—including the author—are quite polarized by sight and the written and spoken word. We identify with these figures (e.g., almost no male can avoid being stirred by any of the sex symbols of the cinema). In this document, the issue presents itself quite strongly.
>
> If I—author—were a Martian noting men (or a human noting mice), their rises and falls, deaths and transfigurations would not touch me. But I am human. Thus life and death, bombing, killing, human despair, poverty, and suffering reach me. However, in the long run, I cannot tell what difference it will make if I try very hard to influence or lead men.
>
> Conversely, there are others like me who hold to tradition. Only by choosing the safe path of what worked yesterday can we stay even.
>
> The fact is that we are a biological species, chemically driven. Too great a degree of objectivity and disinvolvement is not permitted to us. We are driven by our insides. Thus we can only see ourselves and guide ourselves to a very limited degree. A pity!

The dynamic history of civilization is the history of a biological species. In this, it is the story of an evolutionary, adaptive system, which is not fully ergodic but quasi-ergodic. The more things change, the more they remain the same is the ergodic part. However, just as an expanding drop of water in an inhomogeneous medium can illustrate, each new tentacle of fluid creates changing boundary conditions by small quantized jumps as it expands homogeneously in its field. In time-averaged fashion, the motion may even be viewed as continuous.

At this scale, it is more to the point to see the story of the system of nations that comprise civilization as an ecological boundary-value problem. It is the story of the coalescence of civilization in lush river-valley systems, the growth of the means of production, the expansion in space, the rape of the land, the expansion, improved technology, and the further rape of the land. And now—which way?

The historical continuum is not a system that can be described by a long unbroken continuum but by the perturbation epochs, with as many phases or domains as are necessary to describe such piecewise changing dynamic boundary conditions. The study of civilization, in similar light, can be examined in Spengler (22), and Steward (23).

It is clear that anthropologists still distinguish prehistory and history by the beginning of the written record. Yet this is not so important. It simply indicates how close to kindergarten the unraveling of history still remains. The men who established settlements but did not write for 3,500 years were still men. The rulers and others who we can identify during the past 3,500 years were involved in the same basic modes of behavior as their forebears.

Yet we are bounded by an unfolding evolution. At most, speculative investigators like Leakey place the diverging history of man from his ancestors at perhaps 14 million years ago.[1] Man as a species with rapidly changing social characteristics had not come into his own then.

However, the prehistory of early man, perhaps 14 million to $\frac{1}{2}$ million years ago, reveals only a very sparse and dilute (low-density) existence. (Though not immediately applicable, Childe provides an estimate for the number density of the aboriginal population of Australia as never exceeding 1 per 30 square miles. Clark estimates the entire world population in Middle Paleolithic times—30,000 to 40,000 years ago—as one million. With perhaps 30 to 50 million square miles available, a similar density is estimated. However, both of these estimates are for relatively modern man.)

More studied estimates are given by Deevey (24). He estimates prehistoric populations as follows:

1,000,000 years ago	100,000 population
300,000	1,000,000
25,000	3,000,000
10,000	5,000,000
6,000	90,000,000
2,000	130,000,000

Under such circumstances, the compelling drives must have been the roaming patterns for food, the search for adequate shelter and warmth, and sufficient reproductive drive to maintain the species. Thus thermal boundary conditions, and population densities of various species represent the ecological constraints of the human species.

At that the main needs were for techniques or tools to increase the food-gathering productivity to a point that men could live. At most one may surmise that very loose family life existed. It is usual to regard today's patterns among animals (chimpanzees, gorillas, ruminants, rats) and primitive tribes as crude models for early primitive man's behavior.[2]

The temporal bounding or driving conditions are generally taken to be the course of the ice ages. If nothing else, their occurrence has in addition produced sufficient distortions and obliterations of prior remnants of human history that the earlier record of man has not yet been read in any detail.

Thus, the story of modern man begins with the end of the last ice age, nominally 15,000 years ago. In its critical ending, one may trace the opening phases of historically known modern man.

---

[1] A recent study by Wilson and Sarich (1969) suggests divergence of man from ape only took place 5 to 10 million years ago.

[2] Ethologists, of course, request very careful consideration before attempting to transfer inter-species results, particularly species that are biologically apart (e.g., rat and human).

What is to be found at the near-recent period of 15,000 years ago? Nomadic tribes wandering, cavemen communities, small fixed settlements, and a total human population of perhaps a few million. The rudimentary atoms of little tribe-city-state-nations have come into being. Their internal patterns are not well known in detail—this is pre-recorded history—but their tools, their weapons, their vehicles, their buildings, their artifacts, and their burials give us some information. They are organized to solve the patterns of living. It is this key, throughout all of history, that is the story of human civilization. It is designs and patterns for living.

It is too early to detect yet, but one suspects that the rhythms of civilization must already have begun. Over and over again, the internal forces among atoms—these atoms are now like near-ideal-gas atoms with very weak interaction—give rise to molecular spectra representing bound configurations, and their bonding energies begin to present transport coefficients of viscosity (momentum transfer), of thermal conductivity (energy transfer), of mass diffusion (species transfer), and of frozen-out states. The frozen-out degrees of freedom of the solid state are most intriguing in illustrating spatially fixed or bound domains, accommodation at boundaries, temporary coalitions that change only slowly, precipitation, etc. In order to achieve a higher-order, more complex level of organization, a new entity likely must come into play —an impulsivelike nonlinear amplifying element, which can mobilize and entrain considerable amounts of energy. Such amplification is realizable today in communications networks (including speech) and computer networks, and it creates a highly unsymmetric distribution far removed from the Maxwellian. In the social milieu, high-energy mobilizing catalysts or social elites must come into existence. These superenergetic atoms entrain large amounts of "people-energy" (that is the product of the number of people and the individual energy they each can bring to bear). The missing ingredient that is not well documented is the catalytic capability of such atomistic elites among early humans to mobilize their energies. This process, just as in the chain reaction, deals with the question of how much fundamental metabolic "life" stuff the individual deals with in order to live and how can it be mobilized.

The wide-ranging animal, including the example of a human who is occupied all day with food gathering, apparently has no spare energy for other pursuits. For such, he must have leisure and the internal brain capacity to utilize the power made available by leisure.

Perhaps—most speculatively—a relatively "recent" acquisition of speech capability represented the real threshold.[1] Less speculatively, a most

---

[1] For example, Oakley (25) states that the rapidity of cultural change in Europe during Upper Paleolithic times would scarcely have been possible without language and that speech was likely only an invention associated with this late cultural development. We are willing to accept this thought but also prefer to strike the provocative note of calling attention to the issue of handedness.

significant change occurred 5,000 years ago, as "The high yield of about thirty bushels of cereal to four-fifths of a bushel sown revolutionized man's diet."

Once man's food-acquiring productivity exceeded unity[1] (stemming from this most important advance in food-gathering, since such productivity was not first reached on an animal diet in symbiotic relation with grazing animals—although this also involves the agricultural productivity—or in fishing or preserving, both involving greater technological skills), then modern man exploded. One excluded bound in phase space is removed. Man expands ergodically into that space and effectively acts like a new species.

One must then note what the new emergent pattern is, within its new quasi-ergodic limitation. In Marxian terms, the society segregates into two classes (subspecies), those elite who control the new, more effective means of production and those who do not. The societies tend to focus around this central fact. Clan-city-nation begin to form around this. Earlier, only sexually oriented leadership was dominant in primate societies.

Societies now begin to have this dual character of haves and have-nots. The forces of attraction pull the have-nots to the orbits of the haves. Within the scope of the ecological constraints, societies grow to supermolecular size. The problems that face them are now organized food production, organized distribution (process lags begin to show up), organized waste disposal, organized technology (of tools, machines, vehicles, weapons, buildings, artifacts), organized defense, organized propagation of the species, and organized education (26).

Out of this jump in productivity, there begins to emerge the nesting of unstable constellation within superconstellation until stability emerges. The full complexity of human culture begins to emerge. It is no longer the simple Marxian two-class society. Many more elites arise. [See, for example, Mills (19) for its modern structure; or, contrarily, see Lundberg (21).]

Modern civilization has been born. Modern social rhythms have been born. Civilizations rise and fall with time constants of the order of 500 years. Major moral ideas arise with a time constant of the order of 2,000 years. The gestation period of significant ideas is about 200 years.[2] War begins to beat a near-17-year rhythm.

For the last thought, it is not inappropriate to use as modern references (15, 28–30) and Dewey (31).

---

[1] If a man must work all day in order to acquire sufficient sustenance for a day, this may be regarded as unit productivity. It is only when surplus is generated that the productivity exceeds unity.

It may be surprising to some, but animals at both ends of the mammalian size scale—the shrew and the elephant—must devote almost all of their time to foraging.

[2] An interesting overview of technological change in Europe over the past 200 years is presented by Landes (27).

It will be useful to state an initial position to start the dialectic. Although it may be quite dubious that territorial control is a major concern of the individual (the opposite view is expressed at great length in Ardrey's *The Territorial Imperative*—we are certainly willing to concede that it may be of concern to some species, illustratively the dog), the relation of population densities, population needs, and ecological interaction make the problem of the sphere of influence of the nation a primary one. This is no less true for the mobile (in space and time) atom surrounded by its nearest neighbors, or the crystallite, than it is for the nation. The problem starts and may be first viewed in a dilute ecological sense, in which the interactions are very few and far between. It then grows, as we approach the current scene, to the denser ecology in which nations bound each other and have the space fairly divided. However, there remain other boundary conditions. Their own internal structure may be dilute—America today is still quite dilute—or new frontiers (such as our space frontier) can continue to open. Thus it is not a priori clear, although the existence of dynamics and interaction is clear, whether one is dealing with gaslike, liquidlike, or solidlike properties. Actually, as the theory of materials shows, these tend to become matters of relative time scale, of the magnitude of the ratio of time of interest to the relaxation time (which is given by the viscosity or index of momentum transport to the elastic modulus or index of resilience). For biological, in particular human, societies, this becomes a question of time scale of the "bulk" viscosity to "bulk" modulus in which the energy tied up in "elite" degrees of freedom contributes to the "bulk" viscosity, and the totality of acoustic modes (i.e., some measure of the totality of communications channels among humans) contribute to the "bulk" modulus. In general, we believe that the bulk viscosity is quite high in complex societies, so that they tend to act like solid plastic-elastic aggregations or at least supercooled liquids. However, as with some materials, society is extremely inhomogeneous. Thus, the transport is locally facilitated (in analogue) by patches of grease or banana peel and hindered by rubble to speed up and slow down social time.[1]

In treating nations, Morgenthau (28) writes of a balance of power and of alliances. (Obviously, this means balance of forces or of energy.) Aron (29) speaks of power and of constellations. We can little doubt, upon reading their works, that we are discussing dynamic systems in a physical sense. The only thing that is lacking is how to identify the degrees of freedom of these national and international systems.

Morgenthau starts out forthwith with the thought that the history of modern political thought is the story of a contest between two different views of the nature of man, society, and politics. One believes in a rational moral

---

[1] It is the function of social "engineers" of various sorts to learn how to manipulate the society, to "heat" up and direct its interests, to cool off its complaints and frustrations.

political order which can be operatively implemented from principles. (The "ought" of societies.) The other believes in an irrational point of view derived from "forces inherent in human nature," that may be only singled out or implemented empirically. (The "is" of societies.) It appears that Morgenthau, accepting the latter view under the title of "political realism," singles out six fundamental principles for this doctrine:

1. Politics is governed by objective laws rooted in human nature (which has not really changed for thousands of years).
2. The path of political realism in time is governed by the concept of "power" (i.e., force), so that statesmen act in terms of "interest defined as power" (that is, by what forces they can bring to bear in their interest).
3. What constitutes self-interest is mutable.
4. Political realism is aware of moral significance (i.e., of the consequence of one's acts upon others).
5. Political realism refuses to identify the "moral" conduct of any nation (including itself) with an "absolute" moral code.
6. Political realism is a distinctive belief, involving the autonomy of political decision. (Thus contrasting with historical materialism, which proposes the autonomy of economic decision.)

Although we would at one time have accepted such absolute doctrines —categorical imperatives—as representing truth, a quarter of a century of experience in physical measurements and, more recently, the systems science of the biological system (and of other systems) leads to another resolution, whose social consequences will be first stated. Since the individual operates with the patterned constellations of his internal oscillators and within his local interpersonal orbits, the basic conclusion that one draws is that his self-actuated motions (including speech) are externally irrational but internally consistent (32). To this extent the embedding of political behavior in "human nature" is quite real. However, "self-interest" is not the guiding human principle. Instead there is only an internal algorithm which is internally consistent and which may only be marginally useful to the individual.[1] "Successful" societies and species serve self-interest with sufficient threshold to survive, little else. However, it is obvious that no group or entity has done this "forever." On the contrary, they are commonly quite limited. Thus, "self-interest" is a convenient fiction for the short run. We can, therefore, agree with its mutability. The reference to moral "law" is artificial because morality is more nearly what suits elites. Thus the redundant definition proposed is

---

[1] There are so many individuals who commit suicide, literally, or who choose paths—excessive smoking, drinking, eating, drugs, hard-paced high-anxiety lives—destructive of a long-lived pattern that to appeal to "self-interest" for other than the passing instant is imaginary.

that politics is the wielding of force for self-interest, while aware of its self-interest (really in terms of concepts becoming to itself). Politics is, then, actually the wielding of social force for the short-range self-interest of elites. Further, elites (by definition) are those that can tie up the most people-energy into patterned orbits around their own interest.

Now comes the salient feature. Although the elites' vocations[1] may have served as a focus for large segments of the populace, the actual human patterns involve the two handfuls of orbitals that make up its total constellation (32). Their motion "fits" the focal center but has nothing else to do with the center. The human acts through a heterogeneous collection of inputs that have no essential metric. Thus, all sorts of actions, compatible with the orbitals, are possible. This means that people can be incited into orbit or self-actuation by the words of demagoguelike[2] appeals (we are not discussing any value but an objective almost-physiological judgment of facial expressions, gestures, tone level, and sweating, laughing, and breathing responses, etc.)—by Hitler, by Lenin, by politician A, B, or C, by Hal Holbrook, by the local pitchman, by dianetics, by cancer cures, by religion, by almost any theme however sane or outrageous, given time; thus, by the appeal to war. Humans are not externally rational. They get behind the wheel of an auto and, with little provocation, they can kill, or they can kill themselves. Their aggressions are not controllable.

Thus, it is really true that "human nature" governs response. It is only necessary to know the range of human nature, our earlier subject. Thus politics, as conducted, is more a study in manipulation by rules, empirical or otherwise, that will work than a rational unfolding in time. The politician or demagogue must estimate what foci will capture men about him into orbital resonance. Machiavelli's *The Prince* and *Discourses* have always been offered as prototypes of the "political behavior of the human male." To this degree, we agree with Morgenthau.

Further, it is clear that it is not the appeal to reason that moves men. The rules are more in accord with what is known or being learned by advertisers or public relations people.[3]

However, this probes at the question of what time scale—what portion of the spectrum—are the effects being sought. The affairs of nations, obviously, are more nearly to be averaged over (as in the theorem of the virial, or as in

[1] The "elites" in society have most often been politicians, religious figures, performers, governors of production, and military figures.

[2] Note, the use of the term "demagogue" is not used with opprobrium. For example, E. J. Hobsbawn, a historian at the University of London, states, in a review in the *New York Review* regarding some books on the meaning of 1848, "... Louis Napoleon immediately recognized and never forgot that the anti-revolutionary leader must henceforth be a demagogue."

[3] Or confirmed lechers. As a newspaper tidbit reported in the *New Statesman* would have it, a door-to-door salesman reported (memory is not precise): "If they came to the door smiling, I could always talk them into the bedroom."

any attempt to get at continuum laws) the individual atomistic time constant, here the "life" of the man, or at least his effective life time constant. Since he grows for 15 years, is near maturity for 30 years, degrades in "middle age" for the next 15 years, and approaches his final decay state over the remaining 5–10–15–? years, it is perhaps 15 to 40 years that represent his effective "life" time constant. (Today, from 20–25 to 40–60 years of age.) As we attempted to show, history is continuous—by overlap—over a 100-year period.

The discovery of Dewey's study (31) was quite disconcerting to the author. With full understanding that some groups or study directions may be considered controversial and with every effort neither to express opinion for or against the beliefs of the particular group, nevertheless we must express tremendous admiration for the scope of Dewey's study. It covers in detail 2,500 years of man's effort. Skipping the nonsense of three or four significant figure numbers for flexible cycles (in the sense that the econometrician Wald has used it), it appears clear that the data show cycles of 17 to 18-year overt aggressive interactions (i.e., "battles") among intra- and international civilization groupings. If the validity of the data is accepted (the prime investigator's references are given), and this would require extended study to retest, then it requires very serious consideration. To these data must be added Richardson's studies (30), which have provided some background for the course of international quarrels.

These data appear consistent with observations on the physiological nature of "human" behavior and, therefore, are worthy of some preliminary explanation or exploration. We are pleased to credit our colleague, Dr. S. Cardon, with suggesting a salient clue. He pointed out that in most people's lives (males in particular) the most exciting period they may have been exposed to is the war period if it only mildly touched them or its effect could be glossed over in passing.

It has been our biophysical thesis that the physiological-psychological system (in the human) consists basically of a self-actuated motor system made up of large numbers of mediated oscillators[1] or only nonlinearly stable bio-chemical chains, with a plastic command-control and guidance computer (the brain) that develops its own pattern of satisfying the conflicting demands among these oscillators (biologically in the sense of "antagonistic" systems; physically in the sense of force-balance systems). The brain response emerges partially from the near-childhood interpersonal orbital relations and partially from maturation. The net effect of the motor patterning is to maintain a "homeokinetic" (defined as dynamic regulation) balance in the human interior.

Now, while the "postural" sets of the body (i.e., actions of the body on the body) include fundamental physiological ones such as food-seeking,

[1] i.e., oscillators whose operating state may be mediated or changed.

oxygen-seeking, water-seeking, and elimination, they also involve psychological ones. It would appear that the species' psychological characteristics
are not so much evolutionary as emergent; that is, through genetic mutation
and perhaps even ecological pressure, there is generally shifting, perhaps even
branching processes, reminiscent of diffusion (i.e., molecules can move in any
direction; but in summation there is a probability distribution of preferred
path). The major requirement is that what emerges has survival value. (We
are not dealing with a simple linear chain of A psychologically evolving from
B, who evolved from C, etc.) With the emergent evolution of more complex
brains, the patterning texture of behavior increased in complexity. However,
it is not simply a matter of brain size. Certain animals show considerable
patterning. Others do not. However, as is interestingly discussed by Lorenz
(33), aggression is one element quite common in many animals.[1] [Reference
(34) is surprisingly weak in furnishing any theoretic background to its major
topic.] Man, the big-brained animal, has considerable patterning complex
and is very aggressive in societies. In a dilute ecology, he roams in search of
food. With a discovery of the increased productivity of yield of the lands, he
focuses his ranging patterns. He comes into a more fixed spatial orbit. The
family grows into tribe into settlement into state into nation. However, the
underlying patterning texture remains. The question is what happens to the
elements in the patterning texture, now viewed in the complex of nations.

Oxygen seeking (at least until today) posed no problem. (Man had to
confine himself to a particular altitude range.) As far as oxygen demand was
concerned, at low altitudes man was a dilute species in the earth's atmosphere.
His demand did not exceed the supply.

However, water, food, even sex [see such material as (35) for content and
references on development of courting ritual or marriage custom, etc.] have
been organized into orbital constellation.

A posteriori, thus, many of the human hunger cycles are accounted for
in social continua. Now for the usual crucial question—stability.

The question of stability, in a limited sense, is generally whether a signal
circulating around a network will be reinforced in phase or be cancelled out of
phase. It is our thesis that certain orbital patterns act resonancelike; they
reinforce when going around the loop enough times, without damping out.
The characteristic must be regarded as more of a reverberation-type
distributed network than a lumped network.[2] Now with regard to the human,
it is commonly the longer running psychological span, 3 to 4 weeks, that
represents the circulating time constant. (Its atomistic element is likely the

---

[1] In antagonism to these views is Montagu (36). However, once one gets away from the view of
"instincts," the observed social facts speak for themselves. War and aggressive conduct have
been quite persistent in many human societies since Neolithic times. Aggressive behavior is
quite common in many species under conditions of crowding.

[2] Holographic storage, for the brain, possibly even for society, is being considered.

daily pattern of anxiety and complacency, stress and rest, etc.) Over this longer time period, the signals cumulate that lead to an anxiety set for the organism. With the passage of additional circulating time constants, further signals may intervene as a damping "fatigue" set in any anxietylike problem in the individual and is replaced, at least temporarily for the next cycle, by a more relaxed state. Other sets that are found in the human are anger directed at an object (polarized into a friend-foe configuration) and fear (polarized in an extreme state as I-for-I, which can be pictured as I hugging myself, closing everything down on myself).

One man shouting "Fire! Fire!" in the streets may get a certain amount of rational response from passers: "Where? What?" A man shouting so in a crowded room will invariably get a panic. Thus repetition, number, and boundary conditions will govern. However, aggressive tendencies will amplify. A most despicable but human thing happened a few years ago. When a sick person was threatening to jump from a building, a morbid crowd gathered and soon established a chant of "Jump! Jump!" Still more despicable were the Hitler, Mussolini, Stalin spy eras in which so-called mass-hysterias were developed by the big lie, agitprop. What is more common everywhere is the artificial language, the emotive language, the "slogan." "Death to the infidel!" "Down with X!" "Hate Y!" As Walter Lippmann said, "When patriotism is invoked, reason is its first victim." The ground substances of these human responses are the interpersonal communications at biochemical levels. An interesting book on communications in humans is Hall [(37), or more generally a recent review article by McCulloch and Brodey (38)]. Hall is critically received by many of his colleagues.

Applied to the continuum of civilization, the kinetic nature of the order of events leading to national quarrels is discussed by Richardson. Illustratively, his studies indicated that the strength of the opponent served as no deterrent.

Thus, the basic structure of civilization by nations is not stable. At most —as the grains in a continuum—they can only form diffusable aggregate structures. The only possible stabilizing factor, international law, does not exist; only regularizations do.

Morgenthau's book and practically all other political histories before keep saying that the only stable elements which can exist are the changing alliances. The net outcome is what Dewey describes, wars every 18 years (with a wobbling periodicity).

Does this mean that war and peace are the only two states of alteration? No, there is a cluster of events—trade, coalition, confederation, unification, peace, competition, alliance, detente, war. Up the the present however, whatever the cluster sequence, nevertheless peace and war alternate. It seems to us that whenever an internationally (or intercommunity) connected youthful generation exists which has grown up without a war, the aggressive instability is there. The aggressive patterns of their elders can incite the scene and drive it into resonance.[1]

A very speculative "explanation" may be that the presence of the aggressive young adolescent with his forceful sexual drives, who has not tasted the horror of war, possibly creates the unstable situation of egging male elders into seeking another channel for their aggressive expression of frustration at not being able to physically compete with the young. This outlet becomes the less personal, more abstract pattern of nationalistic aggressiveness and war, the vicarious enjoyment of the excitement of aggression.

But what of the small peaceful nation that has not been caught up in war for hundreds of years? War and aggression can only be undertaken without too much thought of consequence by the powerful. The weak, who are not too certain of their strength, can only observe the patterns, note sides, take them, avoid them, or basically involve themselves only to the extent they must. They bounce with a Brownian motion in the wakes and backlash of the large pulsing changes and live the experiences out vicariously. (However, as Richardson pointed out for nations and recent events have shown for individuals, it is possible for the underdog to turn and begin to do battle, irrationally, against whatever odds exist.)

It would be exciting to attempt to show some picture of the changing dynamics, which perhaps could help lend conviction to the 18-year war thesis. (Consider a possible collection of maps of Europe, showing their national groupings at 4-year intervals. If then scaled as a field versus time, any dominant "frequency" of change could be noted by eye.)

Is it possible to live with this kind of nationalistic instability for a long time? No. We may point out that the operational time scales that blanket the lives of nations seem to be:

1 day—the atomistic cueing level for the individual

3–4 weeks—the "Brownian motion" jitter of individuals

1 year—the seasonal cues that drive the human away from intense preoccupation (summer vacation, winter cold in temperate zones)

4–8 years—the national time constants for internal governing

18 years—the periodicity of national aggression

---

[1] Lewis Feuer's recent book, *The Character and Significance of Student Movements*, on the nature of youthful rebellion has been criticized for being quite perceptive in its historical perspective on other times and places but having nothing to say about the current revolution of youth. In our opinion, the problem is that we have had one war too many, the Korean War. The combination of the background threat of nuclear warfare, one war too many, and the instant communications feedback of TV has presented American youth and European youth too intimate a glimpse of an adult world without sugar coating. The disillusionment with all potential father figures has been too complete. This does not mean all youthful society, but a sufficient number. It only takes a small fraction of a society to put it into turmoil. See for example (39).

200 years—the time constant for the emergence of major ideas
500 years—the cycle for nations
2,000 years—the time constant for major moral ideas

At the level of 1 day, the individual faces the repetitive routines of most of living. He eats, works, eliminates, and sleeps. At the level of 1 week, much of the rest of his life cycles make themselves evident. In the range of 3 to 4 weeks, likely the large-scale waves of emotional and intellectual commitment are to be found. In the human, we have proposed that this tends to be entrained at the female's menstrual cycle. At the range of 1 year in most civilizations, the social organization makes itself evident. For example, seasonal celebrations—the harvest, etc.—are marked in so many different ways in different civilizations at different times. It seems clear that this is nearly the border between the individual's consciousness and the awareness of his society.

For the very high-frequency response of nations, we can select a confirming note from Beaufre (40), who points to "... forty-eight hours, the minimum reaction time for international diplomacy."

This has the same sense as McCulloch's observation that it takes 0.3 sec. for a decision by the reticular core as the command-control center in the brain, 0.1 sec. for detecting displacement, 0.2 sec. for detecting velocity, 0.3 sec. for detecting acceleration. Less cryptically, there is 1 day to note, inquire, and receive information and 1 day to compose an answer and recompose thought for action. But then, even though 0.1 sec. is "reaction time" for the body, behavior emerges at perhaps 20 to 60 days. Analogously, national behavior, with its reaction time of days, emerges at 4 to 8 years.

At 4 to 8 years, it is common to find the rudiments of the organized spectral effort of government. Although one casually may think in terms of the rule of kings at a rate of perhaps five per century, nevertheless the unit impulse in politics has always been closer to the shorter time scale named. Three to five times in a century a nation changes its face. At 500 years, the nation-civilization has run its course. It is time for major organization change. Finally, beyond the life of the nation lies the organizing thrust of the whole background civilization. It is this 2,000-year time scale that has given the total moral tone of the civilization.

There may be other elements in the time scale [and more discussion exists in (15) and (17) which we consider interesting but which we cannot accept fully]. Nevertheless, the national components of civilizations do decay. We believe greater stabilization is of some interest to man, particularly if it can improve the quality of life.

SUMMARY

1. The summation of all social atomistic entities covers the two-dimensional world. These entities make up man's civilization.

2. Civilization nominally started at three centers nearly simultaneously—the Tigris-Euphrates, the Nile, and the Indus Valley (about 4000 to 4500 B.C.). It is reasonable to suspect that the earlier agricultural communities of proto-Neolithic and Neolithic times (9000 to 5000 B.C.), having solved the problem of food acquisition, were sufficiently unstable to explode toward the creative patterns of civilization.

3. Civilizations show behavioral modalities. Religion, record-keeping, complex buildings, marked division of endeavor, class distinction, slavery, lawmaking, life and death of nations, science, art, sports, vices (for example, alcoholic beverages and drugs), and warfare mark their conduct.

4. The time scales of their rhythms and tidal oscillations are longer than for the individual. War beats a near-18-year rhythm. Significant ideas have a 200-year time scale. 500 years is the time scale of nations. Major moral ideas have a 2,000-year time scale.

QUESTIONS

1. Probe at the quasi-ergodic hypothesis among nations in the following way. Choose a particular up-to-date period (e.g., a week, or month). Determine as far as you can what were the predominant news media interests reflecting all sides in each nation or major nation during that period all over the world. Note or estimate the pattern of diffusion by which the information reached your collecting point.

2. Probe at the quasi-ergodic hypothesis in another way. Select a pair—sender and receiver—as near or remote as you please. Send messages back and forth between the pair by interpersonal routes passing through specified nations. By choosing and tracing enough routes, estimate the informational relaxation times through various parts of the world.

3. Trace the diffusion of new cultural patterns through the nations of the world (e.g., jazz music, nonrepresentational art, "new" theater, miniskirts, etc.).

4. Depict the largest-scale communities around the earth every 500 years from 5000 B.C. to present.

5. Check in detail the rise and fall of major civilizations (as you can distinctively identify them). Indicate their average time scale and the more common factors that marked their rise and fall.

6. Depict the quality of major wars for the past 1,000 to 2,000 years.

7. Review the history of the most powerful demagogues of the past 2,000 years. Indicate the scale of people-energy they tied up and how extensively they were able to lead these people away from normal life equilibria.

8. Discuss some 100 to 200 year period in well-documented history where the civilization was relatively stable, and compare it with a period in which civilization was maximally unstable.

9. Discuss the temporal spectrum of effects in civilization by pointing up evidence for each periodicity named.

10. Examine the history of some nation in detail. Note the epochs of government change in direction for the past 500 to 1,000 years.

BIBLIOGRAPHY

1. Russell, B.: "A History of Western Philosophy," Simon and Schuster, Inc., New York, 1964.
2. Barnett, L., and *Life*'s Editors: "The Epic of Man," Time-Life Books, New York, 1961.
3. "Everyday Life in Ancient Times," National Geographic Society, Washington, D.C., 1961.
4. Piggott, S. (ed.): "The Dawn of Civilization," McGraw-Hill Book Company, New York, 1961.
5. Linton, R.: "The Tree of Culture," Alfred A. Knopf, Inc., New York, 1957.

6. Mellaart, J.: "Eastern Civilizations of the New East," McGraw-Hill Book Company, New York, 1965.
7. Piggott, S.: "Ancient Europe," Aldine Publishing Company, Chicago, 1965.
8. Wooley, L.: "History of Mankind," vol. 1, part 2, "The Beginnings of Civilization," New American Library, Inc., New York, 1965.
9. Pareti, L., P. Brezzi, and L. Petech: "History of Mankind," vol. 2, "The Ancient World," Harper & Row, Publishers, Incorporated, New York, 1965.
10. Barnes, H.: "An Intellectual and Cultural History of the Western World," 3 vols., 1941 rev. ed., Dover Publications, Inc., New York, 1965.
11. Klein, R.: Mousterian Cultures in European Russia, *Science*, vol. 165, no. 3890, p. 257, 1969.
12. de Lumley, H.: A Paleolithic Camp at Nice, *Sci. Am.*, vol. 220, no. 5, p. 42, 1969.
13. Thom, A.: "Megalithic Sites in Britain," Oxford University Press, New York, 1967.
14. Taton, R.: "Ancient and Medieval Science," Basic Books, Inc., Publishers, New York, 1963.
15. Huntington, E.: "Mainsprings of Civilization," 1965 ed., New American Library, Inc., New York, 1964.
16. Murdock, G.: "Social Structure," The Free Press, New York, 1965.
17. Sorokin, P.: "Social and Cultural Dynamics," 4 vols., Bedminster Press, Totowa, N.J., 1937–1941.
18. Bursk, E., and J. Chapman: "New Decision-making Tools for Managers," 1963 Harvard ed., New American Library, Inc., New York, 1965.
19. Mills, C.: "The Power Elite," Oxford University Press, New York, 1959.
20. Heilbroner, R.: Who's Running This Show?, *N.Y. Review*, January 4, 1968.
21. Lundberg, F.: "The Rich and the Super-Rich," Lyle Stuart, Inc., New York, 1968.
22. Spengler, O.: "The Decline of the West," 2 vols., Alfred A. Knopf, Inc., New York, 1926–1928.
23. Steward, J.: "Theory of Culture Change," University of Illinois Press, Urbana, Ill., 1955.
24. Deevey, E.: The Human Population, *Sci. Am.*, vol. 203, no. 3, p. 194, 1960.
25. Oakley, K.: "Man The Tool-Maker," University of Chicago Press, Chicago, 1961.
26. Goodman, P.: The Present Moment in Education, *N.Y. Review*, vol. 12, no. 7, April 10, 1969.
27. Landes, D.: "The Unbound Prometheus," Cambridge University Press, New York, 1969.
28. Morgenthau, H.: "Politics Among Nations," Alfred A. Knopf, Inc., New York, 1961.
29. Aron, R.: "Peace and War," Doubleday & Company, Inc., New York, 1966.
30. Richardson, L.: "Arms and Insecurity," Boxwood Press, Pittsburgh, Pa., 1960.
——: War Moods, *Psychometrika*, vol. 13, no. 147, p. 197, 1948.
31. Dewey, E.: The 17.7 Year Cycle in War 600 B.C.–A.D. 1957, *Res. Bull. 1964–2*, Foundation for Study of Cycles, Pittsburgh, Pa., August, 1964.
32. Iberall, A., and W. McCulloch: 1967 Behavioral Model of Man, His Chains Revealed, *Currents in Mod. Biol.*, vol. 1, p. 337, 1968.
33. Lorenz, K.: "On Aggression," Bantam Books, Inc., New York, 1967.
34. Fried, M., M. Harris, and R. Murphy (eds.): "War," The Natural History Press, New York, 1968.
35. McGill, T. (ed.): "Readings in Animal Behavior," Holt, Rinehart and Winston, Inc., New York, 1965.
36. Montagu, A.: "Man: His First Million Years," New American Library of World Literature, Inc., New York, 1962.
——: "Man: His First Two Million Years," Columbia University Press, New York, 1969.
37. Hall, E.: "The Silent Language," Fawcett World Library, New York, 1959.
38. McCulloch, W., and W. Brodey: The Biological Sciences, in "The Great Ideas Today 1966," Encyclopaedia Britannica, Inc., Chicago, 1967.
39. *World Health*, American Public Health Association, New York, July–August, 1969.
40. Beaufre, A.: "An Introduction to Strategy," Frederick A. Praeger, Inc., New York, 1965.

part four

Back to the Main Track

15

The Ecological Continuum
in Which Mankind Is Atom

In the continuum of the biosphere, the species that is man is an atom. Had we not taken the excursion through the man sciences after the atomistic biological entities such as man, rat, dog, etc., we would have considered the continuum ensemble of such atomistic species expanding on the surface of the earth. (At the present, we must note that the earth surface limitation—actually the continents and their temperature zones—further tend to quantize the species into groups centered on particular foci such as water or food supply. They now become atomistic entities in the ecology.)

At this level, the civilization of man shrinks into atomic proportions again. His historical civilization represents the story of a single species among a host of other species. He may possess self-pride, but there is no assurance that his position is better than fishes, insects, beetles, bacteria, or viruses. These are all viable species. If man makes one fatal misstep, any one of these might inherit the earth. This is neither morbid nor academic, simply a prognostication of what faces man in the long run. [Examination of the past

experience among various phyla is an indicated exercise for man as a rational
biophysical entity, Homo sapiens, who is capable of paying some attention to
his survivability. The other species are not. See, for example, Simpson (1).]

> Man's pride extends to the arrogant belief that he can mold the milieu to his
> own service. Good luck! But suppose he fails? Is it not prudent that he plans with
> a number of alternates in mind? This, of course, is said with intellectual catalysis in
> mind. Is it possible to catch the attention of the elites, to bring into their focus the
> long 200-year range for ideas or perhaps even the 2,000-year range for moral ideas
> as a requirement for planning? Obviously, man is still a dilute ecological atom. He
> does not feel tremendous pressures. Suppose the loss of oxygen or water was at
> stake; or, his reliable reproducibility; or his viability; would this affect this
> judgment? A first-round answer is not promising, e.g., smoking will affect your
> health; do you, therefore, give up smoking? This generation, bringing saturation
> conditions among a number of ecological factors closer to realization than ever
> before, does not show any biological concern over survival. It seems that there are
> always too many locally patterned issues which carry more weight. Thus elites
> really only have a limited number of directions in which they can mobilize people-
> energy. Yet, the ecological cost of increased technological productivity has con-
> tinued to rise from the beginning of the modern period 7,000 years ago, e.g., man
> does not improve the ecology but is a burden on the ecology. It is this that may
> defeat him through some unfortunate coalition of circumstances.
> The human atomic species thus will continue to be abrasive in the ecology.
> Because of the emergent nature of the evolution (of things and systems), it is not
> certain where the future will settle down. One may suspect that major changes will
> come to pass within the next 4,000 years. This is not meant to be facetious. Some
> three major ethics have come into being in 6,000 years, and it is only the second
> one, the Judaic, which has been accepted in practice, while the third one, the
> Christian, has only current lip service. Yet it is possible to conceive of the next two
> moralities. Note, one cannot guarantee their gestation, only their conception.
> One may thus only guess—as with all emergent cyclic phenomena as opposed to
> deterministic closed simple harmonic-oscillator-type phenomena—some one or
> two future cycles, i.e., it is possible to estimate for one or two cycles in the future.
> Such estimates may be made for a system before the phase information goes
> completely bad.

Some of the salient points to be noted about a viable ecology are the
following:

1. A point made by Pattee is that an essential factor in a viable life cycle is the
   existence of biological degradability. If there is no biological degradability of matter, there cannot be a living temporal cycle, for matter must
   be carried through all the life and death phases.
2. This is not only true of the most fundamental levels but of all operating
   levels. As one species goes down, the others go up, i.e., generalized
   predator-prey cyclic relations always exist.
3. As Kerner has continued to develop (2, 3), Gibbs' ensemble theory can be
   applied to competing ecological species to determine the equations of
   change for the ecological continuum.

4. Some of the rudiments of the energetics of ecology have been developed by Odum (4). A more fundamental view is under development by Morowitz [see, for example, (5)].
5. The foundation of evolution—whether of individuals or species—is that it is not the individual who is preserved by replication. Individuals have various mutational characteristics genetically locked up. Whatever adaption is required will be epigenetically selected. Those individuals with survival value will adapt. The remaining existing individuals, who may otherwise have charming characteristics, are ruthlessly weeded out in time. Thus, it is not the individual who is selected but the line of adaptable characteristics.

A corollary is that regardless of whatever intelligence or learning the human species may have achieved, it does not have to survive. How to assure the development of human elites (Moses-like figures) who can understand that and tie up and lead the human species in our evolving future can easily become a fundamental question for the species.

It is clear that the interpersonal binding forces among men are not great enough to overcome the individual disorienting forces to the point that the preservation of the species is important. Instead conduct patterns must be regularized by moral codes. Consideration for species preservation may arise after the development of a few more moral codes. The codes that have existed and which may yet come into existence are :[1]

1. The "traffic policeman" code. The crush of human traffic requires that an orderly sequencing of human conduct is developed. (Possibly this came into existence 6,000 years ago, in law 4,000 years ago.)
2. The code of justice. The rights of other individuals and groups are to be respected. (Possibly this came into existence 4,000 years ago, in law 2,000 years ago. It is quite generally codified by today, with only a few exceptions. Our own Constitutional Congresses in capturing the intellectual development of that century represented the beginning of its general codification.)
3. The code of love. Beyond respecting other individuals' rights, the individual is fully responsible for the quality of his interpersonal relations with others. (It possibly came into existence 2,000 years ago but has not yet been accepted in law and is only practiced to very limited extent.)

These are the moral codes known to man today. Two more may be projected.

4. The code of responsibility. Beyond being responsible for the quality of his interpersonal relations, the individual is fully responsible for the consequences of his acts. (Barely in existence, it may be considered to be the

---

[1] Moral codes are prescriptions for behavioral patterns that fit the elites (6).

foundation for real professional ethics. Today, we would be considered too ignorant to know how to apply this morality. It is not unreasonable that a growth of all the sciences may make this a feasible code within 2,000 years.)

5. The code of omniscience. Beyond responsibility for the consequences of his acts within his existing capabilities, the individual is responsible for modifying the characteristics of other individuals for the optimal preservation of the species. (This doctrine cannot be said to exist meaningfully today. It would be considered improper meddling. Yet a target 4,000 years hence is not beyond imagination.)[1]

Although the jump from the viable patterns of existence of simpler species to those for man may be violent for the reader, it is one necessary for men to ponder on. They are too recent and too young in the ecology for their future history and role to be assured.

To complete a summary of the status of the entire biophysical continuum—assuming for a moment the subsequent geophysical atom, the earth, with its 4×10^9 years of development—we finally see the development and proliferation of living forms 3×10^9 years ago.[2] Similar to the start-up problem for all other systems, the question arises whether life formed in a single incident or in repeated instances. We surmise, from stability considerations, that a time must have emerged in which the conditions for life formation had been reached. In a vague way, the biochemical conditions are known. The first issue was of feasible sources of energy and kinetic processes that could bridge the barrier of forming amino acids from water, carbon dioxide, oxygen, and nitrogen. Such questions have a history whose tempo is increasing (7). Since Oparin who began the discussion, sunlight, electricity, warmth, water, radiation, and the elements in gaseous state have seemed ample for the first steps.

Explaining the synthesis from amino acids to protein, although a more difficult process to account for, seems to be on the agenda for today. A considerable gap still exists in the step from protein to cells. (Besides the building

[1] As a simpleminded summary, in the code of justice, a mother who wipes her child's nose, who clothes him, feeds him, and cares for his needs can go to heaven. In the code of love, the mother who in addition loves her child can go to heaven, even if her efforts are misdirected. In the code of responsibility, the mother becomes responsible for the consequences of her acts. She is punished for their failures and rewarded only for their success. Her solicitousness must extend to seeing that her child does not catch a cold because of her efforts. In the code of omniscience, there must be a program for making one's fellow man into a better person. One is responsible for the value to the society (or the species) of one's actions concerning those whom one came into contact with. The mother is responsible for making her child over so that he does not catch colds!

[2] Recent estimates (late 1968) have produced evidence (8) for the likelihood of life forms—algae—with an age of 3.2×10^9 years. Thus the gap between the formation of the earth and the start of life has appreciably diminished as compared to the earlier traditional estimates for its beginning.

blocks of amino acids to protein, also needed are the building blocks of nucleotides to nucleic acids, and of carbohydrate polymers. Besides the elements C, O, H, N, also needed are the metallic ions of P, Na, K, and Fe, and action of Cl.)

This problem has revolved around replicating mechanisms. One may not conclude that current molecular biological DNA coding can do it. However, it is clear that biochemical encoding for a living form must be dynamic and electrical. The gross description of the process is catalysis. Catalysis should perhaps be defined as dynamic gating through molecular templates using geometric electrical conformity, according to temporal dynamic patterns. Enzymes seem the most promising catalytic forms for the task of developing dynamic biochemical oscillatory reactions (9).

However, the major element in a cell, besides the dynamic catalysis inside, was the formation of a dynamic surrounding gate, the membrane. It is this gated cell which opened the door to the biological continuum, for by its further specialization more complex organisms evolved, colonies and civilizations of such organisms grew, developed extensions, competed, and, in toto, made up the biophysical continuum.

SUMMARY

1. Broader than man's cultures and civilizations is the sum total of all such atomistic biological species and their "cultures." In toto, they make up the ecological continuum that covers the earth.

2. These systems all have lives and deaths basically starting 3×10^9 years ago (on an earth that started 4×10^9 years ago—that is, the biochemical oscillators that make up life systems have existed for the past 75 percent of earth's history). Man's precursors emerged 10^7 years ago; man (Homo), 10^6 years ago. Modern civilized man only emerged 10^4 years ago.

3. Can man plan for his future? He had better, because he is abrasive in the ecology.

4. What unique contribution has he made? He has contributed ethics, complex adaptive rules of conduct. The pace suggested is that man contributes a new thought every 2,000 years. Six thousand years ago he formulated the traffic policeman code (man must regulate his trafficking). Four thousand years he formulated the code of justice (of legal equity). Two thousand years ago he formulated the code of love (respect for all fellow men). Two more codes can be projected, the code of responsibility (the individual is responsible for the consequences of his acts) and, perhaps 2,000 years in the future, the code of omniscience (man is responsible for the modifications necessary to best preserve his species). These two codes are premature today. Man does not know enough to operate in accord with these codes.

5. What must be understood in all viable systems is that there must be a degradative phase and an emergent phase to keep the system's rhythms beating.

6. Conditions for start-up of such viable systems require "catalysis," a dynamic gating of processes that increase their speed of reaction. The emergence of the biological continuum has likely depended on enzyme reactions and the biological membrane.

QUESTIONS

1. Examine and discuss the history and experience of some living and extinct phyla.

2. Discuss possible means for getting human elites, particularly politicians, to look ahead 100 to 200 to 2,000 years. Is this desirable?

3. What issues might or might not organize mankind to worry about its future?

4. Discuss the thesis of new moral codes each 2,000 years. If you disagree with the two projected codes, make and defend your own suggestions.

5. Sketch out your hypothetical timetable of dynamic events (and evidence)—including a summary of the dynamic surface conditions on earth—for the origin and development of life forms until they are well launched into known fossil history (two- or three-man interdisciplinary team).

6. Propose a dynamic description of catalysis in some well-defined chemical process.

BIBLIOGRAPHY

1. Simpson, G.: "Life of the Past," Yale University Press, New Haven, Conn., 1953.
2. Rashevsky, N. (ed.): Mathematical Theories of Biological Phenomena, *Ann. N.Y. Acad. Sci.*, vol. 96, p. 895, 1962.
3. Waddington, C. (ed.): "Towards a Theoretical Biology, 2. Sketches," Aldine Publishing Company, Chicago, 1969.
4. Odum, E.: "Ecology," Holt, Rinehart and Winston, Inc., New York, 1966.
5. Morowitz, H.: "Energy Flow in Biology," Academic Press, Inc., New York, 1968.
6. Iberall, A., and W. McCulloch: 1967 Behavioral Model of Man, His Chains Revealed, *Currents in Mod. Biol.*, vol. 1, p. 337, 1968.
7. Fox, S. (ed.): "Origins of Prebiological Systems," Academic Press, Inc., New York, 1965.
8. Engel, A., et al.: Alga-like Forms in Onverwacht Series, South Africa: Oldest Recognized Lifelike Forms on Earth, *Science*, vol. 161, no. 3845, p. 1005, September, 1968.
9. Higgins, J.: The Theory of Oscillating Reactions, *Ind. Eng. Chem.*, vol. 59, p. 18, 1967.

16

The Planetary Atom; Geo-, Atmo-, Hydro-, and Biospheres

16-1 THE PLANETARY ATOM

The universe has persisted 10^{10} years, the earth 4×10^9 years, living systems perhaps 3×10^9 years, and man perhaps 10^7 years. The story of the earth, as an atomistic system, does not depend on biological organisms or on man, although the future story may at some remote time. As finite-sized field continua of materials of which the biochemical constituents are only a small fraction, the atomistic earth and other planets came into being.

By itself, the earth is only an atomistic element that has developed, like a hydrodynamic vortex, from a still-large continuum field (although the specific dynamic processes of formation are not certain today). However, at this atomistic level we find the earth to be a remarkably stable example of a solid-state entity. As a planet, this macroscopic grain is assembled in continuum form from a large number of smaller, repetitive solid-state "particles" to a tremendous size. It is only questionable, at the astronomical level, whether the solid state has stability against gravitational self-crushing forces.

It has been popularized by Gamow that self-crushing and thus stripping of atomic and molecular levels down to nuclear particles may take place for condensed planetary material size greater than the diameter of Jupiter.

To illustrate the flavor of possible hydrodynamic stability considerations involved in the formation of the solar system, we can point to an article by Chandrasekhar (1) for a theory of the solar system that was developed by Weizsäcker.

Three excellent books that present a considerable amount of the solid-state story of the geophysical earth are Munk and McDonald (2), Scheidegger (3), and the Jeffreys appreciation volume (4). What appears therein is much of the detailed characterization of the elastic earth. What is not so well detailed there is the story of the plastic earth.

The geophysical earth displays a broad spectral response, varying from the drift of continents[1] to a high-frequency spectrum of earthquakes, tremors, seismic movements, etc. Most of the activity points to an elastic-plastic nature of the earth. This is not so well detailed. We have been quite impressed by the thesis and evidence of Paul Siple[2] that the vector properties of the rotational momenta of the earth provide an excellent point of departure with which to correlate the anelastic-plastic changes of the earth which result from its earthquake and dislocation spectra. Since there are indications that the thesis is being adopted and that geologists are increasingly concerned with a better foundation for a theory of faulting and rock movement, the problem is better followed in the professional geophysical literature.

However, the constituents of the earth that created it upon condensation are not all solid or liquid at operative temperature. (One must note that the earth is in an imperfect "box" which is in radiative equilibrium with the sun. Thus, although imperfectly isothermal, a combination of factors—distance, the solar constant, the earth's rotation, the existence of an atmosphere, the "greenhouse" effect—all tend to create a nearly isothermal surface for the earth. Compared with the extremes of near $0°K$ in space—a common current figure in the scientific literature for the background temperature of space is $3°K$—$6000°K$ on the surface of the sun, and $10^7°K$ internally in the sun, the total earth's surface variation of $180°K$ to $350°K$ is fairly isothermal. A gaseous atmosphere exists, held by gravity. A description of the dynamic characteristics of this atmosphere is the task of the science of meteorology.

The primary characteristics that one finds in the atmosphere are certain near-equilibrium properties such as an elementary law of atmospheres (the pressure and density vary exponentially with height), and a number of

---

[1] Besides a number of physically esthetic arguments for this in the past, which validly did not impress geologists too much, geological evidence for such drift has piled up to a point [even since (3)] that it is now taken quite seriously and is in fact essentially accepted. [See for example (5), (6), and (7).]

[2] Private communication.

nonequilibrium properties such as large-scale tidal oscillations (8), cellular movements of air masses, and wind patterns. A first elementary test for a basic science of meteorology is whether it can develop a theory adequate to account for the average temperature and pressure profile and the average pattern of winds in the atmosphere. The temperature profile is a most serious and difficult test since the profile is the resultant of the tidal oscillations of the atmosphere trapped into nonlinear limit cycles. In a little more detail, it is probably necessary to construct an atmospheric model in two phases: one to account for the law of atmospheres for the entire thermally inhomogeneous atmosphere; and the other to account for the added inhomogeneity introduced by water held as a partial pressure atmosphere. [A third phase is then likely to account for the compositional inhomogeneities at high altitude resulting from chemical decomposition and diffusion. See Mitra (9) for an introduction.] It must be noted that the changing character of weather is dominated by temperature, wind, and water. A novel, perhaps useful, background for the hydrology problem (the interaction of water with the earth and atmosphere) is given in (10). A complete hydrological study requires that the interaction of atmospheric and surface water through rainfall and evaporation be considered. This was a significant portion of what is presented in (10).

It may be pertinent to review the status of some work in meteorology. A valid brief introduction to the history of the development of theoretical meteorology is found in Thompson (11). Thompson notes that the Navier-Stokes equations of motion, equation of continuity, energy equation, and equation of state are a perfectly good mathematical theory of fluid motions in general and atmospheric motion in particular. The reason the meteorologists have not attempted to use them until recently is that they were too difficult to solve and purely numerical methods involved too much computation. Bjerknes attempted a systematic study of idealized mathematical models but failed to find a satisfactory formulation for weather prediction. L. F. Richardson (12) did try by a finite-difference scheme to solve the nonlinear equations but was not successful. He predicted that the large-scale weather disturbances would travel at about the speed of sound and in the wrong direction. The emphasis in the 1940s was on high-speed automatic computing. Thompson cites in this connection the work of Von Neumann and Charney but points out that "it was recognized that a machine would commit the same errors as Richardson." Effort went into remedying various kinds of errors. He notes that the atmosphere is always very close to mechanical equilibrium. Thus, large-scale accelerations of air are very much smaller than the individual forces per unit mass. This implies that the net accelerations are small differences between individually large terms. This difficulty alone would be enough to guarantee failure of the approach. Thompson asks if the hydrodynamic equations could not be modified to leave out all high-speed waves. It is such an incomplete set which is used for meteorology.

Thompson then notes that, in general terms, the problem of dynamical or numerical weather prediction is to project a solution of a limited system of hydrodynamical equations which satisfy boundary conditions. The set of equations contemplated not only involves seven dependent variables and four independent variables but also is nonlinear. One cannot apply the superposition principle to solutions. Its complexity, he feels, far exceeds that of any nonlinear system that has been solved heretofore. [Although still in dispute, the set examined in (13) is a complete hydrodynamic system of comparable complexity.]

He points out further that at the present time, lacking any general and rigorous theory of meteorological approximation, the field is forced to rely heavily on induction from special cases and on physical and mathematical insight gained from the study of relatively simple hydrodynamical systems. In designing improved baroclinic models, for example, they must attempt to decide what parts of the mechanism are essential in order to simulate dominant features of the atmosphere's meteorological behavior.

> We can accept, with Thompson, that the complete Navier–Stokes equations (motion, continuity, energy coupled with the gravitational field of the rotating earth revolving around the sun) are a valid description of meteorological phenomena. The solution of such an equation set is difficult, and it cannot be solved by brute computation force. In the latter point, we disagree with current methodology. In our opinion what are lacking, and have been lacking since efforts in the 1920s, are means to trap the primary tidal oscillations into limit cycles. A path has been outlined for hydrodynamic problems in (13). It must be emphasized that the demonstration in (13), although fundamental, is still in very rudimentary state.

Simplified mathematical equation sets for the atmosphere have been used by Charney and Pekeris (14). The work of Fultz (14, 15) is interesting in furnishing a variety of physical-mathematical model analogues for various details of the atmosphere. However, they are not actual models for the atmosphere. At another extreme, the more abstracted mathematical efforts of Bjerknes (16) do not bring the characteristics of the atmosphere to focus.

Regarding the large structure of the atmosphere, Siebert (15), illustratively, concludes that many results of the treatment of this problem are due to approximations and estimates and that a completely satisfactory explanation of the atmospheric tides does not exist. He hopes that the present incomplete status of the theory and of the observational information may act as a stimulant for further work. [Current work is described in (17) and (18).]

What could be valuable in the field of meteorology is an intermediate science, derived by careful choice or construction, of those scientific elements that are substantial enough to carry the theory across a bridge from first principles to substantial observations. We believe that competence in hydrodynamics and general systems analysis would permit its accomplishment.

The immediate interests would be to model the mean states, cyclic states, stationary stochastic state, and remnant fluctuating states of pressure, temperature, humidity, wind velocity, cloud cover, and precipitation in the atmosphere by a synthetic model, building up from primitive to complex, with emphasis on agreement with known experimental data. The basic hydro-dynamic problem is to first write a complete set of equations and then a consistent theory of approximation. It may be necessary to include electrical and magnetic effects and effects of particular surface features which become important boundary conditions. In this connection, it is perhaps noteworthy that the evaporation model set up in (10) developed a boundary layer 10,000 ft thick rather than the much smaller hydrodynamic boundary layer used by others. This very thick layer was considered relevant to an account of the gross structuring of the atmosphere.

In our opinion, the primary solution cannot be done by computer because it is not an arithmetic task. Instead it requires the development of a hierarchical logic of solutions. Only subsequently, at some stage much farther along in the development, by perturbation methods does the problem degenerate to computation. (Here we differ with the computer-oriented point of view of current meteorology.)[1]

The requirements are to set up a valid set of equations that includes all the necessary hierarchical levels. (The set given by Thompson is not com-pletely valid. It lacks viscosity, thermal conductivity, diffusivity, dissipation, phase change, chemical activity, electrical effects, radiation, and a radiative source and balance. Beyond that, it does not set up a hierarchical solution set. In illustration, whereas an equation of hydrostatic equilibrium may be assumed for one level, it does not correctly model any dynamic time-dependent coupling in the vertical direction.)

It would appear desirable to set up the equations and attempt solution sets that could predict a few fundamental characteristics of the atmosphere ; namely, the vertical mean-pressure distribution, the vertical mean-temperature distribution, the steady-state limit-cycle tidal oscillations, the time-independent temperature distribution at sea level, the mean-velocity distri-bution in the atmosphere, the vertical mean-humidity distribution. These could also provide a rudimentary theory of clouds and the ozone layer and a rudimentary theory of mean precipitation.

After we have modeled a mean state, the characteristics of a stationary limit-cycle spectrum, and a stationary driven spectrum (driven by the earth's daily rotation and the yearly revolution around the sun), then the stochastic characteristics, which may be weakly or strongly coupled with e-m phenomena both in space and on the earth, may be attempted.

---

[1] Currently, there is emphasis toward a 1-year global atmospheric-research program to gather extensive data.

The hydrological cycle is intimately related to meteorology. A scientific introduction to the hydrological cycle was attempted in (10). Extensive hydrology references are given there. It is based on the view of a rotating earth in space revolving around a sun. This constellation produces near-temperature equilibrium (in sufficient time) for the earth. There is thus a segregation of material by condensation. Those materials whose triple-point temperature is lower than the temperature range will be frozen out, and there will be no melting. "Rocks" and "ices," in fact all those components that are regarded as solids, freeze out. To what extent these bond and bind is a complex matter of past thermal history, alloying elements available, and the stress-strain associated with ponderous motion, thermal stresses, and internal inhomogeneity. These materials are not in complete thermodynamic equilibrium.

As a result of all these facts, the wrinkled surface of the solid earth body is not a permanent thing but a thing always in upheaval. Over any long period, we may regard the marginal stability of the body to be sufficient that the earth hangs together globally. (However, no one can state what the future history of earth faults will lead to.) Thus, all the states of the surface making up a set of states are likely to be slowly changing (i.e., "geologic" in time scale) and representative of an ensemble of states that satisfy quasi-ergodic conditions. Hydrology is thus to be regarded as the interaction of a fluid flux, water, gravitationally acted upon with members of this Gibbsian ensemble of surface states.

On the other hand, the temperature range may exceed the critical temperature of some other of the materials of this rotating planet. These will form a permanent gaseous state, an atmosphere bound by gravity. This blanketing layer, of course, may change the temperature conditions on the surface to some extent.

Then, there are materials which may melt and freeze and "boil" or sublime. On the earth, water is the prime example. In fact, water is "condensed" to cover most of the earth's surface, partly as solid ice but mostly as liquid water, because of the earth's temperature range. Jupiter and more remote planets are less concerned with the liquid phase.

The water is not in stable equilibrium. It enters into a limit cycle. The chain is as follows: Water evaporates from the oceans and land (the land cover is not a very effective resistance to evaporation). It recondenses as precipitation directly onto the water surface of oceans or soaks into the porous earth and seeps under a water table to rivers, which run off to the oceans. The problem, somewhat quantitatively, is that about 30 in. of precipitation falls in a year on a surface like the United States, about 22 in. evaporates, and about 8 in. appears in the rivers as runoff. Thus evaporation is nearly the same dominant process over the land as well as over the oceans.

On any global scale, the hydrological cycle may appear to be a minute interface governing a zone perhaps 2 miles thick (e.g., the major cloud-cover zone, and surface waters) or, at most, perhaps involving 60 miles of atmosphere. Nevertheless, it is basic for the many biophysical species, and it is the rate-governing process in determining the changing earth surface.

Although minute, the water material is not in thermodynamic equilibrium. If it were, the atmosphere would be saturated with water vapor. Instead, the average humidity is about two-thirds saturated. It is this moderate difference that is maintained by atmospheric dynamics through the complex of tidal oscillations and cellular air-mass movements.

The corollary of the spectral dynamics that appears in the atmosphere[1] is a spectral dynamics that appears on the earth and in the ground (such as flood epochs, water-table dynamics, etc.).

16-2 CONTINUUM PROPERTIES WITHIN THE PLANETARY ATOM

The planetary superatom thus represents a coalescence of particles by gravitational forces and thereby exhibits, emerging from the liquid, gaseous, and solid states of its many atomic species, the following continua—the atmosphere, hydrosphere, biosphere, and geosphere. The underlying atoms have formed various molecular arrays in essentially chemical thermodynamic equilibrium. The molecular arrays are arranged in various state coalitions which still exhibit enough binding under gravitational forces to hold together for billions of years. These coalitions are not in thermodynamic equilibrium. Their near-ideal gas emanations are. Yet they are close enough to local equilibria that the changes which take place are aptly characterizable as "geological." What are these properties?

Changes are slow enough that the coalitions (crystallite, river, biocell, society, other earth features) are regarded as formed or formal structure. Further, various segments or realizations differ in their "geological" epochs of emergent evolution. Ancient crystal forms, rivers, life forms, social groups or social customs, and volcanoes mingle with the very latest forms and fashions. Yet we propose that in some way the systems remain quasi-ergodic. How? Apparently, regardless of the age there is not that much difference in the structures (i.e., a newly emergent form "quickly" fills out all its many viable mutations or permutations and nearly saturates the available phase space), and there really are not that many discernible steps of difference.

These are the background ideas that must be of concern to GSS whether at the lower atomic level or the intermediate social level.

---

[1] Namely, the dynamic process that maintain mean state parameters such as the 30-in. rainfall per year; fairly regular fluctuating parameters such as 6-, 12-, and 24-hour parametric pressure variation limit-cycle oscillations or driven seasonal temperature cycles; and stochastic epochs, e.g., "Rain this afternoon, fair this evening."

Nuclear particles bind by a few types of force. We defer the details of this binding. Then electrical forces—coulomb forces and valence-type forces—bind these particles together into atoms-ions-molecules.

These particles in ensemble make up the chemosphere—a slowly changing milieu in which the more permanent particles enter into temporary configurations and constellations. This is the resultant of the intermolecular forces, which are electrical. At close molecular distance, these forces are repulsive because the molecule is a stable local structure. At longer distance, there are polar attractions arising from permanent and induced multipoles. If there are no intermolecular (or interatomistic) forces, there is no system.

However, these intermolecular configurations are not everlasting. "Catalysis" and collision change their form at various rates. [The dynamic time scales of these bound configurations have not received much discussion. There is some discussion in (10) of the composition of the solubles in rivers. These result from the interaction with the land.]

For a very oversimplified statement but one with considerable significance, we might say it is the "polar" character of oil and water configurations, plus a few other poorly defined "electrical" properties, that brought organic life systems into being. Again, these configurations have time-changing properties.

When we think at the level of society, we should be conscious of this background.[1] One portion of society does not deal with the same frame of reference as another. If we think that communications and rationality foster rapid information diffusion, this is a mistake. The "molecular" configuration of the human interior is formed at many different "geological" eras.

Thus one must ultimately grasp the mixed-up character of intellectual endeavor.

An "intellectual" is one who verbalizes patterns of description in abstract form. A "nonintellectual" does not to any appreciable extent. However, the act of intellectualization does not guarantee any rationality of description, only the grammatical, sometimes logical arrangement of words.

In planning social patterns, this background too must be kept in mind.

SUMMARY

1. The planetary earth is an atom 4×10^9 years old.

2. In its condensation, there has arisen a stable solid state, the geosphere, a dynamically stable liquid state, the hydrosphere, and a dynamically stable gaseous state, the atmosphere.

3. Beyond these three states of matter is the chemosphere, the totality of all interactions that bring near-equilibrium molecular structures into being.

---

[1] It is not possible to develop GSS by a very logical development proceeding upward step by step. For a first effort, we must take somewhat wide and wild steps to encompass as many phenomena as possible. At all levels, we attempt to be guided by what we know of the physics of the simpler or more elementary systems.

4. The life state emerged from the liquid phase in the form of biochemical oscillators. The primitive materials for life may be said to be oil and water with a little specialized protein, nucleic acids, and sugar thrown in.

5. The dynamics of these states each show complex rhythmic characteristics. The turbulent atmosphere and the hydrological cycle are excellent examples.

QUESTIONS

1. Sketch out a plausible theory of solar-system formation.

2. Sketch out a theory for a planet (or cold star) and its equilibrium state which is subject to self-crushing forces.

3. Having examined the literature on the characteristics of an elastic earth, discuss the forces on the earth that stress it and attempt to sketch out a model of its plastic-elastic deformations.

4. Develop a model for the slowly changing geological evolution from the formation of a solid skin onward. Try to make it roughly conform with known history.

5. Develop as much of an autonomous hydrodynamics theory of atmospheric tides as you are able [start from Wilkes (8)].

6. If the existence of atmospheric tides is assumed, work out as many modeling consequences for the characteristics of the atmosphere as you can from this dynamic instability (e.g., vertical mean-temperature distribution).

7. Describe the mean hydrological cycle and attempt a theory for it (e.g., why the rainfall).

BIBLIOGRAPHY

1. Chandrasekhar, S.: On a New Theory of Weizsäcker on the Origin of the Solar System, *Rev. Mod. Phys.*, vol. 18, p. 94, 1946.
2. Munk, W., and G. McDonald: "The Rotation of the Earth," Cambridge University Press, New York, 1960.
3. Scheidegger, A.: "Principles of Geodynamics," Academic Press, Inc., New York, 1963.
4. Cook, A., and T. Gaskell: "The Earth Today," Royal Astronomical Society, London, 1961.
5. Wegener, A.: "The Origin of Continents and Oceans," Dover Publications, Inc., New York, 1966.
6. Hurley, P.: The Confirmation of Continental Drift, *Sci. Am.*, vol. 218, no. 4, p. 52, 1968.
7. Turcotte, D., and E. Oxburgh: Continental Drift, *Phys. Today*, vol. 22, no. 4, p. 30, 1969.
8. Wilkes, M.: "Oscillations of the Earth's Atmosphere," Cambridge University Press, New York, 1949.
9. Mitra, S.: "The Upper Atmosphere," The Asiatic Society, Calcutta, 1952.
10. Iberall, A., and S. Cardon: "A Study of the Physical Description of the Hydrology of a Large Land Mass Pertinent to Water Supply and Pollution Control," HEW, Contract No. Saph-78640, four qtrly. reports, 1961–1962.
11. Thompson, P.: "Numerical Weather Analysis and Prediction," The Macmillan Company, New York, 1961.
12. Richardson, L.: "Weather Prediction by Numerical Process," Dover Publications, Inc., New York, 1965.
13. Iberall, A.: Contributions Toward Solutions of the Equations of Hydrodynamics, Part B, Primitive Solutions for the Fluctuating Components of Turbulent Flow Between Parallel Plates, *Contractors Rep. Off. Nav. Res.*, Washington, D.C., October, 1965.
　———: A Contribution to the Theory of Turbulent Flow Between Parallel Plates, *Seventh Symposium on Naval Hydrodynamics*, Rome, Italy, 1968, in press.

14. Bolin, B. (ed.): "The Atmosphere and the Sea in Motion," The Rockefeller Institute Press, New York, 1959.
15. Landsberg, H.: "Advances in Geophysics," Academic Press, Inc., New York, 1961.
16. Godske, C., T. Bergeron, J. Bjerknes, and R. Bundgourd: "Dynamic Meteorology and Weather Forecasting," American Meteorological Society, Boston, Mass., 1957.
17. Brown, J.: *J. Atmospheric Sci.*, vol. 26, p. 352, 1969.
18. Smagorinsky, J., K. Miyakoda, R. Strickler, and G. Hembree: *Monthly Weather Rev.*, vol. 97, p. 1, 1969.

17

The Universe as Continuum; From Stars to Galaxies to Cosmos

Matter and radiation in gross amounts, subject to dynamic laws, have now been brought onto the stage of existence and being. At a large-enough dimensional scale, the weak force of gravity binds great amounts of matter into equilibrium. Here, it turns out that the equations of motion are not the Newtonian equations but the relativistic equations of Einstein. In the space-time of sparse collections of matter and energy, one finds remotely separated stellar systems, an e-m radiation field, and a variety of cosmic radiation of various energetic small particles. Within solar-system dimensions, one finds that Newton's theories of motion and of gravitation are accurate descriptions of phenomena (1). As a modification of Newtonian mechanics by using Einstein's special relativity theory with its correction to Lorentz kinematic transformations rather than Galilean transformations, Maxwell's e-m theory and mechanics are quite consistent. We may now consider an indefinite extension throughout space of such particulate systems—of matter bound in stars and possible solar systems, distributed with some sort of density throughout space. Thus in the large, the firmament begins to approach continuum properties. However, different from the previous systems of particles and radiation in equilibrium with the walls of an enclosure, here the discrete star systems have higher temperature (core temperature $10^{7}°K$) than the background radiation ($3°K$). Thus we must question whether we can assume equilibrium or how we may describe their grand ensemble.

We have not touched on the nature of these atomistic stars. Excellent introductions are by Schwarzschild (2) and Dufay (3). The e-m radiation spectra of the stars are continuous and, by comparison, are inferred to represent the same atomic species as on earth, but at high temperature. Materials like hydrogen, helium, metals, nonmetals, and even molecules are indicated. As abstracted in the Harvard classification, it was found that the excitation condition of the ionized atoms or molecules shown by the spectra could be associated with a (essentially blackbody) temperature for the stars. Stars have then been grouped, by their visual magnitude and spectral type, into stars of the main Russell sequence, white dwarfs and red giants; also subdwarfs, subgiants, and blue supergiants [(3), Fig. 21; the presentation is generally known as the Hertzsprung-Russell diagram].

Nuclear processes in the stars' interior are what provide the sustaining energetics for the structure and evolution of stars by the conversion of mass to energy. Equations of change are given in (2). For example, for stars of the main sequence, such as the earth's sun, with an abundance composition as

| Element | Atomic Weight | Abundance* | |
		By No.	By Wt.
H	1	1000.0	1000
He	4	80.0	320
C	12	0.1	1
N	14	0.2	3
O	16	0.5	8
Ne	20	0.5	10
Mg	24	0.06	1
Si	28	0.03	1
S	32	0.02	1
A	40	0.05	2
Fe	56	0.02	1

* Relative to H = 1000

listed in the table, the star "burns" (transmutes) hydrogen into helium either by a proton-proton reaction or a carbon cycle.

Thus, at least for some long period of time, processes for maintaining the high central-stellar temperatures of $10^{7}°K$ have been found. Naturally, such sustaining mechanisms were sought historically, whenever various ways of estimating past geological ages were conceived. The need to account for structural ages of the order of a billion years or greater cannot be answered by an initial hot creation and a passive cooling-off. Ac active "engine" (or limit-cycle) processes had to be found which could be self-formed and by which fuel could be "burned" in thermodynamiclike processes. A number of nuclear processes—according to the Einstein theory of the convertibility of mass and energy—were found and have appeared to be suitable. However, by the very act of finding them, a history of the universe was implied as another systems problem. These references (2, 3) tell at least part of the story of how an evolutionary process could be found for the stars so that their "ages" could be dated—roughly a young population of blue supergiants and galactic clusters younger than 10^8 years; the midrange of the main sequence, like the earth, younger than 5×10^9 years; the oldest stars possibly not older than 1.5×10^{10} years. [The story of how young stars form is interestingly told in an article by Herbig (4).] Furthermore, in some crude way, a process of degradation of nuclear material from a common ancestry is suggested but with a number of other questions that are currently under investigation and in controversy—the size of space, the age of apparently remote bodies such as the quasars [see, for example, (5) or (6)], etc.[1]

Thus the hot atomistic star is capable of existence within the domain of nuclear-fuel processes balancing their outward energy flux against the organizing pressure of a large-scale gravitational force. The question of how the material was brought together in the first place goes back to the formative history of the cosmos, which is not our present concern.

The atomistic stars then are found in near-constant number density distributed through space. As our scale of interest grows, space and its occupants seem to have certain continuumlike properties [even to the extent that a number of hydrodynamicists have begun to concern themselves in recent years with hydrodynamiclike questions; see, typically, a talk by Lin on "Stellar Dynamics and Galactic Spirals" (8) or Arp (9)]. Even though the stellar separations seem great, nevertheless they interact as in a continuum, and the stellar continuum itself then ends in a discrete structure, the galaxy.

As Arp's article (9) indicates, although theoretical modeling of galaxies is still in controversy, only a limited number of forms of organization are found, and these all seem to require specific mechanisms to explain their form. Thus, birth and death of these systems and their dynamic maintenance are required.

Commenting on the issue once again, it is difficult to think of conservative systems as an apt description of these large-scale gravitationally bound systems. This carries the implication that to regard our own "minute" solar

[1] An interesting recent speculation (1969 Physical Society meeting) is that cosmic radiation, one of the remaining mysteries of space, may originate from pulsars or neutron stars. See, for example, (7). Another theory is that cosmic rays originate from supernovas.

system as being independent, isolated, closed, and conservative is illusory. However, for the necessary time scale, we do not have enough examples and experience to discuss the forms of solar systems. [A possible model for the formation of planets is discussed by Chandrasakhar (10).]

Thus, although the use of conservative systems is a useful model when systems are not too lossy per high-frequency epoch of motion, such formulation is not adequate to discuss the long-term stability of a system. This topic must be deferred to systems' theories of the birth and death of systems. At present, we have concerned ourselves only with the conservation and slow evolution of their motion as nonlinear nonconservative systems.

As we view it in this section, now the galaxies spread continuumlike in number density throughout the firmament. We see the cosmos as a continuum of such formed and formless clusters of stars and dust.

SUMMARY

1. The universe is filled with the atomistic stars, some of them holding solar systems. The continuum field of stars clusters into galaxies. The form of galaxies suggests very specific hydrodynamiclike events in their historical formation. Now the galaxies spread out in number density throughout the firmament. The life process of birth and death must go on among these giants.

QUESTIONS

1. Past the early question of start-up, discuss the "hydrodynamics" of stellar and galactic formation.

2. Discuss the theory of changing historical composition of stars and the universe.

BIBLIOGRAPHY

1. Dziobeck, O.: "Mathematical Theories of Planetary Motions," Dover Publications, Inc., New York, 1962.
2. Schwartzschild, M.: "Structure and Evolution of the Stars," 1958 ed., Dover Publications, Inc., New York, 1965.
3. Dufay, J.: "Introduction to Astrophysics: The Stars," Dover Publications, Inc., New York, 1964.
4. Herbig, G.: The Youngest Stars, *Sci. Am.*, vol. 217, no. 2, p. 30, 1967.
5. Burbidge, G., and M. Burbidge: "Quasi-Stellar Objects," W. H. Freeman and Company, San Francisco, 1967.
6. Fowler, W.: "Nuclear Astrophysics," American Philosophical Society, Philadelphia, Pa., 1967.
7. Thomas Gold Talks about Pulsars, *Sci. Res.*, vol. 4, no. 12, p. 32, 1969.
8. *Symposium on the Dynamics of Fluids and Plasmas*, University of Maryland, October, 1965.
9. Arp, H.: The Evolution of Galaxies, *Sci. Am.*, vol. 208, no. 1, p. 70, January, 1963.
10. Chandrasakhar, S.: On a New Theory of Weizsäcker on the Origin of the Solar System, *Rev. Mod. Phys.*, vol. 18, p. 94, 1946.

18

The Nuclear Particles

What is the universe made of? Nuclear particles and radiation. We thus return to the universe of the small. An elementary survey of the history and background of the nuclear particles, as of 1958, may be found in an article by Salam (1); as of 1964 in (2); as of 1968 in (3). Recent textbooks on elementary particles are by Källén (4) and by Bernstein (5). A fairly up-to-date summary of particles may be found in (6).[1] Older texts are by Bethe and Morrison (7) and Blatt and Weisskopf (8). A sketchy account of the particles follows:

Electron (e^-) *and proton* (p^+) J. J. Thomson and Rutherford (1911) were responsible for a solar-system model of the atom, using the same material as in electric currents.

Photon (γ) Planck and Einstein were responsible for the discrete unit description of e-m field that makes up beams of light and other e-m

[1] See footnote on page 326. The casual reader is advised that beyond Beiser (1) and other popular or summarizing sources, the literature on "elementary particles" is difficult, confusing, and in flux. Of course, in time this status may change.

radiation (since all charged particles emit or absorb e-m radiation when accelerated according to Maxwell theory, but only in quanta according to Planck and Einstein, electrons and protons emit or absorb photons).

Positron (e^+) *and antiproton* ($\bar{p}$)   Dirac proposed, on relativistic quantum grounds, the concept that particles must exist in pairs having the same mass and directed-spin (angular momentum) magnitude but opposite electric charge. From conservation, a photon should be able to create a pair, or a pair can annihilate each other and create a photon. Anderson and Blackett soon confirmed pair creation for the positron, and Segre (1955) found the antiproton.

Neutron (*n*)   Proton-proton exchange forces were not sufficient to bind nuclei. A stronger bond was necessary. Chadwick and Curie-Joliot (1932) discovered the neutral neutron. Heisenberg (1932) showed that its binding suited close nuclear ranges. Thus the nucleus was not bound by the excess of protons over electrons but by protons with neutrons— nuclear forces millions of times stronger than electrostatic forces but with short range. Gamow and Condon (1928, 1929) suggested how a quantum tunnel effect could permit a nucleus to remain stable.

Neutrino (v_e, v_μ, $\bar{v}_e$, $\bar{v}_\mu$)   The process of beta-decay (i.e., electron formation) from (quantumly) unstable (radioactive) nuclei and reasons against there being electrons in the nucleus lead to problems in balancing the energy in the process. Pauli (1931) proposed the neutrino, with no mass, to account for a missing piece of energy, momentum, and angular momentum (i.e., if conservation laws were to hold during nuclear reactions). However, this was a weak interaction. (Thus four kinds of interactions became known: strong nuclear interactions; medium e-m interactions; weak decaylike reactions involving an unstable particle like the neutron; and still weaker gravitational interactions. The strengths of their relative forces are 1, 10^{-2}, 10^{-13}, 10^{-38}.) In addition to a neutrino (and antineutrino) emerging from beta-decay processes, there is a neutrino pair that can emerge from the transformation of mesons.

Pi-mesons or pions (π^+, π^-, π^0)   Yukawa (1935) postulated that interacting entities which have a quantized field entity involved in their quantum exchange processes (such as the quantized exchange photons between charged particles) must have a complete analogue in nuclear exchange forces between nucleons (proton and neutron particles). This quantized exchange entity was the meson. Their existence was verified by Powell (1947). There are three π-mesons, depending on charge.

Mu-mesons or muons (μ^+, μ^-)   Observed by Anderson (1936) in cosmic rays [see Heisenberg (9)], they were thought to be Yukawa mesons but later determined to be decay products (π-mesons decay to μ-mesons and neutrinos).

The "strange particles" $(K^+, K^0, K^-, \Lambda^0, \Sigma^+, \Sigma^0, \Sigma^-, \Xi^0, \Xi^-)$ Other decay processes, after Powell demonstrated both the pions and muons, were uncovered as a variety of short-lived particles. Eight particles (and their antiparticles) were at first called strange. (In addition to the quantized "spin" of angular momentum, another vector space is postulated in which "isotopic spin" may be assigned all the strong interacting particles. The concept was advanced by Gell-Mann and Nishijima in 1953.)

For a brief sampler of readings that can be examined by the more ambitious nonspecialist, we might propose Fermi (10), Salam in (1), Gell-Mann and Rosenbaum (11), Hill (12), Chew et al. (2), Fowler and Samios (14), Wigner (15), Heisenberg (16), Barger and Cline (17), Wolfgang (18), and Weisskopf (3). [A more technical summary of particle and states is contained in the review (6) and, still more recently, in the January 1969 issue of *Reviews of Modern Physics*. These would be essentially incomprehensible to the nonspecialist.]

In any case, a number of diverse facts have emerged at this level of the quantization of the smallest (as yet) known particles and states.

1. There appear to be four levels of particle interaction: strong interactions that certainly seem necessary to bind nuclei together; medium e-m interactions to bind atoms together; weak interactions that seem to be concomitants associated with particle instability; and weaker gravitational interactions (1, 18, 3).
2. There are some particles that are quite stable, i.e., they do not transform of themselves, but may enter into strong interaction. These are the photon, electron, positron, proton, antiproton, and the neutrinos. However, these particles will not represent the stable atomic structures that exist. For this the neutron is essential. The neutron is not a stable particle in a free state. It decays in about 1,000 sec.

   These 10 particles are relatively stable. However, there were (as of not-too-ancient-vintage ideas) 20 other "fundamental" particles which decay quite rapidly (all but two in the range 10^{-8} to 10^{-10} sec, one which decays at 10^{-6} sec, and one which decays at 10^{-16} sec). As of 1953 [see for example (11)], there appeared to be numerological selection rules (the Gell-Mann–Nishijima theory) that governed the existence of these 30 elementary particles.
3. What were considered to be "resonance" associations (i.e., in the same sense as we have proposed orbital synchronies as temporary constellations in human behavior or epochs in the hydrodynamic limit cycles) have now enlarged the number of particles to more than 100 particles or states [see for example (12), (2), or (3)] so that there are new

numerological selection rules based on five quantum numbers selected for strong interactions (integral atomic mass number, hyperchange, isotopic spin, angular momentum, parity).

4. The transitory forms of these more than 100 "particles," it was finally realized, were not to be identified as particles but as a new set of excited states, i.e., there exist three quantum spectroscopies. The first spectroscopy derives from the quanta emitted by atoms. The second derives from the quanta emitted by nuclei (e.g., γ rays, electrons, and neutrinos). The third spectroscopy derives from the quantized decay of highly energetic transitory forms.

Part of the proliferation took place with the demonstration that a number of expected "preservations" or symmetries did not exist. (That is, as "strange" particles were discovered and a more elastic set of quantization rules were required, the questions of various kinds of space-time symmetry were raised. Beginning with the question of space-reflection symmetry by Yang and Lee—if right-polarized neutrinos exist, do left-polarized neutrinos exist—and the soon forthcoming answer that it did not exist, anarchy existed for a while in particle physics. It was quickly necessary to experimentally clarify the nature of all possible particles or states.) Just as Dirac hypothesized that for every particle there existed an antiparticle, in the vector domain of spatial geometry and space-time geometry, a number of symmetries were expected—space reflection and time reflection. If a process existed, its mirror image could be expected. Yang and Lee (1956) pointed out that it had not been tested, and it was found to be wrong. Parity conservation (of space-reflection symmetry) did not exist. By 1961, a particular unified system of symmetries was proposed, the "eightfold way" of specifying eight quantum numbers. This has been embedded in the theory of Lie groups. [See for example (13).] Some predictive value to the "theory" may be noted [Fowler and Samios (14). For a more complete discussion of the symmetry conservations viewed today, see Wigner (15) and Weisskopf (3).][1]

It seems clear that a completely satisfying status for the theory of small particles does not exist currently. On the other hand, the physics of stable nuclei is not beset by so many perplexing issues [see (19) and (20)]. It is possibly best to end this section on the following note:

Quantization at the level of nuclear particles exists. Various degrees of stability and instability exist. As a result of "natural processes," we find

[1] It is beyond the scope of this study to systematically reference any of the specialized professional-specialist literature in current particle physics. Unless or until some greater unifications or simplifications emerge, the nonprofessionally involved reader is warned against forays beyond the level of expositions presented in *Scientific American*. This opinion has been derived by examining sources through 1969. One reference that the nonspecialist might examine for a view that summarizes the current position is (25).

"stable" nuclear systems and "unstable" (radioactive) ones. (As a result of highly energetic man-made processes, additional instabilities have been found.) The "unstable" ones suggest a history for the universe. A theory for stellar processes and a cosmological theory have created the rudiments of a historical theory for the nuclear particles [e.g., the Gamow theory (21)].

However, the pot of observations and ideas is stirred in two ways—one, the findings of stellar and cosmic radiation which dredge up a host of spectral lines and small amounts of material derived from complex processes at all stages of cosmological evolution, and two, the more recent ability of man to focus highly energetic hammers upon the problem of smashing of material. This dredges up even more quantized entities.

Although relativity (as of Einstein's early work) would have been satisfied with assuming the static ad hoc existence of mass or energy as a sufficient basis for cosmology, it turned out that dynamic processes were essential. Einstein's own mass-energy interconversion coupling made the process possible. However, he did not adequately resolve the mechanical, electromagnetic coupling problem of mass and radiation and its quantization.

Quantum mechanics provided various schemes for doing this, and a relativistic quantum mechanics as derived by Dirac (22) led to feasible negative-energy states with finite probabilities of transition. Thus, Dirac was forced to postulate that all negative-energy states were filled (to prevent a collapse of the existing particles, another of the unpleasant catastrophies that quantum mechanics has had to be continuously expanded to avoid), e.g., with one electron in each state according to Pauli's exclusion principle. An unoccupied negative-energy state—that is, if the particle could be pried out of that state—Dirac postulated was a positive electron. These assumptions, Dirac states, "require there to be a distribution of electrons of infinite density everywhere in the world," i.e., pair production and Dirac's negative ocean were thus introduced.

The pursuit of developing a quantum mechanics has in fact proved that quantization is a process repeated over and over again. However, its foundations are what we might dispute. The history of quantized particles and states (not necessarily formal quantum mechanics but the observational side) has been marked by requiring such abstract descriptors as:

Quantization
Selection rules
Exclusion principles (or forbidden rules)
Numerological rules
Correspondence principle
Wave-particle dualism
Stability—instability
Pair production and annihilation

Tunnel effect
Negative ocean
Cosmic radiation
Symmetry—antisymmetry

With such properties, we have considered it difficult, since at least 1946, not to see the need for a theory of nonlinear quantization, very much like the views expressed in (23). [To illustrate views even more exotic, one may examine Alfvén (24).]

SUMMARY

1. What is the universe made of? Nuclear particles and radiation. Thus, we return to the small, whence we came. Here lie the details of the viable nuclear-particle systems.

QUESTIONS

1. Discuss what a nonlinear theory of fundamental particles might be like.

2. Describe the character of the ensemble of fundamental particles and states if bound together by gravity as a large homogeneous near-ideal gas at $a. \sim 0°K$, $b. 3°K$, $c. 300°K$, $d. 10^4°K$, $e. 10^6°K$, $f. 10^7°K$, $g. 10^8°K$, $h. 10^9°K$, $i. 10^{10}°K$, $j. 10^{11}°K$, $k. 10^{12}°K$.

BIBLIOGRAPHY

1. Beiser, A. (ed.): "The World of Physics," McGraw-Hill Book Company, New York, 1960.
2. Chew, G., M. Gell-Mann, and A. Rosenfeld: Strongly Interacting Particles, *Sci. Am.*, vol. 210, no. 2, p. 74, 1964.
3. Weisskopf, V.: The Three Spectroscopies, *Sci. Am.*, vol. 218, no. 5, p. 15, 1968.
4. Källén, G.: "Elementary Particle Physics," Addison-Wesley Publishing Company, Inc., Reading, Mass., 1964.
5. Bernstein, J.: "Elementary Particles and Their Currents," W. H. Freeman and Company, San Francisco, 1968.
6. Rosenfeld, A., et al.: Data on Particles and Resonant States, *Rev. Mod. Phys.*, vol. 39, no. 1, p. 1, 1967.
7. Bethe, H., and P. Morrison: "Elementary Nuclear Theory," John Wiley & Sons, Inc., New York, 1956.
8. Blatt, J., and V. Weisskopf: "Theoretical Nuclear Physics," John Wiley & Sons, Inc., New York, 1952.
9. Heisenberg, W.: "Cosmic Radiation," Dover Publications, Inc., New York, 1946.
10. Fermi, E.: "Elementary Particles," Yale University Press, New Haven, Conn., 1951.
11. Gell-Mann, M., and E. Rosenbaum: Elementary Particles, *Sci. Am.*, vol. 197, no. 1, p. 72, 1957.
12. Hill, R.: Resonance Particles, *Sci. Am.*, vol. 208, no. 1, p. 38, 1963.
13. San Fu Tuan: Group Theory in Particle Studies, *Phys. Today*, vol. 21, no. 1, p. 31, 1968.
14. Fowler, W., and N. Samios: The Omega-Minus Experiment, *Sci. Am.*, vol. 211, no. 4, p. 36, 1964.
15. Wigner, E.: Violations of Symmetry in Physics, *Sci. Am.*, vol. 213, no. 6, p. 28, 1965.
16. Heisenberg, W.: "Introduction to the Unified Field Theory of Elementary Particles," John Wiley & Sons, Ltd., London, 1966.

17. Barger, V., and D. Cline: High-Energy Scattering, *Sci. Am.*, vol. 217, no. 6, p. 76, 1967.
18. Wolfgang, R.: Chemistry at High Velocities, *Sci. Am.*, vol. 214, no. 1, p. 82, 1966.
19. Leachman, R.: Nuclear Fission, *Sci. Am.*, vol. 213, no. 2, p. 49, 1965.
20. Feinberg, G.: Ordinary Matter, *Sci. Am.*, vol. 216, no. 5, p. 126, 1967.
21. Gamow, G.: "The Creation of the Universe," The Viking Press, Inc., New York, 1956.
22. Dirac, P.: "The Principles of Quantum Mechanics," Clarendon Press, Oxford, 1958.
23. Heisenberg, W.: Nonlinear Problems in Physics, *Phys. Today*, vol. 20, no. 5, p. 27, 1967.
24. Alfvén, H.: Antimatter and Cosmology, *Sci. Am.*, vol. 216, no. 4, p. 106, 1967.
25. Wick, G.: The Forces of Nature: Testing Their Strength, *Science*, vol. 168, p. 1329, 1970.

19

The Cosmological Atom

The distribution of galaxies throughout the universe might or might not continue indefinitely. Again we face a quantization problem—Einstein (1917) to the fore—by which our very cosmos becomes atomistic. A key observation is the apparent shift to the red of spectral lines of recognized elements (Hubble, 1929) for galactic material at great distances, suggesting that remote galaxies are receding from us (Doppler effect). Such expansion implies a historical development. Thus galaxies, stars, and elements are tied together in their previous history.

One authoritative view of astrophysical origins, from current nuclear physics, of nucleosynthesis is Fowler (E1). However, we must go back to foundations in general relativity. Some sources are Møller (A2), Eddington (A3), and Tolman (A5). Also, see Eddington (A4) and North (A1). [A popular source is (A6). Another popular article is (I1).]

Einstein (1917–1918) applied general relativity to the cosmological problem of a homogeneous fluidlike substance. According to the assumption either of negligible radiation pressure or of negligible mass density, one obtains one of two static universes—an Einstein universe or a deSitter universe (1917). However, Hubble's work showed a red shift increasing linearly with distance. The Einstein model cannot satisfy this requirement, and the deSitter model corresponds to an empty universe. (As Eddington points out, the summary of the day was, "Einstein's universe contains matter but no motion, and deSitter's contains motion but no matter.")

The model of an unstable nonstatic universe with an intrinsic time-dependent spatial metric determined by the time-dependent variation in matter and energy density was begun by Friedmann, Lamaître, Robertson, Tolman, and Eddington. This led to the particular model forms of the expanding universe. The Einstein and deSitter universes represent "momentary" unstable phases of the oscillating universes (although that moment would be extremely long). It would seem that all current models, based on nuclear reactions, in the end really refer back to expanding or oscillating universe models, with our current phase being an expansion. Fowler (E1), for example, presents the Gamow theory for the early status of nuclear material. (The Gamow theory of the evolution of the entire sequence of elements is no longer accepted.)

In more detail, the following position is currently defensible. The past decade (as many eras before have also claimed) has likely illuminated the question of the creation of the universe more sharply than ever before. It is the coalescence of estimates made from three "clocks" (i.e., the processes implied) that provides such confidence. [See Sandage's article in (C2).]

The currently accepted truism in astrophysics is that one can set up the following sequence of temporally ordered events. The creation of protons preceded the expansion of the universe, which preceded the first galaxy, which preceded our own galaxy, which preceded the first stars in our galaxy, which preceded our sun, which preceded the earth, which preceded the crust of the earth. Geochemistry today offers an estimate of $4\frac{1}{2} \times 10^9$ years for the earth's crust [see also (G1)].

The first clock   Hubble's 1929 hypothesis was that the red shift in spectrum is a Doppler (velocity) effect arising from recessional motion. The empirical facts show an increasing red shift (implying velocity) with decreasing apparent galactic luminosity (implying distance). The near-linearity of the relation, whose proportionality is given as Hubble's constant, has been interpreted as evidence for an expanding universe. A current estimate of the reciprocal of Hubble's constant, the "age of the universe," is about 10^{10} years [$7–15 \times 10^9$ years (Fowler, 1965); $10–16 \times 10^9$ (Sandage, 1968)].

The second clock   The Hertzsprung-Russell (H-R) diagram of stellar magnitude, implying luminous or radiant-energy flux, versus stellar color, implying surface temperature, is useful in studying the history of stars. Stars of homogeneous composition fall on a curve called the main sequence, the position of each star being determined by its mass and composition. The diagram has been accounted for on the basis of a hydrogen-helium nuclear ac "engine" cycle. Hydrogen is "burnt" to helium. When the helium concentration near the center of a star reaches a critical magnitude, about 12 percent, the aging star moves off the main sequence. Such stars become red giants. Massive stars (30 sun masses) evolve quickly—in 10^6 years. Very light stars (0.1 sun mass) evolve slowly—say in 10^{11} years. The oldest-known cluster in our galaxy is NGC 188 with an estimated age of about 10^{10} years (globular clusters show ages up to 1.5×10^{10} years—Sandage).

The third clock   As mentioned before, the age of high atomic weight materials in the earth's crust and in the solar system (meteorites) has been estimated from radioactive decay to be about 5×10^9 years. (It is likely that this follows the formation of our galaxy by about 10×10^9 years, i.e., the age of our galaxy is about 15×10^9 years.)

A current evaluation is in progress by A. Sandage. In a January, 1968 lecture at the Franklin Institute, he pointed out his estimates of the mass, age, and curvature of the universe. Due to "self-gravitational" forces, the universe has to decelerate in its expansion. Thus the velocity-distance relation should not remain proportional. Sandage's estimate [see also (I2)] puts the deviations beyond the value for a flat space. Thus he estimates a closed-cycle expansion of about 80×10^9 years, a present age of 10×10^9 years, an end of expansion in about 30×10^9 years, and a closed space [see (I1)].

The controversies which exist in this subject may continue—a current expansion which began with a big bang versus steady-state creation and Einstein's brand versus Brans-Dicke's brand of relativity. Nevertheless there is evidence for creation in the universe and a considerable consistency in describing the processes. The density of atomistic matter and its interconvertibility into energy and the existence of a gravitational field associated with matter make an oscillating universe possible. It is the details of the nuclear mass-energy constituents that represent the "start" of the process. Have there been cycles before? At present, we cannot answer that question. Yet we can now answer that stars—the previous immutable objects appearing in the heavens—are born, live, and die. Thus periodic processes are going on.

Their start and end? Who knows for certain?

In order to leave the reader with some flavor of our present cosmos, let us sketch out the start-up processes, as they are known or hypothesized today, that have taken place within the cosmological atom.

That we have repeatedly confronted real issues in this initial attempt at a GSS can be noted by quick examination of a section in North (A1) on "Cosmology and the Formation of Galaxies." In the context of an expanding universe, the problem exists of explaining the formation of condensations (first of galaxies, then of stars). Lamaître (1933) proposed three stages—fragmentation of the universe in high-energy radiation and velocity; deceleration permitting a gravitational attraction approach to equilibrium, with local inhomogeneity; renewed expansion. He was followed, seriously, first by Gamow and Teller (1939), then by Gamow, Alpher, and Herman (1948–1949). They started with the thought that an equilibrium theory of a single high temperature and density could not produce the abundance of elements observed. They began to develop a nonequilibrium theory of a rapid initial expansion, essentially of a neutron gas. By decay and capture and disintegration, the heavier elements are produced. This has been discredited, except possibly for the residual survival of He which could conceivably have survived the initial high-temperature expansion. The difficulty of maintaining the chain is that it is broken by the short life of mass 5 and mass 8 particles, i.e., it is only possibly light elements that were thus formed. Instead, the later Burbidge-Burbidge-Fowler-Hoyle (B^2FH) model (1957) of the processes of element formation has come to be generally accepted. Von Weizsäcker (1944) introduced into cosmogeny a theory of turbulence as a source for producing gravitational stability.[1]

We can now begin an outline for cosmological processes:

1. *The big bang* (*time* = *0*)   From a previous cycle, the present phase of expansion starts out from a soup of perhaps few or no elementary particles in thermodynamic equilibrium with radiation at extremely high temperature. The primeval fireball, with a temperature in excess of 10^{10} °K (B2) and a diameter of perhaps 100,000,000 miles, starts our present cycle $(1.3 \times) 10^{10}$ years ago. A temperature of 10^{12}°K destroys all heavy elements. The initial material is likely largely radiation (i.e., an intense e-m field). The basic process that will subsequently take place is an adiabatic expansion of this photonlike universe as a near-equilibrium thermodynamic process.

2. *Thermal pair production* (*time* = *a few seconds*)   There is a copious flow outward of leptons (photons, neutrinos) followed by the creation of neutrons. The temperature is about 10^{10}°K.

The neutron reactions are like

$$p + e^- \leftrightarrow n + v$$

$$p + v \leftrightarrow n + e^+$$

3. *Helium production, i.e., birth of the elements* (*time* = *soon thereafter*) As temperature cooled to 10^{9}°K, the beginnings of stable nuclei appear. The

---

[1] We refer to gravitational hydrodynamic instability on a galactic and stellar scale. However, Silk (1968) suggests that radiation damping in the fireball will eliminate density perturbations below 10^{11} solar masses. Thus it is galaxies which are at issue as the first instability.

helium reactions are like

$$n + p \rightarrow D^2 + \gamma$$
$$D^2 + D^2 \rightarrow He^3 + n$$
$$D^2 + D^2 \rightarrow T^3 + p$$
$$He^3 + n \rightarrow T^3 + p$$
$$T^3 + D^2 \rightarrow He^4 + n$$

It appears that helium, a little deuterium, and Li^7 were the only elements produced in the fireball [see Fowler (E1)]. A primordial He abundance of about 27 percent by mass has been estimated (although the initial presence of helium in the fireball is one of the greatest of current controversies in astrophysics).

Evidence for the primeval fireball and the big bang was recently forthcoming [see Peebles and Wilkinson (B3), Partridge (B4)] when the background radiation in space was found to be about 3°K and isotropic. This was quite close to (and found without knowledge of) the estimate of Gamow's colleagues of about 5°K (1949) as the residue from the big bang, i.e., that the radiation has been maintained in thermodynamic equilibrium during its lifetime of expansion.

4. *Formation of the ionized elements*   It is vague whether elements were formed in the initial fireball, although there is little discussion of the passing transient phase, as the fireball expanded and cooled off.[1] Current matter, outside of hydrogen and the lesser amount of helium (and other light traces), seems to have emerged later in high-temperature stars. Yet the reactions which brought the other elements into being had to be a sequence of reactions taking place at rising temperature. One must surmise that a falling temperature of the fireball preceded a subsequent inhomogeneous segregation into spots that became hot spots and produced elements, i.e., an instability problem emerging from e-m radiation.

5. *Formation of nonionized matter such as galaxies* (*time = few* × 10^5 *years*–10^6 *years*)   Until the temperature of the universe drops to about 3000 to 4000°K, matter is ionized. The free electrons keep radiation and matter tightly coupled by means of Thompson scattering. Radiation exerts a viscous damping which prevents any large-density inhomogeneities. Thus galaxies cannot form until the temperature drops to perhaps 4000°K and recombination of the plasma begins.

6. *Formation of the galaxies*   Once small-density irregularities exist in this gas, those inhomogeneities greater than a critical dimension grow for

---

[1] As an example, it is not known for certain at the present time whether all the atomic species heavier than hydrogen have been produced in stars without the necessity of element synthesis in a primordial explosive stage of the universe.

stability reasons. These are the first gravitational-bound systems in the universe. The relative density of matter for the first time exceeds that of radiation.

The somewhat speculative evolution of galaxies may be followed in an article by Arp (C4). Based on the original proposed classification of Hubble more than 20 years earlier (1936), galactic forms are viewed as evolving from spherical forms of "gas" that become elliptic as they spin, flattening out to discs and then branching off in two directions, one of normal spiral galaxies, the other of barred spirals (occasionally there are ring galaxies). For example, ours is a spiral galaxy which has a dense, fattened, luminous nucleus with a diameter of about 50,000 light-years and spiral extensions extending to about 200,000 light-years in diameter (our sun is in one spiral arm about 33,000 light-years from the center, the other spiral arms are 5,000 light-years distant). Our galaxy has about 10^{11} stars. Distributed around the nucleus are more than 100 globular clusters of stars, each containing a few thousand to a million stars.

E-m energy is radiated by the stars and radiated or absorbed by the gases. Magnetic fields organize the pattern of radiation. The spiral as a whole rotates (we have a speed of about 260 km/sec around the galactic center). Thus gravity, electricity, and magnetism organize the motion and state of the galaxy.

In the spiral arms, there are very hot blue young stars and also great clouds and filaments of dust. The form of the spiral arms of our galaxy is outlined by hydrogen. This gas and dust must furnish the material for the formation of the bright new stars that are found to illuminate the spirals.

It has been proposed by Chandrasekhar and Fermi (1953) that the magnetic field is responsible for the spiral arms, balanced against the gravitational force. Further, it appears that there may be enough migration of gas material from the galactic center to create about one star (of sun mass) per year. However, there are still difficulties in developing a valid (i.e., consistent) theory of the spiral arms.

In Arp's view, instead of the Hubble concept of galactic evolution, he favors the idea that more than one "species" or type of galaxy emerged from the instability conditions of formation. He uses the criteria of mass and angular momentum per unit mass. He finds a line extending from giant spheres with little angular momentum per unit mass to more and more flattened and lighter galaxies with increasing angular momentum per unit mass. A second line of more nearly constant and small mass (irregular and dwarf systems) increases in angular momentum from near-spherical form to ellipses to irregular masses to the patch of Magellanic clouds. The independence of the determinants of mass and angular momentum suggest independent "birth" for these various "species."

Thus he believes that out of the essentially uniform expanding and cooling gas of near 10^{10} years ago various inhomogeneities or irregularities formed[1] of different size and momentum. These slowly contracted under the force of gravity. They were galaxies in the process of formation.

Further, he points out, galaxies frequently come in clusters. He shows that the dense clusters of galaxies have high intergalactic density (near 10^{-26} g/cm^3) and are made up of the large mass elliptic and disc-shaped galaxies, whereas in clusters approaching the mean density of the universe (10^{-29} g/cm^3) the "clusters" are single spiral galaxies.

Some view of peculiar galaxies is given in an earlier article (C6). It suggests that galaxies may still be in process of forming; i.e., these peculiar galaxies seem to be young because those masses of gas could not persist in their present form "and because the natural forces acting in a rotating mass of gas always tend to produce order from disorder—symmetry from dissymmetry."

For a hierarchical ordering of the units of cosmology, the Burbidges point out for us that the diametral unit of stars typically is of the order of 10^6 miles (the solar diameter); star clusters of the order of 100 light-years; spiral galaxies of the order of 10^5 light-years; galactic clusters from a few to many thousands of galaxies with diameters of the order of 10^7 light-years. Beyond lies the universe with a diameter of 20×10^9 light-years.

We can abstract a few additional ideas (and issues) from an article by Dicke (C3). Our galaxy (most galaxies) was formed originally as an enormous mass of highly turbulent hydrogen gas. There is little reason to believe that much He or any heavier elements were there. The age of the universe, from its beginning expansion, is of the order of 10^{10} years. (However, Dicke has a problem with the existence of old globular clusters, stars with a possible age of 2.5×10^{10} years. Sandage regards such issues as easily buried in the uncertainties.)

The evolutionary age of a galaxy is determined by the rate at which hydrogen gas is converted into long-lived stars.

The composition of our spiral galaxy is about 2 percent H gas. It possesses three classes of stars—a halo of very old stars spherically surrounding the nucleus; a disc population of old stars; and a patchy spiral population of young and older stars.

Dicke traces the theoretical evidence for the evolutionary age of a galaxy. After the "rapid" formation of the halo population, the galactic gas decreases exponentially with time. The mean life is about 2×10^9 years. Thus with the present concentration of about 2 percent H in our galaxy, this leads to a galactic age of about 5–12×10^9 years.

With regard to the evolutionary age of the universe as compared to the galaxies, Dicke presents the following table. (We present only his relativistic finite start-up, and closed universe data.)

[1] One cannot avoid the inference of hydrodynamic forces and thus hydrodynamic criteria of stability, such as in turbulence or thermal convection, organizing the field. However, see (C5).

| Object | Type of Age | Ages (in 10^9 years) | |
		Per general relativity	Per Brans-Dicke
Our sun	Radioactivity	4.5	4.5
Our sun	Stellar evolution	4–15	2.5–7.4
Globular cluster*	Stellar evolution	16	7.8–8.1
Old galactic cluster	Stellar evolution	16	7–8
Galactic system	From depletion of H	5–12	5–12
Elliptic galaxies	Stellar evolution (mean age)	10–16	5.5–7.5
Uranium production ("prompt")	Time of first formation	7.5–11.1	7.5–11.1
Universe	Hubble (galactic) expansion	13	13–15
Closed universe	Based on Hubble age	< 8.6	8

* Near-spherical clusters of 10^5 or more stars. Iben (C8) dates them from a big bang start-up of 13×10^9 years ago.

One may say that stars are $0-8 \times 10^9$ years old; galaxies are $8-10 \times 10^9$ years old; and the universe is $9-13 \times 10^9$ years old within a framework of current dating.

With regard to Dicke's proposal that galactic gas decreases exponentially in time and his thesis regarding galactic evolution, Roberts (C7) found that the amount of H in a galaxy seems correlated with its appearance and structure. Spiral galaxies have about 1 percent mass in interstellar H. More open spirals run as high as 14 percent. Elliptic galaxies have no detectable H. However, the author declines to infer that this supports an evolution of galaxies. In any case, it supports a potential age for galaxies of the order of 10^{10} years.

7. *Formation of stars* (*time* $= 7 \times 10^9$ *years*; *i.e.*, 6×10^9 *years ago*) The oldest stars in our galaxy, Halo Population II stars in a more or less spherical halo around the nucleus, and globular clusters of 10^5 or more stars have this age. These are metal-poor stars.

How a star forms is briefly touched on in an article by Abt (E6). Stars form out of gaseous nebulae such as the Great Nebula in Orion. Initially that gas is turbulent, but it gradually will contract into stars. The angular momentum of the turbulent gas cloud tends to be conserved in the resulting stellar cluster. It shows up in two principal forms, the rotational spin of stars and the orbital motion in binary star systems. With single stars, all the angular momentum goes into rapid rotation. In binaries, it shows up mostly in the orbital motion. (If the star possesses planets, nearly all of its angular momentum goes into the orbital motion of the planets.)

The evolution of stars may be followed in Gamow (B1), Schwarzschild (E3), Dicke (C3), Fowler (E1), Herbig (F1), and Unsöld (D3).

It has now become reasonably clear that our galaxy was formed originally as an enormous mass of highly turbulent hydrogen gas [Dicke (C3)].

Of course, the simplest assumption to make is that the primordial material of the galaxy was hydrogen. This hydrogen could have originated as the material "created" to sustain the universe in the steady-state cosmology.[1] It could also have originated in a primeval explosive phase of the universe in which nucleosynthesis beyond hydrogen may or may not have taken place. Again the simplest assumption is that the syntheses of the elements heavier than hydrogen did not take place at that time; the "creation" of a neutron "ball" was followed during expansion only by neutron decay to protons and electrons and the subsequent formation of neutral hydrogen atoms. However, this simple assumption may not be correct, and as we have seen, there may have been some helium, but little or no heavier elements, produced in the big bang. The arguments go round and round [Fowler (E1)].

However, the inhomogeneities leading to galaxies could not begin until low temperatures near the magnitude of 4000°K are reached. Now galaxies can form.

But the galactic gas must be comparably unstable, after starting its rotation. Such additional hydrodynamic instability must lead to the inhomogeneities of stars. As these inhomogeneities tear off and become isolated, they may expand somewhat further. However, ultimately they start to contract. The process that takes place depends on the size of the inhomogeneity.

The stars make their appearance on the visual magnitude-color index H-R diagram as a sequence of points by which their evolutionary history can be traced. The visual magnitude of a star may be transformed by theory into the star's luminosity (its radiated power). The color index may be transformed into an effective blackbody temperature. Thus the H-R diagram may be expressed in terms of the quantities luminosity and effective temperature.

On this diagram, the star first appears as a large-radius, low-surface-temperature object. As it contracts, the stars move in at near-constant luminosity and increasing temperature until they meet the "main sequence" line. At this point, the central temperature has reached near 20×10^{6}°K. Nuclear hydrogen burning reaches a significant rate. The cooking of nuclear elements begins.

---

[1] In our opinion, it would seem that the steady-state cosmology has subsequently lost almost all of its ground, particularly since the confirmation of "initial fireball" background blackbody radiation temperature. However, one may note the remarks of the editors in *New Scientist*, p. 623, June 19, 1969.

Heavy stars run through their life histories in a surprisingly short time (4×10^7 years for a star 10 times the sun's mass), first forming helium from hydrogen, then forming heavier elements from helium in a variety of processes (Dicke).

The detailed evolutionary tracks of stars of various masses, how they spend most of their life in their main sequence position, how they depart from the main sequence position, how they depart from the main sequence at a critical helium density and become red giants and evolve further is part of the detailed story that can be followed in Gamow, Schwarzschild, Dicke, Herbig, and Iben.

8. *Formation of the sun* (*time* $= 8.5 \times 10^9$ *years*, 4.5×10^9 *years ago*) Our sun is a "typical" star on the main sequence.

9. *Formation of the heavy elements* Since it was clear that elements much past helium were uncertain of formation in the initial high-temperature big bang and that the heavier elements could not be formed till they were "cooked" as intermediates in the high-temperature "burning" of hydrogen to helium in the hot cores of stars, a theory is necessary for the subsequent formation of the heavier elements. Gamow pointed out that Eddington found that main sequence stars had central temperatures of about $20 \times 10^{6\circ}$K and that by the quantum theory for nuclear reactions it was possible to show that the observed stellar energy releases originated from the reactions between hydrogen and the nuclei of light elements (Bethe and von Weizsäcker in 1938 —the carbon-nitrogen cycle, plus a small fraction of the Critchfield hydrogen-helium process).

However, Gamow's scheme for the formation of the elements (i.e., the entire process that leads from the initial formation of elements to account for their present detected abundance) from an early neutron gas was found to be defective. It has been superseded by the more specialized theses that may be found in the classic B^2FH paper (D1). Elements were formed in the contracted hot cores of stars by a series of different processes.

It seems probable that all elements evolved from hydrogen, since the proton is stable while the neutron is not. Hydrogen is the most abundant element in the universe, and helium, the immediate product of hydrogen burning by the *p-p* chain and the C-N cycle, is the next most abundant element. As an approximation, the atomic abundance curve has an exponential decline to atomic weight $= 100$ and is approximately constant thereafter. However, abundance diminishes in a series of cascades, varying with atomic weight, as one descends along the exponential decline.

The star exhibits a regulating mechanism in which the temperature is adjusted so that the outflow of energy through its envelope is balanced by nuclear-energy generation. The temperature depends on the fuel available. Hydrogen burning takes place at a lower temperature than helium; helium, a lower temperature than carbon; followed then by a further increasing temperature sequence ending at iron. As hydrogen becomes exhausted during the stellar evolution, the temperature rises by contraction until helium becomes effective as a fuel, and so on.

B^2FH list a sequence of processes from hydrogen burning (the cycle by which helium is synthesized from hydrogen, and also isotopes of carbon, nitrogen, oxygen, fluorine, neon, and sodium are synthesized); to helium burning; to an alpha particle process which when added to Ne^{20} synthesizes Mg^{24}, Si^{28}, S^{32}, A^{36}, Ca^{40}; then an equilibrium process that synthesizes the iron peak elements; and then others. From these processes, some estimates of abundance curves can be made.

As far as where we stand now, at least 99 percent of the heavy elements were produced during the first $1-3 \times 10^8$ years,[1] whereas our galaxy is about 7×10^9 years old. Although the elements tend toward equilibrium, we are still far removed from heavy-element equilibrium. The older stars still show the original early abundances of material in their outer mantle. It is only recently formed stars that are "dirty" with heavy elements, with roughly 4 percent of elements heavier than helium.

In all stars, the relative abundance ratios of all heavy elements from carbon to barium and beyond, and probably the H/He ratio, are the same.

Another paper on the origin of the elements that discusses the cosmological significance of the helium abundance in stars is by Clayton (D2). We may start with the observation that in the sun the H/He ratio is about 16:20. (With regard to the serious beginnings of nucleosynthesis, Clayton points to the historic review paper of H. Seuss and H. Urey, *Reviews of Modern Physics*, vol. 28, p. 53, 1956; later H. Urey, *Review of Geophysics*, vol. 2, p. 1, 1964.) However, samples for the other elements from the solar photosphere, the corona wind, cosmic rays, the earth, and meteorites are not completely compatible in the abundances found. Also, there are stars different from the sun. (Actually they are similar in composition but not identical.)

---

[1] As a crude frame of reference for the reader, we can offer the following extrapolated schedule for a uniform homogeneous post-big-bang ideal adiabatic expansion of radiation.

Now: $R = 5 \times 10^{22}$ miles $T = 3°K$ $t = 1.5 \times 10^{10}$ years
    $= 10^{10}$ light-years $= 5 \times 10^{17}$ sec

Harrison (B2): $T \propto t^{-1/2}$ $V \propto T^{-3}$

Thus $RT = $ const.

$T, °K$	t, years	R, miles
3	1.5×10^{10}	5×10^{22}
10^2	1.5×10^7	10^{21}
10^4	1.5×10^3	10^{19}
10^6	1.5×10^{-1}	10^{17}
10^8	1.5×10^{-5}	10^{15}
10^{10}	$1.5 \times 10^{-9} (= .05 \text{ sec.})$	10^{13}

These should only be regarded as hypothetical "bar-talk" numbers.

Clayton makes the following points:

a. The oldest known stars are formed from material less rich in elements heavier than He than were stars formed recently. Thus elements were not always with us but were manufactured.

b. Population II objects (primarily globular clusters and high-velocity stars) are less abundant in heavy metals relative to H by a factor of 100 and 1,000 as compared to the sun.

c. The first low-mass stars, in big bang cosmology, are supposed to be formed from material in which heavy elements are lacking, so this is not surprising.

d. If the universe is metal-poor, then most (99 percent) of the heavy elements were formed between the early formation of first Population II objects and the formation of young stars.

e. The logical site for this synthesis of heavy elements is in the thermonuclear epochs encountered in the thermal evolution within stellar interiors.

f. We now have a photon gas of $3°K$. Assume it to be the same photon gas that existed when the original dense universe expanded sufficiently to be transparent to an average thermal photon (perhaps $40,000°K$).

g. Detailed calculation based on current abundance makes only H, D, He^3, He^4, and perhaps Li^7 possible then. This checks with the finding that the old globular clusters are underabundant in heavy elements.

h. Although the ratios in these stars are difficult to check, one can estimate in Halo Population II stars a He/H ratio of $1:10$ by number or a mass fraction equal to 0.27. This is near the value for He for the simplest big bang theory. It lends support to the big bang.

i. However, there is a low He abundance in old horizontal-branch B stars, although young stars of the same branch show normal He abundance. Therefore, He has also been synthesized within the interiors of stars or like objects. Thus the simplest big bang theory is not necessarily correct.

j. The question of element formation is not settled by astronomy today. The He mystery is perhaps the biggest current mystery in astronomy. If the He concentration were universally low when first stars were formed, we must abandon the big bang cosmology unless we either have a low initial He theory—e.g., according to the scalar-tensor theory of Dicke or some other exotic conditions that Clayton specifies.

k. Even if He is primarily a remnant of the cosmic fireball, p-p and C-N chains are the internal power source of main-sequence stars and will convert several percent of galactic mass of H into He.

l. After H burning, the next process is then He burning. It is the later thermonuclear epochs that are responsible for most element synthesis. Carbon (or oxygen) burning is then next.

m. Rapid cooling of burning carbon as in an explosive ignition (W. Arnett, *Nature*, vol. 216, p. 1344, 1968) and rapid cooling from $2 \times 10^{9}\,°K$ to $0.5 \times 10^{9}\,°K$ produce an abundance like the solar abundance of nuclei. Ne^{20}, Na^{23}, and Mg^{24} may be so yielded (and others up to Si^{29}, by explosive ignition near $10^{9}\,°K$).

n. Si burning (at temperatures above $3 \times 10^{9}\,°K$) is responsible for nucleosynthesis between $A = 28$ and $A = 57$.

o. Beyond iron, synthesis results from free-neutron capture. The presence of heavy radioactive nuclei in substantial numbers is proof that elements have not been a permanent part of the universe (i.e., originally, no heavy atoms; now there are. Thus they have been formed).

Even more recent is Unsöld's paper (D3) on the origin of elements.

a. In normal stars the outer layers indicate the original composition (except for Li, Be, and B which are proton sensitive) when formed. There are rare stars which show higher abundances of He, Ba, C, and others, where a spectral line is strengthened. However, with regard to normal stars:

b. In old stars, the heavier elements C to Ba are less abundant relative to H by factors up to 200. The problem is whether this lack and the abundance of He (the end product of H burning) for old stars differs from the corresponding ratio for the sun and younger stars by the same factor.

c. All normal stars have the same abundance (ϵ — by number of atoms), even young stars (source: Müller and others, 1960, 1961), as shown in Fig. 19-1.

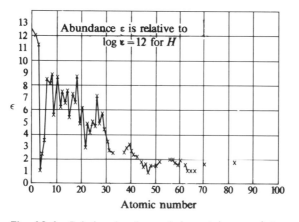

Fig. 19-1   Relative abundance of elements in normal stars.

These abundances are for main sequence stars, old or young, even for young supergiants, red giants, and planetary nebulae.

d. Therefore, most of the life of our galaxy has been spent practically with the same composition.

e. Most stars (more than 95 percent) classified by two parameters—spectral type and luminosity class—thus have the same composition on their surface. (Unaffected by internal nuclear processes.)

f. In internal nuclear reactions at low central temperature, H burns into He first by fusion, then at higher temperature by C-N-O cycle; with increasing temperature, He burns to C^{12}; then carbon burns to heavier elements. At very high temperature, iron group elements burn by the so-called e process (a kind of frozen thermodynamic equilibrium); and finally the heavier nuclei by neutrons—the s (slow) and r (rapid) process[1] (whereas in the Alpher-Bethe-Gamow paper, all nuclei were formed by rapid neutron capture; thus elements are not all made by a process of rapid capture). All of these reactions were studied in the B^2FH paper (D1). (This still remains the fundamental paper for the origin of elements.)

10. *Formation of the earth* (*time* $= 8.5 \times 10^9$ *years*; i.e., 4.5×10^9 *years ago*) The solar system's age is also about $4.5–5 \times 10^9$ years.

11. *Formation of the earth's crust* (*time* $= 10 \times 10^9$ *years*, i.e., $3^+ \times 10^9$ *years ago*) Earliest known rocks date from 3×10^9 years ago. The origin of earth took place 4.2×10^9 years ago. Temperatures 4×10^9 years ago were [modeled by Allan (1956)] 450 to 700°C at 50 km depth and 4200 to 4300°C at 2,900 km (temperatures today are essentially the same).[2]

12. *Formation of life* (*time* $= 10 \times 10^9$ *years*; i.e., 3.2×10^9 *years ago*)

13. *Formation of young stars* Population 1 stars are young metal-rich stars found in spiral arms, strongly concentrated in the galactic plane, associated with interstellar dust and gas (Baade 1944). These are less than 10^8 years old.

19-1 A SCENARIO FOR START-UP

Since those not trained in physics may not have gotten the picture completely, nor is the picture actually complete, we will sketch out a loose scenario for the start-up of the cosmos. It is not meant to be perfectly valid, but it will fill in steps to make the full story graphic. Present and future experts can correct what is wrong.

Either as a single act of creation or more likely as the remnant of a previous cycle, the entire universe had collapsed to a hot, nearly pure photon gas (at 10^{12}°K all matter has disappeared). At this point it starts its expansion.

---

[1] The s process takes place in massive stars; the r process after supernova explosions.

[2] J. T. Wilson et al. (1956) in Scheidegger (G1).

This hot gas is essentially an ideal gas. It is going to expand throughout its entire "life" as an adiabatic ideal gas expansion

$$pV^\gamma = C$$

$$TV^{\gamma-1} = C'$$

where γ typically may be 4/3, a value for a photon gas.

However, during its expansion, various epochs of instability will arise, and some detailed elements of form will emerge—nuclei (stable nucleons and other unstable particles), galaxies, stars, heavy nuclei. The expansion will continue until a uniform "final" distribution of matter will exist. All the nuclear flames will be out. All matter and radiation will be "cold." Then gravitation will begin to draw all the matter together. Again it will heat up. All matter will disappear, "boiling off" into radiation. The history ends as the next cycle begins.

Start back from the initial expansion. As the temperature drops, the near-pure photons "condense" quantum mechanically (or nonlinearly as one might prefer) to form the stable nucleons—protons and electrons. There is a passing phase of neutron formation. To various uncertain extents, some more light particles and nuclei appear. However, as long as the gas is quite hot, it is a "gemisch" of ionized particles and radiation in near-thermodynamic (but complex) equilibrium.

The gas expands homogeneously until the temperature drops to perhaps 4000°K (not very far above the boiling point of all "atomic" and "molecular" material that can exist on earth). At this point, nonionized material can exist. The gas becomes hydrodynamically unstable. (If $TD = $ const, $D = $ diameter of universe. $D_{then}/D_{now} = T_{now}/T_{then} = 4/4000 = 1/1000$. This, crudely, is the "diameter" of the universe at that point, compared to now. Its dimensions are more compact.)

Inhomogeneities separate away from the expanding gas like "droplets." What this likely has to mean is that the homogeneous expansion runs away from the inhomogeneous droplets. They continue to expand more slowly. But these droplet galaxies contain momentum. They form from unstable hydrodynamic eddy lines.

Now the expanding droplet goes through its complex early history. It expands and builds up its spin and early form. However, as it expands, it too is unstable hydrodynamically. It pulls apart (still mainly as a hydrogen gas) into stabilizing "droplets." These will form stars. Again the galaxy likely rushes on in its expansion, leaving the expansion of the star droplets to fall behind.

Inevitably the gravitational forces take hold, and the large star mass, with its diameter many times its subsequent diameter, begins to contract. (All of this may take place as a complex series of events going below and finally above 4000°K.) Internal stellar temperature rises and rises. Finally when $10^{6}°$K is reached, nuclear reactions can begin their "fires." The star begins to crawl through its long quasi-equilibrium, in which it burns its hydrogen fuel, forming heavier elements as by-products. The first phase involves core temperatures of the order of $10^{7}°$K. Subsequent phases go higher and higher, until supernova states of explosion ensue. The star grows, ejects its material, and is then finally "dead." The temperature cools off to the background of space.

In the case of the youth of our star, and likely of many others, one more phase of hydrodynamic instability is reached. Material is shed and our solar system is born. As it cools, one additional chemical instability is found. Chemical compounds including the biochemical oscillators come into being. Its early start more than 3 billion years ago suggests that the formation of life is ubiquitous—not too hard to obtain. (A little specialization in condition but not too much. Man thou art not more rare than a day in June!)

The only fault in the scenario is where or when the elements formed. Gamow's original scenario (B1) formed them quickly. Everyone, apparently, would like to have them formed within the first 10^{8} years. However, they have to be formed on a rising temperature scale. This might fictitiously have happened 1) if after the initial expansion there were a second jog in temperature, nuclear fires lighting and running through a sequence, 2) if there has been sufficient convection to turn over internal nuclear constituents in the stars, or 3), if Gamow's hypothesis were correct, with some other early nuclear process common throughout the cosmos. These causal ideas in

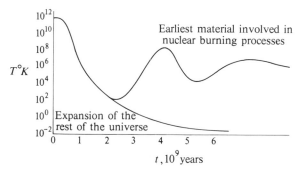

Fig. 19-2 A purely hypothetical temperature diagram for the start-up oscillations.

combination or some other as yet undisclosed thesis may straighten out the issues of element formation in the near future.

An extremely speculative picture is presented in Fig. 19-2, showing some extreme processes that cosmological material undergoes. Part enters into an initial expansion, a subsequent heating, a relatively rapid burning out by near-supernova explosions to form all the elements, a subsequent more gentle gravitational contraction and reheating, and the beginning of a long burning-out process of a hydrogen type. This is not all taking place with one fixed quantity of gas. The picture represents an upper-contour limit for energetic processes.

The bottom limit is of material that has simply undergone an adiabatic expansion to the present background of space.

Such an "oscillating" model, even if only conceptual at this point, should not be looked at with disdain. If we imagine it superimposed on the Sandage closed-universe model as depicted below, then we may see that a few "minor" oscillations that enrich the pot of cosmological events, in the temporal transformation of mass energy, are just tempests in a teapot, other "minor" manifestations of other limit cycles. (See Fig. 19-3.)

Thus, essentially, we are back to the possible thought that large-scale processes, through a conversion of mass and radiation density (perhaps open, perhaps closed), govern the size and curvature of our atomistic cosmos. The details of whether space is open or closed, whether time is uniform throughout or not, whether gravity varies, of the nature of systems like novas and quasars, and of the sources of all cosmic radiation are still open for investigation. However, the possibility of indefinite limit-cycle oscillations of the cosmological atom still exists. The details of the nuclear processes that follow the expansion lead us back to the other end. [Examples of more recent commentary are by Wagoner (D4), Peebles and Wilkinson (B3), Maran and Cameron (B5), and Chew (B6).]

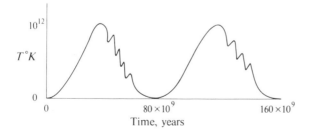

Fig. 19-3 The limit cycle of cosmology enriched by a few start-up oscillations.

SUMMARY

1. The cosmological atom ends our line of systems. The accepted model is the dynamic universe, at present undergoing an expansion. The first real estimate of the cycle-time has recently been made. It is about 80×10^9 years.

2. The cosmological "atom" indeed furnishes a complex "atomistic" level.

3. From a finite total magnitude of matter-energy, a primeval fireball at about $10^{10°}$K began an expansion by its radiation pressure about 10^{10} years ago. The fireball may have been left over as the gravitationally contracted stage of a previous cycle.

4. Except for some "minor" nonlinear issues in the formative character of mass particles, the expansion would have been a simple "adiabatic" expansion of self-contained electromagnetic radiation (i.e., a photon gas) to its current temperature of 4°K. However, these minor issues involve the question of the interconversion of energy into quasi-static stable states of matter, in general consistent with the Einstein relation.

5. However, matter quantizes—perhaps by quantum theory, perhaps by a nonlinear theory. The history of this quantization of form out of the atomistic fireball becomes the story of the formation of nuclear particles, of galaxies, of the stars, and then ultimately of all other autonomous systems.

6. The sequence is first the formation of a few simple nuclear particles in the cooling expansion, protons, neutrons, electrons; the further cooling of the universe in its ionized form down to where the ball was hydrodynamically unstable; the formation of galactic "eddies"; their subsequent additional instability to form stars; the contraction of stars to light up their nuclear fires; and the "cooking" of the heavier nuclei.

QUESTION

Examine the scenario for cosmological start-up, and write your own scenario with as much detail as you can add. Review all available material.

BIBLIOGRAPHY

A. Expanding universe—time dependent mass-energy

1. North, J.: "The Measure of the Universe," Oxford University Press, New York, 1965.
2. Møller, C.: "The Theory of Relativity," Oxford University Press, New York, 1965.
3. Eddington, A.: "The Mathematical Theory of Relativity," Cambridge University Press, New York, 1938.
4. Eddington, A.: "The Expanding Universe," Cambridge University Press, New York, 1933.
5. Tolman, R.: "Relativity, Thermodynamics, and Cosmology," Oxford University Press, New York, 1934.
6. *Scientific American* Editors: "The Universe," Simon & Schuster, Inc., New York, 1957.

B. Initial start-up and formation of first "particles"

1. Gamow, G.: "The Creation of the Universe," The Viking Press, Inc., New York, 1952.
2. Harrison, E.: The Early Universe, *Phys. Today*, vol. 21, no. 6, p. 31, 1968.
3. Peebles, P., and D. Wilkinson: The Primeval Fireball, *Sci. Am.*, vol. 216, no. 6, p. 28, 1967.
4. Partridge, R.: The Primeval Fireball Today, *Am. Scientist*, vol. 57, no. 1, p. 37, 1969.
5. Maran, S., and A. Cameron: Relativistic Astrophysics, *Science*, vol. 157, no. 3796, p. 1517, 1967.
6. Chew, G.: Hadron Bootstrap: Triumph or Frustration? *Phys. Today*, vol. 23, no. 10, p. 23, 1970.

C. Formation of galaxies

1. Mihalas, D.: "Galactic Astronomy," W. H. Freeman and Company, San Francisco, 1968.
2. Woltjer, L. (ed.): "Galaxies and the Universe," Columbia University Press, New York, 1968.
3. Dicke, R.: Implications for Cosmology of Stellar and Galactic Evolution Rates, *Rev. Mod. Phys.*, vol. 34, no. 1, p. 110, 1962.
4. Arp, H.: The Evolution of Galaxies, *Sci. Am.*, vol. 208, no. 1, p. 70, 1963.
5. Burgers, J., and R. Thomas (eds.): Proc. Third Symp. On Cosmical Gas Dynamics, *Rev. Mod. Phys.*, vol. 30, no. 2, p. 905, 1958.
6. Burbidge, M., and G. Burbidge: Peculiar Galaxies, *Sci. Am.*, vol. 204, no. 2, p. 50, 1961.
7. Roberts, M.: Hydrogen in Galaxies, *Sci. Am.*, vol. 208, no. 6, p. 94, 1963.
8. Iben, I.: Global-cluster Stars, *Sci. Am.*, vol. 223, no. 1, p. 26, 1970.

D. Formation of the elements—the remaining "particles"

If, as for example Fowler (E1) and others are willing to state, "in the majority of stars there has been no mixing between surface and interior over the lifetime of the star," and as Unsöld (D3) and others point out, "Throughout most of the lifetime of our galaxy, the interstellar matter (out of which the stars originated) has had practically the same composition.... [T]he chemical composition of the outer parts of the stars in our galaxy remained mostly unaffected by nuclear processes which...must have considerably altered the chemical composition of their interiors. The well-known fact that the vast majority of stars (> 95 percent) can be classified by two parameters, spectral type and luminosity class...clearly shows that all these stars have the same composition, too, and that the foregoing conclusions apply to them"; one must draw the inference that one major point of view in current cosmological beliefs would require the formation of the "elements," with near-current cosmological distribution, before the nuclear-fueled stars.

As Clayton (D2) ends, "The quantitative success of (his) theory leaves little doubt that the elements have had a primarily thermonuclear origin. The question is where and when."

1. Burbidge, E., G. Burbidge, W. Fowler, and F. Hóyle: Synthesis of the Elements in Stars, *Rev. Mod. Phys.*, vol. 29, no. 4, p. 547, 1957.
2. Clayton, D.: The Origin of the Elements, *Phys. Today*, vol. 22, no. 5, p. 28, 1969.
3. Unsöld, A.: Stellar Abundances and the Origin of Elements, *Science*, vol. 163, no. 3871, p. 1015, 1969.
4. Wagoner, R.: Cosmological Element Production, *Science*, vol. 155, no. 3768, p. 1369, 1967.
5. Starr, V., and P. Gilman: The Circulation of the Sun's Atmosphere, *Sci. Am.*, vol. 218, no. 1, p. 100, 1968.

E. Formation of stars—the early history

1. Fowler, W.: "Nuclear Astrophysics," American Philosophical Society, Philadelphia, Pa., 1967.
2. Dufay, J.: "Introduction to Astrophysics: The Stars," Dover Publications, Inc., New York, 1964.
3. Schwarzschild, M.: "Structure and Evolution of the Stars," 1958 edition, Dover Publications, Inc., New York, 1965.
4. Saltpeter, E.: Stellar Energy Sources, *Rev. Mod. Phys.*, vol. 29, no. 2, p. 224, 1957.
5. Iben, I.: Stellar Evolution, *Astrophys. J.*, pts. I–V; vol. 141, p. 993, 1965; vol. 142, p. 1447, 1965, vol. 143, pp. 483, 505, 516, 1966.
6. Abt, H.: The Rotation of Stars, *Sci. Am.*, vol. 208, no. 2, p. 46, 1963.

F. Later stars and elements and remaining mysteries

1. Herbig, G.: The Youngest Stars, *Sci. Am.*, vol. 217, no. 2, p. 30, 1967.
2. Limber, D.: The Pleiades, *Sci. Am.*, vol. 207, no. 5, p. 58, 1962.
3. Burbidge, G.: The Origin of Cosmic Rays, *Sci. Am.*, vol. 215, no. 2, p. 32, 1966.
4. Sandage, A.: Exploding Galaxies, *Sci. Am.*, vol. 211, no. 5, p. 38, 1964.
5. Ginzburg, V.: The Astrophysics of Cosmic Rays, *Sci. Am.*, vol. 220, no. 2, p. 50, 1969.
6. Burbidge, G., and M. Burbidge: "Quasi-Stellar Objects," W. H. Freeman and Company, San Francisco, 1967.
7. Alfvén, H.: "Worlds-Antiworlds," W. H. Freeman and Company, San Francisco, 1966.
——: Antimatter and Cosmology, *Sci. Am.*, vol. 216, no. 4, p. 106, 1967.
—— and A. Elvius: Antimatter, Quasi-Stellar Objects, and the Evolution of Galaxies, *Science*, vol. 164, no. 3882, p. 911, 1969.
8. Hewish, A.: Pulsars, *Sci. Am.*, vol. 219, no. 4, p. 25, 1968.

G. The solar system and earth

1. Scheidegger, A.: "Principles of Geodynamics," Academic Press, Inc., New York, 1963.

H. Life

1. Eglington, G., and M. Calvin: Chemical Fossils, *Sci. Am.*, vol. 216, no. 1, p. 32, 1967.
2. Echlin, P.: The Origin of Plants, *New Scientist*, vol. 47, no. 638, p. 286, 1969.

I. The end of the universe

1. Thorne, N.: Gravitational Collapse, *Sci. Am.*, vol. 217, no. 5, p. 88, 1967.
2. Sandage, A.: Observational Cosmology, *Observatory*, vol. 88, no. 964, p. 91, 1968.

part five

Speculation

20

The Giant Pinball Process
for Systems

It is interesting to speculate on a grand picture of the development of systems. If we regard our cosmological atom as starting from its "origin" in time (as the current phase of cosmology), then we have an expansion which we can depict as shown in Fig. 20-1.

We can adopt the thermodynamic principle (for this phase) that as a result of natural processes the order in the universe is decreasing; i.e., processes run down. Thus we have powered the pinball machine.

However, its "surface" is not plane and homogeneous. It is pebbled, particularly near the start. See Fig. 20-2.

These pebbled regions are actual pockets.

As our matter-energy descends the thermal ladder and cools, various parts of it reach local space-time-temperature stability regimes. The first is a "rainfall" of nuclear particles (protons).

Think, reader, what this likely means. It suggests that the formation of "quantized" matter emerges not much differently from its "continuum"

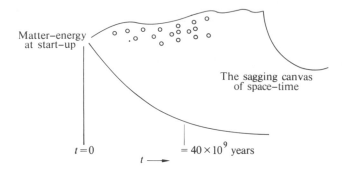

Matter–energy
at start–up

The sagging canvas
of space–time

$t = 0$ $= 40 \times 10^9$ years

$t \longrightarrow$

Fig. 20-1 The giant pinball of space-time in which matter-energy runs downhill (i.e., losing order).

matrix than did the hydrodynamic turbulence "vortices" and that both emerge as nonlinear stability processes. The fact that ultimately these "particles" (at the moment we would prefer processes, representing function becoming form, i.e., functional degrees of freedom appearing as form) may be described by quantum mechanics does not negate their being quantized particles emerging nonlinearly from a near-continuum field.

We find it peculiar in many discussions of irreversible thermodynamics to find many writers fixed with a strange obsession that there is no capability for "natural" processes as opposed to "man-made" processes to form viable systems or for active systems to form and operate near but not at thermodynamic equilibrium.[1] This is particularly true with regard to living systems and their start-up.

[1] We likely have a different view of "near-thermodynamic equilibrium." For irreversible processes, we do not regard $dS = 0$ but that neighboring regions do not differ markedly in their irreversible entropy production. For example, we find ourselves in essential agreement with K. Mendelssohn in *Physics Today*, p. 46, April, 1969.

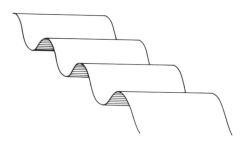

Fig. 20-2 However near the start, the canvas of space-time is pebbled. Processes get caught for awhile.

People seem to worry about how "negentropy" manages to get pumped. We find the issues no different in any of the systems. Our views are contained in the following theses:

1. Irreversible thermodynamics is satisfied if locally at every point "as a result of natural processes" (better still if there is no local engine feeding the region) there is an irreversible production of entropy; that is, that degradative processes exist. This will assure one that locally it is not possible to build perpetual motion machines of the second kind.

2. The apparent assumption is that if a medium is locally degradative in each infinitesmal volume, by summation it is totally degradative throughout the entire field. This is not true. It is only true if the field is homogeneous and there are no convective mechanisms or sources located elsewhere in the field.

3. If homogeneous, the field must be degradative, so that the final state of such an open system is one that comes to a static equilibrium in which entropy is maximized.

4. However, if not homogeneous, the broader field may form a nonlinear limit cycle. It may in fact transfer potential energy from one higher level to a lower level (dc) so as to convert it into an active ac form. It forms a thermodynamic engine. This engine can pump "negentropy." It cannot do it at a rest state.

5. The only static rest state is the degradative maximum entropy state. However, there is a second possible steady state. This is an oscillatory state, which transforms without violating the first law (nor does it really violate the second law) by converting dc energy to ac energy. It charges the overall universe for this service, but it does it by not running the system at its maximum level of entropy production. As long as the system can persist (i.e., for its life), this will continue as a metastable state on the pinball machine surface. Ultimately, the system will degrade, but "temporally" it may remain locked up.

6. Thus several books on nonequilibrium thermodynamics speak validly about the results for stationary states having only degradative equilibria. However, dynamic states can be near thermodynamic equilibrium at every point and yet operate far away from the maximum entropy state.

7. We have sketched this in the hydrodynamic field. We can specifically point to "convection" as the mechanism by which the turbulent field can propagate energy into the core to maintain the turbulent field far removed from its degradative laminar flow level.

8. In parallel fashion, we can point to "algorithmic convection" by which the living system biochemically stuffs its mouth full of food and similarly periodically counters the flow of entropy. We found it quite amusing to note that comparable mechanisms stuff the gut of the galaxy and the star to run these similarly viable processes.

Ultimately—it is true—they will all run down, but in the interim a short life and a merry one!

Then the field continues its expansion of mixed energy and quantized nuclear particles. With some theoretical difficulties, the rain of the first nuclear elements begins. The evolution of the periodic table has begun. As Gamow imagined, the elements have been cooked.

However, the matter is not uniform. It was formed inhomogeneously. At every level, including the one before, lossy processes take place. Viscosity, thermal conductivity, and diffusivity represent them formally. Here we had to descend to 4000°K. Ionization, which stirred these nuclear particle-radiation fields violently, ends for the first time. Because of the local inhomogeneities, "condensation" takes place. Larger portions grow and ultimately pull together. How does this happen? It means that the expanding fireball starts to outrun the local expansions. The continuum of nuclear particles and radiation begins to pull apart into "droplets," the larger ones predominating (i.e., there is a critical size). Thus we have a "mist" of galaxies forming in the expanding fireball. This misty rainfall is now the "countless" number of galaxies, all at near 4000°K. In these galaxies, coalitions of a few atoms and molecules can form. Such hot material you can nearly visualize in the laboratory.

Why did these droplets of particles form? Again, to provide nonlinear stability.

As the fireball races in its further expansion, what of the galaxies? They too are expanding, but more slowly. Why are they not hot? Because there is a large amount of radiation energy that has still been stored as radiation, confined by the previous confining fireball cavity. Now for the first time that radiation is not supported, it drops down to matter.

The galaxies proceed in their expansion. However, they too are inhomogeneous and locally unstable. Coalitions of atoms (above a critical size) begin to pull apart once again. The stars are forming. In fact, once again it "rains," now stars, for instability reasons. Now the gravitational forces begin to take hold. The galaxies throw material out. Gravity begins to dominate the stars. They begin their contraction. They tear apart and strip their simple coalition atoms and molecules back down to nuclear particles. However, now they discover that they have fuel—protons. The fuel ignites with increasing temperature. The subprocess of hydrogen burning into helium begins. Once more, another instability shows itself. The stars locally heat up. The main sequence of stars has begun.

Incidental to the formation of stars, some small drops shake off, possibly for local stability conditions. A solar system forms. Again it may simply be an excessive chunk of star material which shakes off vortices in a critical early step in its ignition process as an instability.

These particles of the solar system condense, cooling off back into atoms and molecules.

Consider now an added hypothetical thought. Here we have hot material (planets) cooling back down toward 4000°K, where ionization will end and atoms and molecules will form. The material is boiling. The abundance distribution will be much like the parent stars. What will move toward the surface as the gases condense and form a melt? The light gases may boil off. The heavier metallic atoms may remain. Down, down, comes temperature toward a few thousand degrees. Materials solidify. The atmosphere is still filled with heavier atoms. The convection is tremendous. A fantastic greenhouse effect can hold a large temperature gradient. We come down to a state of the order of 500°K. An atmosphere of the lower-melting-point metals—vapors of lead, mercuric compounds—can exist. The gradients in the atmosphere are sufficient to cover the whole temperature range of near-zero to 500°K. Water vapor–ice can exist on the top, going down to these heavy molecules.

As the temperature drops, consider now the following possibility. As the smaller molecules finally approach the capability to provide appreciable partial pressure in the atmosphere, they can sweep through extensive temperature gradients. Thus once again, an instability makes itself evident. Is it not quite conceivable that hydrodynamic-thermal-chemical oscillators may arise from stability consideration; i.e., it is conceivable that they can arise from the coupling in the gas-vapor-surface interaction without having to have a full liquid or solid surface to play within. Thus the precursors to biochemical oscillators may very well have started in the atmosphere, in which the liquid-solid earth may have only been used as a substrate for a number of boundary layers or "film" processes. The convection would be a useful hydrodynamic stirrer.

Thus we can conceive of prelife as emerging in the vapor phase and then of some stage in life being taken over in the liquid phase on earth. In either case, we conceive of an instability process. However, the hydrodynamic violence of convective turbulence appeals to us to speed up the process. Electrical discharge, etc., may augment the process rather than providing its sole source.

Meanwhile the primeval fireball continues to expand towards its current ghostlike 4°K state. In another 30×10^9 years, material having reached the bottom of the pinball machine, all processes having nearly cooled off, and matter having extended itself, then the gravitation contraction and rise in temperature may start again.

Now the purpose of this scenario was not to poorly review present cosmology once again but to present the reader with enough of a picture to point out that near-equilibrium thermodynamics always governs, that atom keeps alternating with continuum, and that as the giant pinball process tumbles downhill there are always pockets to get caught in as "temperature" permits new instabilities to be uncovered. "We shall always stumble into gopher holes."

Thus systems—temporary lockups, whether for a man-made day or a cosmological datum—will always occur. Further it is the lossy processes, when they exist and when the "temperature" drops sufficiently in value that a local process can take hold, that make the lockup possible. (See our hydro-dynamics paper on why a selection of "near-resonant" reinforced or "high

Q" waves out of the mix may take place in that specialized case.) These are the local inhomogeneities on the giant inclined pinball surface where the local material can play games and tie itself up for a lingering moment—even gaining and feeding on the local energy—before it again continues its drop to the bottom.

This has been our last chance to convince the reader that the formation of a life phase of limit cycles is not an analogue. If it had not been clear before, it would seem clear, from the issue of nucleosynthesis, that the history of a complex formation and changing abundance of nonvolatile elemental nuclei "proves" (in the sense of probing at) the thesis that these quantized particles arise from nonlinear stability criteria, i.e., that the cosmological universe (of discourse) has loss mechanisms, and that there are specific thresholds at which elements form and disappear (Unsöld).[1]

We cannot adduce like evidence for the case of atoms. However, it still seems clear that the mixed m/e-m field and the surrounding spatial "cavity" and walls (i.e., the matter-radiation equilibrium) similarly likely provide non-linear thresholds for their formation. That the ground state of the atom is not inviolate is indicated by the fact that a rather moderate wrench can in fact violate it. A nonlinear stability state is certainly suggested.

Thus the only two apparent "conservative" fields (i.e., that seem to be describable by a Hamiltonian, without losses) narrow to the case of the ground state of the atom and the isolated near-point-mass solar system, and this defense nearly vanishes or essentially vanishes when historical cosmology and the interactions of the entire field are brought in. Thus in our opinion, the point of view of the Hamiltonian of conservative systems, except for limited periods and for elementary tutorial purpose, must be given up. Instead, the formation of systems under the action of nonconservative forces must simply emerge as an essential concomitant of stability due to size, intrinsic inhomo-geneity, and an intrinsic process where increased energy density locally is always preferably wrapped up into atomistic local "structures" emerging from these functions. The model, generalized, may be based on the factors that make up the Reynolds number, i.e., a ratio, as comparison between competing organizing forces.

For example, with regard to the ground state of the atom (take the hydrogen atom of proton and electron as the simple illustration), the thought occurs to us that a "classical" limit cycle might be demonstrated if only the proton be permitted to be deformable, perhaps even elastic, and the lossy damping of the self force be allowed. A wave-guide nonlinear "resonance" result might be imagined that would lock up the ground state. Whether this classic description would account for the Bohr state is not known. However, it is at that point that other modifications of classical e-m theory are possible

[1] We find support for our hierarchical concept in a recent article by de Vaucouleurs [Chap. 21, (6)], who attempts to make out a case for a hierarchical cosmology.

to take advantage of the combination of lossy mechanisms and energy-restoring mechanisms that are required to form limit cycles.

[The possibility of classical derivation of the blackbody radiation laws and other quantum mechanical results is being vigorously explored by Boyer [Chap. 21, (1)], whose work has recently come to our attention. Boyer shows that there must be temperature-independent e-m energy in the universe, and that Planck's constant can enter theory solely as a constant setting the scale of the zero-point e-m radiation spectrum. He then shows that he can use this result in a cavity to derive the blackbody thermal spectrum. For the average energy of a nonmassive oscillator in an e-m field he obtains the quantum result indicating a zero-point energy.]

21

Subfields Beneath
and Universes Beyond?

Thus, in this maiden venture, theses have been sketched out for a general systems science, and hints have been left how one might gradually enlarge and deepen its scope. At this point, some speculation is proferred on the question of levels beyond those covered.

From man's self-centered view of reality, there seems little doubt that mass-energy defines space-time. However, it is an interesting speculation of Bohm's (2) that it is in the timeless hierarchical levels of order itself from which time itself may emerge. It may be that the closed-loop nature of appropriately ordered chains can eliminate time and the problem of quantization. However, it is beyond our competence and responsibility to pursue these metaphysical foundations except to note them.

What has been more appealing to us was a thought that occurred to us in the early forties during graduate study, that, in the extremely small space within the nucleus, instead of there being exchange forces and a nuclear binding that might be described, in extremes, as a liquid droplet model or as an ideal-gas model, the particles actually dissolved their boundary and became a "fluid." The nominal objection was that it might not be possible to obtain a stable fluid, the self-force objection. However, our vague feeling was that there might be "hydrodynamiclike" means to achieve a stable configuration. In our present terms of today, perhaps one might visualize an internal vortical structure, involving inhomogeneous-appearing "elite" elements— smaller internal nonlinearly quantized structures.

This thought, applied at the small nuclear end of systems, really was companion to a later thought, that a continuumlike dense universe might be a much more acceptable structure for a physically comprehensible universe than a universe sparsely filled by quantized particles. Inspired then by Dirac's negative ocean, we thought that here was a possible concept by which such a dense universe could be understood. We proposed an infinite coextensive continuous plastic-elastic-electric medium—the negative ocean (or anti-matter as it has become called). The properties and magnitude of the medium were not a priori known but were to be discovered by inference. We then imagined a primary cataclysm (which became referred to as a "big bang") which produced intense stresses in that negative ocean. If these stresses could result in an extremely regular vortex system hydrodynamically developed in the negative ocean, these vortices might appear in our coextensive positive space as a sparse universe of quantized particles. In particular, while the large vortex system represented or carried the disturbing energy, it was the very smallest regular local eddies that represented the "fundamental" particles.

It was clear that we were attempting to revive a form of ether theory. However, here the ether was meant to be real and continuous. We were consciously trying to forge a link between elasticity in one medium and e-m communications in another. We sought competent advice on the merit of such modeling in about 1947, since it appeared to us to be a very difficult problem to tackle and we wanted at least some encouragement before spending what appeared to us to require at least 10 to 20 years of study. When this was not forthcoming, we had to put away such hydrodynamic universe ideas for a more appropriate time.

Had we pursued it, we would have tried to develop a relativistically classical mechanical, e-m, gravitational model of a continuous unquantized universe—the negative ocean—before perturbing it with disturbances to see what nonlinear quantizations might emerge and only then seeking modifications of these classic laws to try to fit mass particle quantization. The program seems so horribly complex that one is not tempted to undertake it lightly. The only encouragement we could see was to seek some phenomenological bridge between the elastic modes of propagation in the continuous dense universe and e-m propagation in our sparse universe (i.e., how could one interpret the velocity of light as a velocity of "sound"). This may be possible even without an ether if some elastance is permitted in the mass and charge distribution that makes up such bodies as the proton.

Thus instead we waited until we had the occasion to tackle in classical hydrodynamics some quantization problems in such continuum fields. In 1948, we found our first problem, which we completed in 1950, namely, the complete homogeneous small-amplitude pressure propagation in a fluid field—the propagation of pressure in a tube (3). This is the problem, begun by Helmholz, Kirchhoff, and Rayleigh, which they felt provided a real test for the applicability of statistical mechanics. (That is, to see whether a theory of the transport coefficients—represented by the static values of viscosity and thermal conductivity—could account for the dynamic dispersion and absorption of a wave propagating in the long Kundt tube.) In collaboration with Dr. D. P. Johnson while at the National Bureau of Standards, we were able to show that the complete Navier-Stokes equations very nearly factored in the continuum region, for both gases and liquids possessing real (not ideal) molecular properties, into three modes—an elastic propagative mode, a viscous (transport of vorticity) diffusive mode, and a thermal (transport of energy) diffusive mode. Furthermore, there was a parameter z associated with frequency, which governed the damping. At low values, the system was overdamped; at high values, quite underdamped. Our solution thus went over continuously from the Rayleigh damping low-frequency solution to the Helmholz-Kirchhoff low-damping–organ-pipe solution.

(The parameter $z = D^2\omega/v$, where D = characteristic dimension, commonly the diameter, ω = angular frequency of any driven wave, and v = kinematic viscosity. The parameter $\beta = v/cD$ is a spatial continuum parameter. It is the ratio λ/D. λ = mean free path = v/c, where c = velocity of elastic propagation, i.e., of sound. It is small for spatially continuous media. The parameter $\Gamma = \omega v/c^2$ is a temporal continuum parameter. It is equal to the product $\omega\tau$, τ = relaxation time, $\tau = \lambda/c$. It is small for temporally continuous processes. Thus z is the ratio Γ/β^2. If the process is temporally very continuous, it is overdamped, but if spatially very continuous, it is underdamped. We can loosely suggest that our universe is spatially very continuous —that is, the square of the ratio of the mean free path associated with nuclear particle dimensions to cosmological diameter is so much smaller than the ratio of nuclear relaxation times compared to the cosmological life. Thus the hydrodynamic properties are not very lossy.)

It was clear (4) that it was possible to achieve modes capable of propagation over extremely long distances (under the continuum assumption). For example, because of the nature of the cross-coupling, whereas temperature, density, the scalar velocity potential, entropy all had two propagation modes, one elastic and the other thermally diffusive, the pressure wave only had one, the elastic mode. With suitable boundary conditions, the pressure wave could propagate "indefinitely" if Γ were small. (This same result is erroneously given in many acoustics books. The pressure wave within a tube is correctly given very nearly by the pure elastic equation. Written by itself, it creates the impression that the wave system has no damping. However, when the complete boundary conditions are written, attenuation is discovered. It is just misleading to infer that the other field components are purely elastic. They are not. They retain an attenuation component.)

An interesting property that was found in the solutions was that as the damping parameter was changed, imagined in particular along some logarithmic scale, (say by driving successive members of a sequence of tubes of increasing diameter, each at the same frequency), the propagation characteristics changed somewhat abruptly from overdamped to underdamped. (Namely, from a region in which the viscous damping dominated to one in which elastic propagation dominated.) It was this property that we found taken over in the larger-amplitude problem.

A decade later (in about 1961) we took up the problem of larger-amplitude signal. (In the interim, it was not possible to find hydrodynamic interest in combined acoustic-hydrodynamic attacks on turbulence. Since other duties intervened, the problem was dropped.) By tackling the nonlinear problem of stability of propagative solutions for the fluctuating components of a field, we showed that nondissipative modes could exist—using the example of turbulent flow in a long tube. In fact we can essentially show that the same condition which made propagative modes persist in the small-amplitude field acts in the large-amplitude field. The tube acts as a selective filter, providing a "resonant amplification" for elastic modes for large pressure gradient, making the small-gradient laminar field less stable. (This elastic mode does not have to be the source of instability for all steady drives or boundary conditions. However, selection rules according to some secular equation are what determine the choice of governing modality. The nonlinear problem contains within it much of the same character as the eigenvector solutions of linear spaces.) Furthermore, we believe that we can demonstrate the existence of conditions for a continuous spectrum. ("Intensity" conditions still escape us.) Whether our argument has flaws requires considerable study.

Beyond that, we were able to suggest an explanation for how stochastic field properties (the stochastic spectrum of turbulence) might arise from deterministic field equations, in which the nonlinear limit cycle has been demonstrated essentially in the form of deterministic traveling waves.

The first aspect of this problem is that the assumption of a traveling-wave system upon which the linear instability was demonstrated[1] and which was proposed as presenting the limit cycles is indeterminate in phase. Any

---

[1] A function of limited variation can always be written as a sum or integral of Fourier components. If these waves, individually indexed by frequency, can be found to satisfy the equation set as self-sustained oscillations, one may expect that they can be called into existence.

"disturbance"—whether in inlet generating conditions[1] or elastic or thermal wall fluctuations—can produce phase fluctuations.

The more compelling aspect is that there are more than one set of solutions which satisfy the equations for the fluctuating components. In satisfying boundary conditions (of zero velocity and isothermal temperature at the walls), we only attempted to isolate the set which could satisfy the boundary conditions as a nondissipative solution. All others were thus discarded as being unexcited. The result was that a self-propagating set of vortices and eddies of different frequencies which acoustically propagated themselves as fluctuations into the core seemed to be called into existence within a boundary layer near the wall.

However, a second possibility presented itself. These eddies had a scale. Was it not possible that this grainy scale, forced upon the field near the wall, could act as a source of nonself-propagative waves according to Huygen's principle? These might produce damped and slightly built-up disturbances, before being damped, to stir the hydrodynamic field in a tube into its horribly complex picture of turbulence.

Thus, we had located a potential source for the epochs of stability, a possible reason for the inexact kinds of self-disturbed systems often found, the warbling, intermittency, noiselike character, or flexible cycles in nonlinear fields that one suspects could exhibit limit cycles and yet also exhibit stochastic properties. The field is to be decomposed into three types of phenomena: the mean (time-independent) field; the stationary time-dependent field (with moderate amplitudes, an imaginary form for the characteristic exponents of Poincaré permits decomposition, to first order, in a harmonic spectrum) for which its modalities are to be separated out and from which secular equations must be derived to determine the combination of modalities which will satisfy boundary conditions and still persist; and a third set which is excited by the scale parameters of the stationary disturbance. These represent a beginning theory of the bandwidth of fluctuation around each stationary limit cycle that the unstable field can generate. This possibly brings nonlinear field quantization closer to a satisfactory theory.

Thus, nonlinear quantization in continuum-field equations of a classical type can be demonstrated in a qualified sense. We suggest that such an analogous problem (in fact quite similar) can be set up for a electrohydrodynamic negative ocean. We have long thought that this might have some merit in unifying the "primitive" particles stemming from the intensities associated with a possible sharp line spectrum of propagative modalities (the ether vortices sought in the nineteenth century), particularly as vortical structures, with an entire time-varying unstable cosmological model.

---

[1] The turbulent field is formed in a long line leading from a laminar entrance reservoir through an entrance zone before ultimate long-line conditions come into existence.

Alternately elastic properties associated with charge in the positive field of material might do. A varying structure of quantization conditions with time (or sparse mass density in the expansions) would emerge as a natural possibility.

The question of universes beyond has received some exploration in cosmology in the question of whether the space is open or closed at its existing mass density. At the present time, science begins to suspect that our space is closed. We cannot answer, at present, whether there are similar closed spaces beyond. There may be.[1]

As for universes below (that is, are nuclear particles made up of continua, etc.), we found the remarks of Davis (5) intriguing. He raises the question—by postulating an elementary time element, essentially as a delay time—whether stability criteria for space-time cannot be considerably upset. In the linear form he postulates, he regards the time-delay element that he introduces as a dissipative element (namely, one out-of-phase with inertia and compliance forces). However, more generally it can represent a graininess or intermittency in time. Thus, there still remain a number of ways to possibly crawl around the persistent problem of quantizing things.

What is most striking, within the domain of "fundamental" particles and states, is that apparently space (or space-time) is not a simple manifold or even simply the manifold that comes into existence as a result of a distribution of point masses or energies. This seems to be suggested by current high-energy physics. It begins to be tasteful to imagine a structure beyond what we see, which with high localized energy sources can become interactive, even if only for very brief times. One dimly suspects an "ether" beyond. Others prefer hunting the quark.

SUMMARY

1. From their start-up, cosmological processes wax and wane. In our present phase, ordered energy tends toward disorder. Our cosmos is expanding toward a cold state of matter. However, as it rushes down this great inclined plane of energy, like a giant pinball machine, there are local pockets. All our ordered systems and processes form for a while in these pockets.

2. Are there universes beyond the little and the big? Perhaps.

BIBLIOGRAPHY

1. Boyer, T.: Derivation of the Blackbody Radiation Spectrum without Quantum Assumptions, *Phys. Rev.*, vol. 182, p. 1374, 1969; also vol. 186, p. 1304, 1969.
2. Waddington, C.: "Towards a Theoretical Biology, 2, Sketches," Aldine Publishing Company, Chicago, 1969.

[1] At present, there is some conjecture that there are dark spots in our space that might represent such closed regions.

3. Iberall, A.: Attenuation of Oscillatory Pressures in Instrument Lines, *Natl. Bur. Std. J. Res.* vol. 45, p. 85, 1950.
4. Mitra, S.: "The Upper Atmosphere," The Asiatic Society, Calcutta, India, 1952.
5. Fischer, R. (ed.): Interdisciplinary Perspectives of Time, *Ann. N.Y. Acad. Sci.*, vol. 138, p. 467, 1967.
6. de Vaucouleurs, G.: The Case for a Hierarchical Cosmology, *Science*, vol. 167, no. 3922, p. 1203, 1970.

part six

Summary

22
The Principles of General Systems Science

The following is an introductory formalization for a general systems science:

1. There exist ac active atomistic systems that are capable of absorbing and emitting energy. Their quantization in space encompasses their structure; their quantization in time represents their orbits. Such elements will be regarded as nonlinear limit-cycle oscillators.

2. However, such open thermodynamic systems with active properties cannot come into existence unless the fields are locally degradative. In other words, the first and second laws of thermodynamics cannot be locally violated. There must be a local irreversible production of entropy.

3. Furthermore, such active systems could not come into existence in a field which is or can remain homogeneous. The question arises whether there is a motional path, made by man or by natural associations, by which a homogeneous system which has decayed or nearly decayed (there may be disturbing noise sources always present) may become inhomogeneous.

4. The necessary theoretical foundation for active elements to appear is linear instability; that is, a combination of such physical forces acting on discrete entities such that the system cannot come to an equilibrium rest state. Instead there are causal chains that will then keep them in motion.

5. There are conditions, whose sufficiency cannot be presently assured, under which these systems instead can change their state in alternation from A to B (e.g., from warm spell to cold spell, from anxious to calm, from radiating to nonradiating). When such rhythmic alternations in time long persist in taking place, we may view them as limit cycles. They need not be rigid cycles, only long persistent.

6. The important necessary condition for the persistence of rhythms is that the system possess dissipative properties. The occurrence of long persistent limit cycles, paradoxically put, is associated with the occurrence of losses. These "shave" the system down to the nearest possible limit cycle rather than permitting an indefinite wandering through phases that might only change if only conservative collision epochs were available.

7. Because of the current inadequate theoretical foundation, we cannot state more completely the internal conditions in systems that ensure motion by limit cycles but simply investigate their existence. However, we postulate that they occur, hierarchically, at all levels of organization of physical entities. It is very possible that out of the very hierarchical ordering of systems, time itself emerges cycle by cycle, chain by chain, pulse by pulse.

8. We have touched on illustrations, with various degrees of ignorance, of how the instability that leads to quantization comes about.

  a. In a hydrodynamic system, for some simple turbulent field with a spectrum of turbulence, we can illustrate deterministically how such instability comes about.

  b. In a meteorological system, it is possible to illustrate with some self-consistency how tidal oscillations come about and thus how the dynamics of weather may begin.

  c. In the cosmological system, we can show semitheoretically the stellar cycles, the origins of the universe, and the "beginnings" of the (current) cosmological cycle. (Here it is not yet possible to demonstrate persistent cycles. Attention is focused on demonstrating one. However, it is not inconceivable that the source of this cycle lies in the instability of a larger system.)

  d. In the social system, we have traced man's civilized past to likely roots, with cultural beginnings under *Australopithecus* and the beginnings of civilization in Neolithic times. The period 5000 B.C. to 1500 B.C. showed a considerable number of cyclic examples—of growth, life, and decline of civilizations and their replacement by successive cycles. Thus, the second more familiar half of such history, 1500 B.C. to 2000 A.D., presents no essential new characteristics.

The "start" in Neolithic times of the cycles of civilization is well defined. The "causality" of the linear instability is not.

e. In the atomic and nuclear case, the quantization is describable, but there is no theory for origins. Theories are suggested but without any probing foundation (nonlinearly coupled mechanical-electromagnetic quantization in the combined field of atomic particles, radiation, and the atoms making up any enclosure in the case of atoms; vortices in the "ether" of a continuous negative ocean of matter, in which a primary cataclysmic unstable explosion quantized stress fields in the negative ocean, which make up "particles" in our positive sparse ocean of "discrete" particles).

9. These atomistic oscillators do not arise in isolation in the large vistas of space-time but are found occurring over an extended field, most often with boundaries. It is generally in interaction with the boundaries and each other that the atomistic properties—of limited extension and quantized fluctuation—arise. Generally there is an interaction between these entities. However, not all regions of the space nor all interactions are freely accessible. Yet the fact that these entities were formed out of common substance and common causality roughly makes them alike. When these are nearly alike and interacting, then their dynamic patterns resemble each other. This is generally described by Gibbs' ergodic hypothesis (preferably in quasi-ergodic form). All portions of accessible phase space are occupied with roughly equal probabilities by the members of the ensemble. Averages in space and time approach the same limit, upon sufficient sampling.

10. The extensive collection of such ac active atomistic particles makes up a continuum. The conditions for continuumlike properties are the following:

a. The ratio of the mean free path (between relaxations) of an atomistic particle inserted into the collection of such particles to the dimensions of the field should be small. A likely estimate, from a number of fields, is smallness in the ratio of 1 part in 1,000 or less. This highly restricted value has been chosen so that even the boundary layers of the system appear continuous. Else a much less stringent value would have been permitted. (The boundary layer is likely a transition layer that may appear in many diverse fields of phenomena. For example, in a city, it may be the suburban boundary between city and country. In a turbulent hydrodynamic field, it is the so-called viscous boundary layer between the stationary impervious wall and the busy hurly-burly of the potential flow core.)

b. The ratio of the relaxation time (both for external relaxations and relaxations of internal degrees of freedom that are not frozen out) of an atomistic particle inserted into the field to the shortest period of interest (e.g., the period of the highest-frequency phenomena of interest) should be short. An estimate is the ratio of 1:10 or less.

These relaxational properties, indicating physical degrees of freedom associated with the atomistic elements, create transport coefficients in the continuumlike ensemble. The types of transport may be identified as delays in the transport of momentum (making up the relaxation property of viscosity—of shear viscosity associated with translational momentum and bulk viscosity associated with all other forms of internal momentum, e.g., vibrational and rotational); delays in the transport of energy (making up the relaxation property of finite thermal conductivity or electrical conductivity); delays in the transport of mass species (making up the relaxation property of finite diffusivity—both of self-diffusion and interspecies diffusion).

The collection is then described by continuum equations of change (e.g., the equations of hydrodynamics, the equations of elasticity-plasticity, the equations of chemical interaction, the equations of magnetohydrodynamics).

11. A well-defined formalism, based on Onsager's linear theory, exists for nonequilibrium thermodynamics. It indicates the foundations from which various kinds of interatomistic coupling forces can result in transport coefficients.

12. One formal result which comes out of this formulation is the significance to be attached to the bulk viscosity and to the tensile modulus of elasticity (as opposed to the shear viscosity and bulk modulus). The shear viscosity represents the result of delay in the transport of momentum associated with translation. However, if there are interatomistic coupling forces and internal degrees of freedom (by definition, ac atomistic elements have internal modes for absorbing energy), then another delay to transport occurs, identified by the bulk viscosity. The bulk viscosity reflects energy absorbed by all means other than through translational momentum.

The existence of the property of high bulk modulus is the organizing gateway toward rigid structure (i.e., to the solid state). That is, at low bulk modulus a collection of atomistic elements acts fluidlike, with rapid lively relaxation (through shear viscosity). The ratio of viscosity (in particular now the bulk) to elastic modulus (say, also the bulk) defines a time, the relaxation time. (In this case, the rapidity of internal relaxations.)

If now this relaxation time is "high" (for example, for high-quality elastic materials it may be nearly infinite, namely, thousands to millions of years), then these internal degrees of freedom are "frozen out."

In addition, with increased bulk modulus, it may happen that the tensile modulus may grow. From the greater proximity of ensemble elements to each other, increased interelement coupling may result. If the coupling is favorable, then the elastic reliability of the continuum field can grow.

13. The results of interatomistic association are particularly accentuated by atomistic density itself. Formally, the laws of association of ensembles, groups, populations, and civilizations begin to emerge.

14. The basic distinction in characteristics between continuum systems of low bulk modulus and of high bulk modulus is the difference between (structural) form and function. The mobility of ensemble elements creates function. As their mobility is impaired by densification and increased interatomistic forces, function is frozen out and becomes formed structure.

15. Fundamentally, the specific nature of the transport properties depends on the interelement forces. These forces may depend on interelement distance or be independent of distance. The classification of such forces is one of the tasks of physical science. However, one salient force, which is apparently fundamental to quantization, should be stressed. This is the exchange force. In quantum physics, it is said to have no classical physical analogue. We propose one. It is a binding force of configuration. In an ensemble arrangement in which element A can hardly be distinguished from element B, an exchange can be imagined by which they interchange position (somewhat independent of their location, if not too far apart). These two arrangements, somewhat degenerate, somehow have an energy associated with the exchange, representing an instability within which the system can fluctuate. This binding energy and fluctuation is proposed as the foundation for quantization. What is likely required is coupling of more than one kind of force field, e.g., the mechanical and electromagnetic.

Likely examples of such binding are:

Nuclear binding
Atomic quantization
Chemical (Heitler-London) binding
Nonideal gas binding (van der Waals forces are often used as illustration)
Catalysis
Enzyme-substrate interaction
Biological replication
Interpersonal forces
Coalescence of societies

16. Such ingredients for interelement exchange are sufficient for the constructure of near-equilibrium statistical mechanics and thermodynamics of continuum systems. They suggest the behavior of an ensemble placed within a boundary milieu. In time, the system may approach a statistical mechanical equilibrium. Interactions—"collisions"—will take place until whatever energy is associated with boundary exchange will be distributed among the available degrees of freedom.

17. A system should be examined with a spatial and temporal "box," that is, with a scale extending from a minimum time to a maximum time, from

a minimum distance to a maximum distance. Unless its relaxation spectrum too extensively overlaps the ranges of this box, a bounded system will be found to "quickly" relax, among its degrees of freedom, to where the statistical mechanical motion of all the atomistic constituents as viewed in a phase space of all its degrees of freedom and the associated momenta freely wanders within the phase space available to it. Frozen-out degrees of freedom will not be invoked.

Such a free motion describes an ergodic (really a quasi-ergodic) system. Any collection will behave more or less like any other collection. The averages from any one collection (or epoch) will agree with any other collection or epoch.

18. The characteristic motions that these near-continuum ensembles will display is a motion in modalities. These modalities are of two distinct types—diffusive and wavelike—and of combinations of these types. There may be many more than one kind of diffusion or wave motion.

19. As the continuumlike collection increases in size, it ultimately becomes linearly unstable again. The characteristic motion is a sustained combination of the modalities permitted by the continuum. A new society appears composed of a space-time motion of larger entities, of superatomistic elements. Each of these elements is itself made up of a near-continuum of subatomic elements.

20. These large-scale atomistic elements also display lumped motional characteristics by modes. The modes now are associated with internal degrees of freedom of the subatomistic elements.

21. Thus, the hierarchy of general systems consists of a line of . . . atomistic element—continuum ensemble—atomistic element— . . .

22. The continuum is unstable and becomes transformed into a collection of superatoms at a Reynolds-numberlike criterion. The Reynolds number is the product of two ratios. One is the ratio of the velocity to the propagation velocity in the medium, the other is the ratio of dimensions of the field to the mean free path. The propagation velocity measures the rigidity of the system, being determined by the ratio of the elastic modulus of the medium to its density. A continuum thus breaks down if the velocity of any of the dynamic processes that may be developed in the medium gets too large, if the field becomes too large, if the medium is insufficiently rigidly coupled element to element, or if the mean free path gets too short. As a corollary for each governing dynamic process, there is an atomistic size (of continuum) that will fit.

The continuum characteristics of an ensemble of atoms break down or become vague in accordance with the previously stated parameters, when the mean free path of the atoms to dimensions of an enclosure is greater than 10 to 1 and when driven periodicities to relaxation time are less than 10 to 1. (In the range of mean free path to dimensions of 1 to 10 to 10 to 1 lies the domain of slip flow. An element is still within the social range of "a hoot and a holler" of its neighbor.)

23. There are transitional configurations, for example statistically numerous ensembles, in which relaxations do not take place rapidly enough to qualify the states as being fully ergodic (example, plastically yielding materials), or grainy fields in which some degree of condensation may take place (example, slip flow). Such systems thus tend to relax much slower and exhibit a much more restrictive kind of equilibrium. Yet this does not prevent their description. Only the complexity of description is affected.

The important property which is difficult to deal with in such systems is that they exhibit emergent evolution. One is hard put to estimate when a new relaxation process may take place from a temporarily frozen-out degree of freedom (or in particular, when a catalyst may appear for the process).

The emergent evolution generally takes place through a number of steps toward more equilibrium configurations of continua or atomism. It is as if a very slow-motional relaxation process is in progress.

The time scale of relaxation is then set by the bulk viscosity (i.e., internal relaxations) or the time scale of atomistic clustering or diffusive evaporation.

24. With a little more detail, the sequence -A-C-A-C- is not a simple one, for the following reason. Stability, in a nonlinear sense, is not necessarily sharply defined. For example, the stabilization may not be fully completed in just one particular domain. Thus, there is a hierarchy of efforts at stabilization.

Stabilization begins around a focal center. Stabilization takes place into space-time orbits. Most often, there is an individual element which captures or directs a considerable amount of energy. As such, we shall refer to it as an elite, or key element.

Thus, for example, the condensation of an ensemble of atomistic particles may take place into an array possessing some kind of symmetry (e.g., crystal symmetries). At first glance, this array might be thought of as being indefinitely repeated. However, for dynamic reasons, it is unstable. An individual element, a dislocation, "keys" the structure. The dislocation, which may be the site of an "impurity" (i.e., slightly different composition or characteristic), acts as an elite to key or fix that level of stability of the rudimentary crystallite. This structure now begins to extend toward continuum size. However, such a structure is not fully stable. A collection of such like crystallites extends outward till it is stabilized in the grain as a more homogeneous structure than its neighboring material. Again, there are border "elites" who make the mating configuration possible.

25. Having a universe of space and time to play around in, a discrete density of material, ac active elements, and radiative "action-at-a-distance" energy transfer, the only potential motional states are a uniform distribution or singular inhomogeneous condensations. Apparently, the uniform

distribution is less stable. Thus motion tends toward the condensations, i.e., expansion away from one region and condensation around an "elite" focus. Elements tend to cluster.

From a more general point of view, the grand scheme of things appears to be cosmic "dust" condensing to form new stars; the stars clustering in galaxies which also may go through a birth and life process; the galaxies tending to form clusters, typically of a few thousand galaxies (similar in number to atoms in a solid forming a crystallite). These galactic clusters then "fill up" space, and—according to the latest information—thereby close space. Thus, under the action of interatomistic forces—in this case gravitational (the stellar "atom" does not form except at stellar dimensions such as are associated with 0.1 to 30 sun masses)—a sequence of levels is required to stabilize the structure. However, the hierarchy passed through condensed ac active forms.

In any local domain (the argument repeats) the only potential motional states are a uniform distribution and inhomogeneous condensations. The uniform distribution appears to be less stable. Thus motion tends toward the condensations. The dynamic stellar processes continue—main sequence to red giant to white dwarf. Condensed material such as planetary objects may go through a life phase—such as on earth. "Unstable" dynamic processes lead to a geophysical solid, liquid, and gas state and a fourth category of chemically reactive change. The argument repeats again. Apparently, some chemistry is self-replicational, i.e., unstable in the same sense, so that it creates dynamic limit-cycle processes—"life." Life evolves, becomes unstable, forms new patterned structure, etc. This appears to be the continuing chain of thought in the science of general systems.

26. In recapitulation and summary, a system has three phases:

The first phase is a start-up from the local milieu. This may arise by design or by an initial inhomogeneous growth or organization. It does not represent a negentropic contradiction of thermodynamics, just the development of a local inhomogeneity which is locally unstable. The common negentropic pump is "convection," whether the convection of a turbulent hydrodynamic field in which organization is propagated from unstable eddy formation (the eddies are local discrete rotational components that have drawn energy from the constant potential field) or the algorithmic "convection" in which the living system stuffs discrete energetic chunks of "food" into its gut.

The second phase is the normal life phase. This is what this book was about.

The third phase is the degradation phase.

However, that the birth, life, and death of limit-cycle systems emerging from the large-scale degradation promised by thermodynamics be understood, regard the matter-energy-space-time cosmological universe as a giant pinball machine. As a result of natural processes, energy runs down toward the disordered form of heat—in this cosmological phase. However, the cosmological matter-space-time milieu is not homogeneous. There may always form local pockets in which a relatively long-term "life" process will form and lock up. Galaxies will thus form, stars will thus form, elements may thus form, solar systems will thus form, geophysical and geochemical processes will thus form, life will thus form, social organization will thus likely form.

Are each of these foreordained? This is not clear. If we could see the entire inhomogeneous continuum, we might guess at the density of pockets, or if we had experience with 10 or 100 or 1,000 such cosmologically expansive, thermodynamically degradative grand process phases, we might see a more general answer. But we have not the experience, we are puny man. Thus we can see the processes as essentially stochastic. Whatever grand design is apparent is not available to us. Nevertheless, we can be certain that pockets exist and systems form—for a time! These are not in a strict sense "evolutionary" as much as they are temporary "stases" of a dynamic sort.

27. What makes the systems colorful are the many different kinds of internal modalities within which they may operate.

23

Metaphysical Foundations

This book was written as an extension of an earlier unpublished philosophic essay, "Philosophy for Mid-Twentieth Century Man" (1957). The essay contained the rudiments of a working scientific philosophy that has now been more fully developed in this study. Readers have said that there was nothing significant in that essay which is not now in this book. Thus we may put that earlier document out of mind.

However, there was a brief introduction to metaphysics which may be of use to the reader. Although it represented the author's attempt to outline his personal metaphysics, it may help the reader by forcing him to assess his own metaphysics. Scientific paradigms are only meaningful when projected against the underlying structure of metaphysics by which the viewer beholds.

The basic implication of a system of metaphysics is that it lies outside of the realm of the observable. It is, therefore, a set of principles that connect our observable sciences with concepts and relationships that increase their palatability.

We hope that for a structure of metaphysics the following principles or axioms are sufficient.

1. The principle of Occam's razor. This principle, the parsimony of assumption, becomes the starting point of all "thought." A medieval principle, it states that of all explanations that fit a given set of phenomena and their correlations the simplest explanation is best. Note its implications: it is consistent with human "need" for thought and "explanation" (i.e., the need for the organized thought that leads to science); it is consistent with the idea of the human brain as a computer; it ends arguments as to the nature of "ultimate truth." By its very nature, a denial of this assumption simply leads to endless argument.

2. The assumption of reality. There is a unique one-to-one relationship between objects in reality and our perceptions of these objects. Thus non-material ghosts do not exist. This is the assumption of the pragmatist. The Cartesian statement "I think, therefore I am" comes closest to the pragmatic statement. We prefer "I perceive, therefore I am" as a better statement. Solipsistic arguments basically contradict this metaphysical assumption. It does not appear fruitful to debate the issue. The pragmatic fact is that there is a common experience which we share with other men. We all are posed with a common range of daily problems that we must deal with before we die. Thus assuming that to a reasonable extent there is a reality corresponding to our common observations gets us along with the task of living.

3. The assumption of existence. This metaphysical principle premises our existence. It thus premises the class of all phenomena by which we are aware of or sense our existence. One may argue whether this premise or the previous premise comes first.

4. The assumption of identifiability. If we exist because we perceive and if our perceptions are "known" to us in some unidentified way, we again might be at an impasse. However, if we assume that in some undefined way we can create a third element in the chain of one-to-one identification —namely, (as the first element) the symbolization, whether on paper or verbally or by signs, (as the second element) of perceptions (as the third element) of real entities—then we can proceed to exchange information (whether we mutter to ourself or our neighbor). The exchange of information is the whole crux of the matter of philosophy.

5. The assumption of mass-space-time. Whereas in earlier times the philosophers, having agreed in some way as to "reality", then proceeded to argue out the "nature of reality" and spoke of extension, space, time, and their attributes, we may avoid today these needless and lengthy conceptualizations and choose a very simple operational (or pragmatic) point of view. If we exist, because we perceive, and we can identify, there arises the question as to what are the undefined perceivables and thus identifiable[1] elements. We may

[1] Identifiable—able to be brought into one-to-one correspondence with perception.

choose today to take the mass of a particle (operationally defined through the concept of its resistance to motion), the distance between particles (operationally defined by one-to-one correspondence yardstick judgments), and duration of motion (operationally defined by one-to-one yardstick comparisons between the motion, i.e., change in perception, of a mass particle with some other reference mass). Taking the view of a physical scientist of the day, we refuse to admit anything more about the nature of time than that our heart beats, the earth turns, and entropy increases. We are willing to admit that there is a parametric correspondence between the states of these systems, and we will call the parameter time and even, if the reader wishes, say time "flows," but we will not endow it with any more extensive metaphysical nature. It is undefined, except for the parametric character that we have briefly alluded to.

6. The assumption of uniqueness. There is a considerable range of phenomena that are in one-to-one correspondence with our and other human beings' perceptions. We regard it that these agreed upon perceptions represent unique entities. In fact, coupled with the first assumption, we have regarded it the task of science to bring all of our perceptions into such correspondence.

7. The assumption of connection. We may also assume that our persons are part of the real universe, made up of its elements, and that the act of perception is commutative—we can view the world outside of us and be viewed. Thus, just as mass particles can affect other mass particles, we can affect mass particles.

8. The assumption of self-actuation. Just as some systems move, we may note that our human system moves. Although we recognize outside effects whose space-time coincidence is accompanied by our motion, we may assume that there is actuation of ourselves which is not so accompanied by such coincidence. We, therefore, assume as the simplest hypothesis that we have some degree of self-actuation (i.e., we are not always being pulled by outside strings).

9. The assumption of causality. Our perceptions occur in sequences that we have parametrically identified with time. We also note a one-to-one correspondence of sequence in arrangements outside of ourselves. Certain of these arrangements repeat. If in "psychological" time (both for ourselves and others) there is an invariable sequence of before and after in our perception of these repeated arrangements, then we assume that they are causally connected. We may note that we are sometimes fooled. Although strictly we might require invariable reproducibility, we most often settle for lesser reproducibility.

10. The assumption of attribute. The act of identification requires some added assumptions. On the basis of our scheme so far, we could name perceived entities as A, B, C, etc., and probably achieve common identification. However, consider large rock and small rock. Since our perceptions differ

between these two entities, we gradually learn to name entities other than rock. Since finally it dawns on us that we are not naming the entity but the perception of the entity and that this encompasses a broad class of identifiable perceptions, we may begin to recognize that our name consists of the class of all attributes that we can perceive for that entity. We give up, thus, the idea of absolute identification. (Is the rock we saw yesterday the same one we saw today? Are you sure?) By putting names into sequence, big rock (note this is not the only scheme that could be used), we identify the narrower class of entities that have the mutual attributes—so we can get to "the little red sequin with the broken hole that was attached yesterday to the bottom of my purple dress." In addition to passive attributes (like red), we have attributes of action. These are symbols that represent complex arrangements of mass-space-time that we can identify either by illustrating with our own bodies or others —John runs, John loves Mary, John is under a haystack.

11. The assumption of continuity in space-time. We postulate that there is a continuity in space-time of the arrangements that we perceive. This metaphysical assumption, one supposes, is selected to calm our fears.

12. The assumption of the continuity of reality beyond us. Recognizing that humans die, it is of great convenience to postulate that reality persists beyond our limits.

13. The assumption of the operational test of reality. It is more than likely that the last metaphysical assumption that made possible a simple picture of the world around us was Dr. Bridgman's concept of an operational test for "truth" (i.e., explanation—namely, the one-to-one correspondence with "reality"). We only know reality through our instrument perceptions.

Glossary

Abduction   In logic, a syllogism in which the major premise is evident but the minor premise, and therefore the conclusion, is only probable. To illustrate in a common diagnostic use, if A is a (diseased) state marked by a set $[x_i]$ of symptoms and if B is a system that exhibits a subset $[y_i]$ of such symptoms or exhibits a subset with various significant probabilities attached, then B is a system that is probably in that (diseased) state (abductive—involving abduction).

Ac—alternating current   Within the popular engineering technique of identifying physical analogues or "equivalent networks," an ac network would be one containing active sources in which the fluxes varied in periodic or transient fashion.

Active network (or element)   Within the engineering technique of identifying physical analogues or "equivalent networks," an active network (or element) contains (or is) a potential source from which energy can be withdrawn for an appreciable period of time (i.e., it is a source of power). If the potential source is constant, it is known as a dc (direct current) source. Otherwise it is an ac source or element.

Analytic   The general connotation, as used here, is of a representation by the mathematical techniques of analysis. Roughly, this is the representation by means of continuous or piecewise continuous mathematical functions (i.e., "equations"). Analysis includes representation involving both algebraic relations and the relations of the differential and integral calculus.

Atomistic   Atomic has the technical meaning of the smallest discrete particle of an element that is not capable of further subdivision without loss of identity as an element. Similarly, atomistic refers to the smallest discrete unit (if one exists) of any organized system which is not subdivisible without loss of identity as an element of the system.

Autonomous   Independent in governing; mathematically when no factor in an equation contains time, as an explicit variable.

Burnett equation   The equation of motion of a fluid, that is, a simple gas or liquid, is represented by the Navier–Stokes equation of motion, named in honor of its first enunciators. Loosely speaking, it states that a region of fluid accelerates under the action of pressure forces but is retarded by viscous forces developed by adjacent parts of the fluid. At high vacuum, for the flow of gases—as the fluctuations of the molecular constituents make themselves evident—the Navier–Stokes equation must be modified. The pertinent equation is called the Burnett equation in honor of its enunciator. Loosely, there are additional terms in the equation.

Calorimetric, calorimetry   Pertaining to calorimetry; the measurement of thermal heat energy; more generally, the measurement of energy by converting it to a thermal form.

Chain or causal chain   The electrical engineer might have used the word "network." We imply both the causal chain of mechanisms and events that are connected together (e.g., the descriptions of a Rube Goldberg system—mouse A nibbles cheese B, ignites string C, etc.). It denotes both the logical-descriptive connections and the mechanistic connections. Why chain instead of network? Because the pieces may be spread out in field form and not perfectly compartmentalized: e.g., in biology, the stepwise process of activation of precursors, such as in intrinsic factor coagulation, represents an open chain; the SA nodal potassium leak and subsequent reabsorption, as a determinant of depolarization rate, represents a closed chain.

Constellation   Any cluster or gathering of entities that are tied or connected in their orbital motions by some unifying theme.

Continuum   A space or field whose elemental parts cannot be separately discerned at the scale of observation.

Cosmology   The large-scale theory of the nature and principles of the universe.

Coupled   When mechanisms or functional subsystems are connected causally to influence each other, they are said to be coupled. If A is causally connected to B, the connection is often described by coupling coefficients or influence coefficients.

Culture   The unifying concepts, habits, skills, arts, instruments, tools, etc., of a given people in a given time epoch. It is the common elements of a way of life that a people occupying a given physical environment create.

Degradation   In physics, any process by which available energy becomes unavailable, as by conversion into heat.

Degree of freedom   In a discrete entity, the independent kinds of state within which the entity can be transformed without losing its identity. For example, a point particle can be transformed within three translational displacements; an extended body such as a cylinder has three additional rotational displacements; an extended body such as a spring has an additional positive or negative compressional displacement. In an extended entity such as a string or a plate or a whole continuum, each local region may have independent kinds of state within which each segment can be transformed. Each point of a string, for example, can have independent displacements, subject to connectivity limitations in the string.

Deterministic mechanics   The detailed description of the motion of particles or a system of particles by Newtonian mechanics or Einstein mechanics; i.e., their laws of motion, such that the detailed motion at every instant of time, within a given epoch, of every particle is achieved.

Dialectic   The Hegelian method of logic, based on the concept of advancing contradictory arguments, of thesis and antithesis, and seeking their resolution by synthesis.

Displacement   In physics, the physical states, all associated with each degree of freedom, through which an entity can transform. Illustratively, these are the different positions reached by physical motion, the different concentrations that some constituent in a solution may achieve, the amounts of electric charge in a condenser, or the amount of fluid in a container, etc.

Domain   Generally a limited region or field marked by some specific property. In mathematics, it can have a somewhat more specialized meaning.

Dynamics   In physics, the study of the motion of entities in space and time under the action of influences and forces that change the motion.

Elite   A selected part of a group. The selection is based on some specialized property that makes the subgroup rare.

Emergent evolution   That development of the state of a system which issues, flows, or proceeds from its past states. It implies that a system can change its life pattern and that what evolves is related to the internal operating characteristics of the system and the external environment in which the system operates. In biology, more restrictively, the systems are species, organisms, or organs.

Energetics   The science that deals with the laws of energy and the transformations of energy.

Ensemble   An aggregation or collection of elements connected by a series of relations.

Entrainment   From the concept of the act of boarding a train. The technical meaning is the act of one system becoming trapped, enmeshed, or drawn into a track, orbit, or path with other systems.

Entropy   A technical concept in physics with many complex, but essentially equivalent, definitions. The dictionary states that entropy is the theoretical measure of energy which cannot be transformed into mechanical work in a system. More colorfully, if an ensemble of energetic particles displays a considerable degree of ordering upon its release from bounding into a less-confining space, the ensemble loses its specialized ordering. Then entropy is a measure of the ordering in the state of the ensemble as it changes.

Epigenetic   Related to the doctrine that the entity that will develop into a viable system (e.g., the germ cell developing into an organism) is acted upon and depends both on the conditions in its environment as well as its internal coding (i.e., it is both the phenotype and genotype that determines the emergence of the living organism).

Epoch   A period of time marked by noteworthy or characteristic events.

Equilibrium or steady state   Systems can have one of two types of equilibria or steady states. They can have the static steady state of rest, in which all motion ceases, or they can have the dynamic steady state of endlessly repeated motion, which is marked by a stationary spectrum. Equilibrium is regarded by many as a static steady state achieved when all forces come to a balance. However, it is used here in the broader sense of the balance achieved when a system becomes autonomous or isolated.

Equipollent   Equivalent in force, weight, validity, significance, and measure.

Ergodic   A collection of systems forms an ergodic ensemble if the modes of behavior found in any one system from time to time resemble its behavior at other temporal periods and if the behavior of any other system when chosen at random also is like the one system. We do not require identical performance, only quite similar time averages and number averages. (If you cannot tell one youth from another or one adult from another, they belong to an ergodic ensemble.) In an ergodic population, any single individual is representative of the entire population. The salient

characteristics of this individual are essentially identical with any other member of the group.

Ethology   The newer definition relates to the study of animal behavior, founded on a comparative zoological and physiological base.

Extension   In physics, that property of a body by which it occupies space; in logic the extent, denotation, or scope of a term.

Eyring's formalism   One of the most creative views of the interaction of dense collections of molecules has been developed by Eyring. Loosely speaking, his theory of rate processes suggests how to view and isolate a chain of relaxation steps by which dense collections of molecules are brought into motion.

Field   An area of observation, a space in which phenomena are to be noted, a realm of knowledge.

Fluctuation   The change in some physical quantity in time, particularly if it varies around some average value for the quantity.

Flux   Generalized flow, whether of heat, electricity, fluid, or chemical constituents.

Form, formed, formal structure   A field of entities that does not change in time; the invariant character and nature of a particular field; the essential character and nature by which a field is patterned or organized.

Function   The normal or characteristic action of a system of entities, generally in time; the variation of some magnitude that depends upon the variation of some other magnitude.

Functional transformation   The act of changing the form or condition of a system from one functional state to another.

Gain at zero frequency   The gain of a system generally refers to the ratio of the output or some regulated variable to the input. Strictly speaking, it is the influence coefficient for a power amplifier. Systems are tested by their frequency response. One notes what "gain" occurs as signals of different frequencies are applied to a system. In the case of zero frequency, the signal put on the system is just a set of constant variables—its environment. Now we may characterize the gain at this zero frequency state. If statically stable, it would have a fixed gain. If dynamically stable, namely, characterized by limit cycles, the gain would represent these fixed limit cycles. (In linear theory, the system would be said to be unstable because it possessed limit cycles. In nonlinear theory, it would be dynamically stable.) However, a system which produced indeterminate limit cycles (from time to time, or case to case) would be said to have indeterminate gain at zero frequency.

Gestalt   The organized structure or pattern that makes up all of a man's experience of some field system. This integrated view is more than the sum of the individual elements by which the field can be described.

Gibbs' ergodic theory   The doctrine of ergodic theory as originally proposed by Gibbs to denote an ensemble of energetic entities sufficiently constrained in a phase space of displacement and momenta so that time averages and space averages for the ensemble agree.

Heuristic   Narrowly, learning by self-discovery; we interpret this more broadly to denote the first most primitive act of discovery of the entities that make up any field.

Hierarchical   Relating to or constituting a related series arranged in connected levels. A complex network which is coupled by very comparable coupling coefficients at connected levels (such as a bank of electrical components arranged in a computer or the members of a chain of command) is not meant to be hierarchical. We reserve the term for the coupling that represents a drastic change in type yet leaves the system related and connected. (For example, nuclei are connected as atoms, atoms are connected as molecules, molecules are connected as biological cells, biological cells are connected as organs, organs are connected as a complete organism.)

Hierarchy   Narrowly, a group arranged in order of rank or class; we interpret it to denote a rank arrangement in which the nature of function at each higher level becomes more broadly embracive than at the lower level.

High mobile   A technical term in advertising used to denote the elite group in a technological society who are socially mobile because of their managerial position and who by their actions tend to create style.

Hunger   Narrowly, any strong desire or craving; we interpret this more broadly to denote any flux that a living organism needs must pass through its covering. It includes the common conceived-of foostuffs—water, food, oxygen—but also energetic fluxes—light, heat—and information fluxes—verbal exchanges and others.

Hypothesis   A supposition or system of suppositions which are taken for granted for the purpose of an argument; hopefully they would be made explicit.

Isomorphic   Having the same or similar form; we have interpreted this more broadly to represent similarity in both form and function.

Jump phenomena   In many fields, there are surfaces of discontinuity on both sides at which the field phenomena change drastically. The change in conditions measured between the two sides is said to be described as a jump and represents jump phenomena.

Kinematics   Motion in the abstract without relation to force or mass, i.e., displacements as functions of time.

Kinetics   The motion of masses or mass particles in relation to the forces acting on them.

Knudsen flow Many of the laws of flow of gases at high vacuum were investigated and enunciated by Knudsen at the turn of the century. In his honor, flow at high vacuum is commonly referred to as Knudsen flow. An appropriate general force law is commonly the Burnett equation.

Limit cycle In a linear system (such as a vibrating string, or a pendulum), if the system is displaced (pluck the string), it will start to vibrate or oscillate. However, by the second law of thermodynamics (loosely, that as a result of natural processes systems lose energy into lower forms; that is, disorder increases in the universe), the system will decay to rest. In a nonlinear system (examples: a watch, a human, a working engine) supplied with a constant source of fuel or energy, it is possible to obtain configurations such that if the system is started vibrating, oscillating, or "running," it will continue. If the cycle thus formed operates independent of the precise initial starting conditions, in spite of the fact that the system is lossy and in spite of moderate disturbances that try to slow the process down or speed it up, then it is said to be a limit cycle.

Loop In the electrical theory of networks, any one complete circuit. In the theory of analogues for other physical phenomena that has developed from network analysis, it still remains any one complete circuit.

Lossy system We visualize that for many systems if their power supply (e.g., a source of electric power, or chemical energy, or a flame) is turned off the processes will gradually decay in their operating amplitude. This is a lossy process or system.

Marginal stability (or instability) A system which is either operating stably (or unstably) is said to be marginally stable if a small change in the constant operating conditions (e.g., if the fuel supply is increased a little or if the environmental conditions, such as temperature, are changed a little) will change the system operation from stable to unstable or from static to dynamic steady state or vice versa.

Mediate To be the intermediate mechanism for bringing about change or providing communication; i.e., to manipulate to new operating conditions.

Metaphysics The branch of philosophy that deals with first principles; in the Aristotelian sense, the causal essence, the matter, the sources of change, and the purpose. More loosely we denote by it the first principles that go beyond (i.e., before) physics which can be discovered by the senses. Metaphysics holds the structure of our hypotheses about what we are sensing.

Milieu Surroundings, environment, in the sense of a fluidlike environment in which many diverse species (of things or biota) are immersed.

Mode   A complex rhythmic scheme in a system or a basic form or pattern in a system (e.g., a flow field can exhibit a diffusion mode or a wave mode; a child can exhibit a tantrum mode).

Momentum   In mechanics, the measure of motion of a moving body, as far as forces are concerned that can impart motion. It is measured as the product of mass and velocity.

Navier–Stokes equation   See **Burnett equation.**

Network   In electrical engineering, the connected series of loops that make up a system of electric-current-carrying conduits. With the development of analogues to electrical network analysis, any system of flux carriers is referred to and analyzed as a network.

Noise   A spectrum which is continuously filled with all frequencies or whose frequencies vary over a very large band and whose phases are not stationary. More loosely, it is any extensive group of frequencies whose causality is not understood and whose occurrence is not well predicted by the observer.

Nonlinear mechanics or mathematics   Narrowly, the dynamic equations of motion of a particle or system of particles or a field, visualized as describable by systems of differential equations, which represent how a set of a variety of dependent variables, displacements, or degrees of freedom change in time. If these displacements all change in proportion— that is, if they only differ by a scale factor—then the mechanics or mathematics is said to be linear. The major indicator of linearity, as defined here, is the superimposability of independent inputs or causes in creating change. If this property does not exist, then the description or the equations are nonlinear. (Beyond this rudimentary characterization, the description is somewhat hopeless for those with no mathematical training. The concepts are too extensively linked with the entire logical structure of mathematics to provide more detailed insights.)

Operational   Processes capable of change or transformation in a quantity; more restrictively, physical processes changing or transforming in space or time.

Orbit   The closed path or track of a body in space that repeats the path periodically.

Ordered time scales   The periods of spectral frequencies imagined arranged in temporal order.

Oscillator, oscillation   An apparatus producing oscillations; periodic variations of a physical parameter or displacement between limits.

Pair production   The action of an energetic photon or quantum of electromagnetic energy when brought to bear on some point in space (this requires a marker, generally a heavy nuclear particle at the point) can create (by producing them) a pair of particles—an electron and a positron.

Periodic   Occurring at regular intervals, usually of time.

Phase space   Imagine a system having one or more displacements: for example, x which will vary with time t so that it can be presented in a Cartesian graph of x plotted against t; imagine that the velocity v which is the rate of change of displacement x is also plotted against t on a separate Cartesian graph. If now a third Cartesian graph is plotted of velocity v against displacement x for the same values of t, this graph is said to represent the phase space of that system. The concept is generalized for all of the displacements into a more complex Cartesian hyperspace.

Phenomenological   The technical description of phenomena without any attempt at explanation.

Potential   That which can, but has not yet, come into being; technically, energy that is stored in a constant state by virtue of displacement rather than energy in motion (e.g., the potential, or potential energy stored in a compressed spring).

Power engineering   That branch of engineering concerned with the chain of mechanisms and functions from the combustion of fuel to the engines and prime movers in which such energy transformation is made available to do work or provide heat; e.g., in biology, it would represent the dynamics of intermediary metabolism, including catabolism and anabolism.

Protocol   Ceremonial form and procedure; technically, the form and procedure by which some complex task (e.g., a surgical operation or biological experiment) is to be conducted.

Quantize   Technically, to establish conditions such that a system dynamically operating will produce discrete units or half-units of one or more of its physical parameters (e.g., the wavelengths or frequencies associated with a vibrating column of air in an organ pipe are quantized).

Quantum   A discrete elemental unit of energy; any discrete element of a set that has been formed by a dynamic physical process.

Quantum mechanics   Whereas classical or Newtonian mechanics represents the motion of particles or systems of particles under the action of forces by a few elementary laws, the motion of very small particles (atoms and nuclear particles) are described in probabilistic terms by a logical structure that is built on top of classical mechanics. This system is known as quantum mechanics. Its essential feature is that the motion of atomic particles emerges as quantized, even though the deterministic position of the particles is replaced by a probabilistic distribution.

Quasi-static   A static description of a process is the description that does not vary with time, as opposed to a dynamic description. A quasi-static description is one that takes place essentially at such extremely slow motion that any accelerations can be virtually neglected.

Singularities of motion   In nonlinear mechanics, particularly as described by differential equations, one allows all the derivatives of displacement to vanish except the lowest. If then the remaining algebraic equation is solved for one or more roots of that lowest derivative, these roots represent the singularities. Typically, the lowest derivative is the displacement. Then the singularities represent those positions at which the displacements can instantaneously linger.

Spatial or temporal domain   Since a domain generally represents a region which is dominated by some particular physical feature (e.g., the landed domain of a lord), we use the more abstract notion of a region which we can characterize geometrically, by a graph, or by algebraic definition, in which some particular physical concept or function is dominant over a particular spatial region or for a particular temporal region or frequency region. Examples are the domain of sleep (temporal), anxiety (temporal), hairiness (spatial).

Spectrum   The series of frequencies or wavelengths, arranged in increasing or decreasing order, associated with a system (e.g., the spectrum of light waves emitted by some radiating source as detected in a spectroscope which disperses or spreads out the wavelengths). A system may show characteristic cycles of function (e.g., the atomic spectrum, the electromagnetic spectrum, the spectrum of colors in a white-light signal, the mechanical vibrational spectrum of a complex test object such as an entire airplane). When presented in a frequency or wavelength domain, this is called a spectrum. If during observation the frequencies do not change or do not change much, this is called a stationary spectrum. A particular frequency is often called a spectral line; if particular lines in the spectrum change in time, they are said to warble or wow. The repetition of a function at fixed frequency, such as the sleep-wakefulness cycle, is a spectral line. If the duration of each cycle changes from day to day, then the line warbles or wows.

Stability   The capacity of an object or system to return to equilibrium after having been displaced. Note with two possible kinds of equilibrium one may have a static (linear) stability of rest or a dynamic (nonlinear) stability of an endlessly repeated motion.

Static or dynamic regulation or feedback regulation (control)   If an environmental variable (such as temperature) or an input or output variable (such as the flow demand on a system) changes and the system can nearly compensate for those changes in some other variable (such as outlet pressure), then the system is said to be regulated or regulated for that variable. If the regulation is obtained by a static compensation in which some linkage or component is introduced that diminishes the sensitivity to change, then this is static regulation (e.g., a spring scale is designed with materials that thermally compensate the spring against temperature change; a dc motor is designed by the choice of its field windings to give a speed regulation against changes in the load put on the motor; a chemical buffer shifts the operating point of chemical equilibrium to hold the pH of a solution constant). In dynamic regulation, two different switch states (an on and an off state) are arranged so that the system switches from one state to the other when the regulated parameter rises to an upper limit and vice versa when the regulated parameter drops to a lower limit (an on-off thermostat). In feedback regulation (or control as it is technically referred to), an error signal is produced between the existing state of a system and the desired regulated level. This error signal is operationally acted upon, amplified in power, and fed to an actuator to operate a network which can influence the regulated variable so as to reduce the error, e.g., in biology the Na^+-angiotension system. The signal is the sodium concentration. When this concentration decreases, aldosterone is liberated from the adrenal cortex. This agent acts on the kidney distal tubules to increase the reabsorption of sodium ions and reestablish the proper concentration of sodium.

Statistical mechanics   In a large collection of particles whose motion is subject to the laws of mechanics, it has proven impossible to follow the detailed effect of the countless numbers of collisions and interactions. Instead, a methodology has been developed for treating the statistical effects of all of these interactions in determining various measures, particularly average measures to associate with the collection. This methodology is known as statistical mechanics.

System   A set or arrangement of entities so related or connected so as to form a unity or organic whole.

Systems engineering   The systematic application of engineering to solutions of a complete problem in its full environment by systematic assembly and matching of parts in the context of the lifetime use of the system.

Theory   An imaginative formulation of apparent relationships or underlying principles of certain observed phenomena. It may have been verified to some extent, or it may be pure hypothesis or conjecture.

Thermodynamics   The science that deals with the relationship of heat and mechanical energy or other forms of energy and the conversion of the one into the others.

Tizsa's formalism   Newtonian mechanics treats the motion of a system of particles by decomposing it into two motions, (1) the motion of a single equivalent particle moving with the center of mass of the system under the action solely of external forces, and (2) all other motions of the particles relative to the center of mass. Tizsa (Massachusetts Institute of Technology) has formulated a far-reaching explicit means of characterizing the momentum associated with all of the internal degrees of freedom of a molecule as differentiated from the momentum associated with the translational degrees of freedom of the center of mass of the molecule. The translational motion is associated with the shear viscosity of the molecular ensemble. Tizsa associated the remaining motion with the bulk viscosity.

Turbulence Disturbance; technically, in a fluid flow field the state of motion which is continuously disturbed in time without any apparent temporal source for the disturbance.

Vortex A whirling or eddying mass of fluid.

Index

Précis and key words from each chapter are offered as a basis for an index more suited to this book than a disconnected alphabetic list. This thesis was tested by examining philosophic texts with the same conclusion. The reason it is suggested is that such books are not histories or accounts but discourses. Pagination is provided for each chapter and in the body of each précis or for key words whenever the page range covered is too extensive to be grasped at a glance.

Preface (vii–xi). Development of a systems science–the purpose and scope of the book. Apologia for nonlinear quantization and nonconservative systems physics.

Part One. Physical Preliminaries

Introduction (3–9). Principles for a general science of systems. The chief methods
of discourse are dialectic and mystical arguments, as aspects of "scientific method."
Isomorphic modeling of systems as an ideal; formal models as substitute. Key
thesis–alternation of atomistic and continuum levels of active organization, hierarchically nested to form many levels of systems. Definition of a system. Other tool
subjects required for a systems science.

science
unifying principles
explanation–formal,
   isomorphic
scientific methods
induction
deduction
dialectics
mysticism
pragmatism
abduction
active systems
atomism
continua
energetics

quantization
ergodic theory
general systems science
hierarchical organization
equilibrium distributions
viable system
metaphysics
observation
communication
mathematics
philosophy
physical sciences
orbits
constellations

1. *Dynamics of Historical Speculation and of Physical Systems* (14–24). Cyclic
change in history and physical systems (14). Time and scale of human change
(15). Physical ideas traced in Mach up to current thinking (14–18). Form and
function emerge from background noise (15, 18). Scientific logic can be patterned
on alternation of atomistic idea and continuum idea (19). Energetics, nonlinear and
statistical mechanics as foundational working physical tools (18, 20). Energetics as
outlined by Ostwald (20–24). Rejection of his view of nonatomistic foundation for
phenomena (24).

kinematic processes
mechanics
"noise"
"elite" entities
classical physics
statistical physics
energetics
quantum concepts

theory of matter
atomism
continuum
nonlinear mechanics
thermodynamics
energy transformation
equilibrium
inhomogeneity

2. *Some Preliminary Comments on Nonlinear Mechanics* (26–37). Nonlinear
mechanics–a physical tool subject. A rudimentary sketch of its protocol and principles (26–30). Application to physical problems–ex. the three-body gravitational
problem (30–31). Stability of orbits (28–29, 31). Complex coupled phenomena
with degeneracies (31–32). Energetic sources plus losses as foundation for nonlinear
systems quantization (29, 31–32). A mystique of the bounding walls. Quantum
mechanics and nonlinear quantization (33–37). The complex problem of coupled

mechanical and electromagnetic forces (35–36). Coupling in the hydrodynamic field (36–37).

3. *Some Preliminary Comments on Quantum Mechanics* (39–58). Nonlinear stability (implying linear instability) may be the basis of quantization in systems (39). How quantum mechanics is built up from classical mechanics, starting from the conservative Hamiltonian (40). Can the lossy requirement for nonlinear limit cycles and a nonlossy Hamiltonian of quantum mechanics be reconciled? (40–42, 57–58). History of quantum mechanics (43–47, 56). Heisenberg's (1958) view of quantum mechanics (47–52). Derivation of Planck's law (52–56). In defense of quantum mechanics; is it equivalent to a nonlinear process of quantization? (57).

4. *Some Preliminary Comments on Statistical Mechanics* (61–104). Statistical mechanics of active interacting entities (61). From Newton's laws to Boltzmann's equation (62). Radiation coupling to the mechanical motion (62). Blackbody

radiation in a metallic cavity (62–63). Thermal and electrical resistivity and losses (63–64). A conceivable classical quantization (64–65). Classical mechanics; the Hamiltonian (65–67). Rudiments of statistical mechanics–microcanonical, canonical ensemble, Liouville's theorem, ergodic hypothesis, Maxwell distribution, temperature as a tie from mechanics to thermodynamics, ensemble averages, averaging examples (equipartition of energy, virial theorem, equation of state) (67–87). Some rudimentary statistical thermodynamics, per Caratheodory's formulation (87–96). The existence of a temperature function (87–88); the first law of thermodynamics (energy, work, heat) (88–89); the second law of thermodynamics (temperature, entropy) (89–93). The second law emerges from statistical mechanics (93–96). Irreversible production of entropy (disorder increases in the universe because of ongoing processes) (93–94). Boltzmann's result–probability distributions as a bridge from statistical mechanics to thermodynamics (94–95). Introducing statistical mechanics for radiation (96). Some remarks on the quantization issue in the radiation field (97–101). Some dialectics regarding quantization, conservative nature of systems, collision cycles in statistical mechanical systems, and sustained cyclic patterning of behavior in nonlinear systems (101–105).

statistical mechanics
ac active systems
limit cycles
Hamiltonian
quantum mechanics
nonlinear mechanics
Newton's laws
Lagrange's equations
Liouville's theorem
Boltzmann's equation
Planck's radiation law
classical electrodynamics
thermodynamics
temperature
cavity oscillations
Drude-Lorentz theory of
    conduction
dislocations
electrical and thermal
    resistivity
superconductivity
absolute zero of temperature
Wiedemann-Franz law
Fermi-Dirac statistics
atomic lattice
crystallite
nonlinear quantization
Bohr hydrogen atom
Hamilton's principle
canonical equations
generalized coordinates
    and momenta

phase space
ensemble
microcanonical ensemble
wall conditions
holonomic and nonholonomic
    constraints
isothermal and adiabatic
    boundaries
harmonic oscillator
background noise
mass and energy fluxes
open systems
canonical ensemble
conservation of density-in-
    phase
nondecaying motions
spectral response
intermittency or wobbling
conservation of extension-
    in-phase
ergodic hypothesis
a priori probability in
    phase space
Maxwell-Boltzmann distribution
    law
summational invariance
distribution function
ideal gas
thermodynamic temperature
system averages
molecular velocity
virial theorem

Part Two.   Introduction to Viable Systems

5. *The Many-body Problem Exhibited by Atoms and Molecules* (111–114). A
starting level of atoms and molecules. The atomistic building blocks–electron, proton,
neutron, and photon. Localizing their properties. Their characteristics. Intermolecular
forces bind molecules.

6. *The Physical-Chemical Continuum of Molecules* (116–132). Transport coefficients and equations of change for collections of atoms and molecules (116).
Interatomistic force laws (117). Equilibrium and disequilibrium–equations of state
and change (118). The transport coefficients–of viscosity, thermal conductivity, and
diffusivity (118). Significance of bulk viscosity in establishing form (118). Derivation of hydrodynamic equations of change (118–122). States of matter (122–123).
Fluid relaxational properties as bridge from functional change to formed state
(123–124). Elasticity-plasticity, changes in moduli and viscosity with temperature
and time (124–126). Elastic and inelastic effects (125). Lattice structure (126).
Crystallites, dislocations (126). Complex stabilization in the continuum (127).
Chemical unit processes–their categories (128–129). Biochemical milieu–replicating
chemistry (129–130). Chemical oscillator (130–132). Hereditary process (131–132).
General concept of catalysis (132).

7. *Continuum Instability: Physical and Biological Examples* (136–147). The continuum becomes unstable and forms superatoms (136–137). Examples–hydrodynamic turbulence, surface phenomena in fluids, point and surface phenomena in solids, biochemical cellular processes (137–142). The membrane, a basic biochemical active element (142–145). Water and oily polar bonds as foundation for the cellular process (145). Molecular continua do not persist, hydrodynamic instability breaks up the field (146–147).

8. *The Cellular Colony as Continuum* (151–154). Biochemical cells form colonies. They bind by "exchange" forces. Example of a slime mold.

Part Three. Detour–The Line of "Man" Systems

9. *Functional Organization of Simple Organisms, Organs, Parts* (159–163). A detour among biological systems. The cooperative cellular colony extends out to form specialized organs. Introducing the dynamics of organ systems. Similarity of organs and functional parts–both require informational coupling for sustained operation. Self-organization of a system requires nonlinear oscillators. As competitive systems, they require communications feedback to produce their sustained motion.

10. *The Internal Organs and Brain as a Biochemical Continuum* (165–167). Cellular constituents form discrete organs; the extensive organization of an ensemble of organs. Homeostatic regulation; its dynamic form, homeokinesis. Instability–the need for a bound between interior and exterior, limiting size for the organism, and a steersman. Stabilizing information from outside. The brain.

11. *The Complex Biological System in the External Milieu* (170–184). The homeokinetic individual, its bank of oscillators, and modes of operation (170). Stabilization from outside (171). Operation by modes (172). Nature of behavior and the brain (172–173). Some ethological findings (173–174). Long-term behavior emerging from the endocrines and neuroendocrines (174–176). Chemical chains and Markov processes in behavior (175–176). Man's quantized extensions–clothing, tools, machines, mechanisms, instruments, feedback communications, epigenetic coding (learning) (176). Communications and information (176–180). Further development of the thesis of an endocrine-chemical basis of behavior (180–182). Man's engineering extensions (182). Extensions of internal dramas–dreams and creative foci, morality (183). Creativity (183–184). Esthetics (184).

12. *Steps to a Social Continuum; from Person to Culture* (188–264). Individuals are bound, social history begins (188). Physical ideas behind social history–interparticulate forces, effect of density, molecular flow at low density, configurational (i.e., form or posture) fluctuations at medium density, increasing bulk viscosity, frozen-out degrees of freedom, emergent evolution (188–194). Ideas behind a theory of social history–the tool subjects of physiology, psychology, ethology, anthropology, history, sociology, economics (194–195). Dynamic interaction within the individual and between individuals in ensembles (195–196). Human wants as required flux patterns of matter and energy (196–197). Sustained motion within an ecology (196–197). Long past history of a family unit in man and hominid ancestors (196–197). Maturational phases in the human orbit–mother-infant, peer-peer, association with opposite sex, life constellation (197–198). Interpersonal relations and constellations (198). Kinship to *n*-body problem, in particular liquid-plastic state of matter (198). Social organizing effects of density, configurational and exchange forces (199–202). Condensation forces and association among mammals, primates, and humans (202). History of a human ensemble as ethological entity in ecological milieu (203). Describing their wants from the ecology leads to a possible dynamic history of action (203–204). Customary history (205–206). Alternate history of physiological-psychological "hungers"; connection to ecological science (206–207). Quantitative change with innovation of tools and food production (207). Emergence of elites and division of effort (207–208). Emergence of tools and speech (207–209). Limit cycle orbits, stabilization and dynamic "catalysis" or dynamic gating by elites as organizers of social systems (208–209). Quantum mechanical, thermodynamic character in all human societies with densification (209–210). Purpose of society (210). Complex systems character (210). Plastic-solid fixed nature of civilization (211). Comparison in highly industrialized civilizations–society in USSR and in USA (212–213). The marketplace to provide commodities that satisfy hungers (213). Flow fields to satisfy traffic among hungers (214–215). Elites as regularizers of flow (215–216). Contrast of physical view with classical economics (216–217). Idealistic and materialistic view of economics (217). Dialectic with a materialistic thesis–did human culture emerge from barbaric animal life through man's enslavement by man? A systems start-up problem (218). How did human culture start? (218–222). Culture (218–220). Old, Middle, New Stone Age men (221–222). Culture began among hominids with development of tools and speech (222–223). History of hominids (224–228). Cultural evolution in man, neurological-endocrine basis? (227–229). Man's tools, speech, aggressiveness (228–230). Beginning of food producing and fixed settlements–Near East, Europe. Civilization begins (232–235). Possible relations between emergent character of civilizations and neural-endocrine foundations (235–238). Introducing socioeconomic theory via Durkheim's thesis regarding division of labor (238–248). Agricultural industry (239). Social solidarity

13. *The Tribe-Settlement-City-State-Nation; Ethnological Atomism* (274–277). Men, bound in ensemble, spread over the earlth. Horde, tribe, settlements, regional galaxy. Communications limitation on size. Interaction of settlements as *n*-body problem. History.

territory
atomistic nations
population density
cultures
political subdivisions
continuum of civilization
Neolithic beginnings
moral principles
communications
statistical mechanics
translational mobility
glandular response
emotional rhythms
war
power
words
chemical drives
dynamic history of
  civilization
ecological constraints
anthropology
prehistory
bulk viscosity
history
Near East
ancient and classical Greeks
Crete
population
Europe
Sumerian culture
Akkadian
Babylonians
Egypt
Mesopotamia
Indus Valley
Chinese civilization
Mencius
codification

Hammurabi
Justinian
English common law
Code Napoleon
Hohenzollern Code
Code of Ur-Nammu of Ur
Code of Eshnunnu
Code of Lipit-Ishtar
Hebraic
Buddhist
Christian
Islamic
elites
Western history
quality of life
quasi-ergodic hypothesis
neocortex
organs
compelling drives
modern man
ice age
atomistic settlements
rhythms of civilization
atomistic spectra
catalysis
ranging animals
speech
food productivity
means of production
social growth
stability
social rhythms
nations
balance-of-power
alliance
society
politics
political realism

17. *The Universe as Continuum: From Stars to Galaxies to Cosmos* (319–322). Gravity binds matter (319). The physics of atomistic stars (320–321). Stars are bound into galaxies (321). Galaxies spread throughout the cosmos (322).

18. *The Nuclear Particles* (323–328). Nuclear particles and radiation (323–325). The particles and the four levels of particle interaction (325–326). Quantization issues (327–328).

19. *The Cosmological Atom* (330–347). From the galaxies as continuum to cosmological atom (330). Cosmological modelling–general relativity and its origins (331–333). Outline of cosmological processes–big bang, thermal pair production, birth of elements, birth of galaxies, birth of stars, birth of earth, birth of life (333–343). A story for start-up (343–347).

Part Five.   Speculation

20.   *The Giant Pinball Process for Systems* (353–359). All matter-energy tumbles down the canvas of space–time (353–354). But there are inhomogeneous pockets–systems are trapped (355–358). Recapitulation of the story of start-up (356–358). Analogy or "true" description? (358). Support for classical quantization (359).

21.   *Subfields Beneath and Universes Beyond?* (360–365). Mass-energy defines space-time (360). Is nucleus fluidlike? (361). Is the negative ocean a dense continuum? (361). Hydrodynamic insights (362–364). Application to universe. The universe is spatially continuous and not very lossy (362). Quantization in hydrodynamics (364–365). Space-time is not a simple manifold (365).

Part Six.   Summary

shear viscosity
tensile modulus
atomistic density
laws of association
form
function
mobility
interelement force
exchange force
degeneracy
binding forces
statistical mechanics
time and space scale
modalities
diffusion
wave motion

Reynolds number
plastic yield
emergent evolution
stabilization
space-time orbits
dislocations
inhomogeneous condensation
stars
galaxies
main sequence
planets
earth
life
normal life phase
degradation

23. *Metaphysical Foundations* (378–381). Some formal principles are offered–parsimony of assumption, assumption of reality, assumption of existence, assumption of identifiability, assumption of mass-space-time, (379); assumption of uniqueness, assumption of connection, assumption of self-actuation, assumption of causality, assumption of attribute (380); assumption of space-time continuity, assumption of reality beyond us, assumption of an operational test of reality (381).

philosophy
metaphysics
Occam's razor
explanation
reality
relationship
perception
pragmatism
existence
identifiability
symbolization
information
mass-space-time
extension

attribute
parametric correspondence
uniqueness
connection
commutation
self-actuation
causality
invariable sequence
name
continuity
arrangement
operational test
Bridgman
instrumental perceptions